Animal Cell Culture

The Practical Approach Series

Related **Practical Approach** Series Titles

Membrane Transport
Flow Cytometry [3e]
Apoptosis
Cell Growth, Differentiation and Senescence
Virus Culture
Eukaryotic DNA Replication
Growth Factors and Receptors
Cell Separation

Affinity Separations
Subcellular Fractionation
Platelets
Epithelial Cell Culture
Neural Cell Culture
Basic Cell Culture
Plant Cell Culture [2e]
Plant Cell Biology
The Cell Culture

Please see the **Practical Approach** series website at
http://www.oup.co.uk/pas
for full contents lists of all Practical Approach titles.

Animal Cell Culture
Third Edition
A Practical Approach

Edited by
John R. W. Masters
3rd Floor Research Laboratories,
University College London

OXFORD
UNIVERSITY PRESS

OXFORD
UNIVERSITY PRESS

Great Clarendon Street, Oxford OX2 6DP

Oxford University Press is a department of the University of Oxford.
It furthers the University's objective of excellence in research, scholarship,
and education by publishing worldwide in

Oxford New York

Athens Auckland Bangkok Bogotá Buenos Aires Calcutta Cape Town
Chennai Dar es Salaam Delhi Florence Hong Kong Istanbul Karachi Kuala
Lumpur Madrid Melbourne Mexico City Mumbai Nairobi Paris São Paulo
Singapore Taipei Tokyo Toronto Warsaw

with associated companies in Berlin Ibadan

Oxford is a registered trade mark of Oxford University Press in the UK and in
certain other countries

Published in the United States by Oxford University Press Inc., New York

© Oxford University Press, 2000

The moral rights of the author have been asserted

Database right Oxford University Press (maker)

First Edition First Published 1989
Second Edition First Published 1992
Reprinted 1994, 1995
Third Edition First Published 2000

All rights reserved. No part of this publication may be reproduced, stored in a
retrieval system, or transmitted, in any form or by any means, without the
prior permission in writing of Oxford University Press, or as expressly
permitted by law, or under terms agreed with the appropriate reprographics
rights organization. Enquiries concerning reproduction outside the scope of
the above should be sent to the Rights Department, Oxford University Press,
at the address above

You must not circulate this book in any other binding or cover and you must
impose this same condition on any acquirer

British Library Cataloguing in Publication Data
Data available

Library of Congress Cataloguing in Publication Data

Animal cell culture : a practical approach / edited by
John R. W. Masters.—3rd ed.
p. cm.—(Practical approach series ; 232)
Includes bibliographical references and index.
1. Cell culture. 2. Tissue culture. I. Masters, John R. W. II. Series.

QH585.2 A55 2000   571.6′381—dc21   00-026267

1 3 5 7 9 10 8 6 4 2

ISBN 0 19 963797 0 (Hbk.)
ISBN 0 19 963796 2 (Pbk.)

Typeset in Swift by Footnote Graphics, Warminster, Wilts
Printed in Great Britain on acid-free paper
by The Bath Press, Avon

Preface

The use of cell culture is increasing exponentially, and is an essential tool in nearly all biological and medical research laboratories. New techniques are constantly being applied, and this new third edition of Animal Cell Culture builds on the superb foundation laid by Ian Freshney. Stem cell assays, apoptosis, genetic manipulation, and tissue engineering are areas of current interest that have been introduced, while all the other chapters from previous editions have been completely rewritten or updated.

There is one message that has not yet penetrated the minds of the whole scientific community. Cross-contamination of cell lines continues to occur at a high rate and continues to be ignored. The use of cell lines which have not been checked and which are not what they are claimed to be amounts to fraud, and is not something which cannot be ignored as lightly as it has been in the past. Scientists must now authenticate their lines (see chapter by Robert Hay and colleagues from the American Type Culture Collection) and editors should demand that cell lines are authenticated before publication.

J.M.

May 2000

Contents

List of protocols *page* xiii
Abbreviations xvii

1 Introduction to basic principles
R. Ian Freshney

1 Background 1
2 Biology of cells in culture 2
    Origin and characterization 2
    Differentiation 3
3 Choice of materials 3
    Cell type 3
    Source of tissue 4
    Subculture 5
    Selection of medium 7
    Gas phase 8
    Culture system 9
4 Procedures 10
    Substrate 10
    Medium 11
    Cell culture 11
References 16

2 Scaling-up of animal cell cultures
Bryan Griffiths

1 Introduction 19
2 General methods and culture parameters 20
    Cell quantification 20
    Equipment and reagents 21
    Practical considerations 22
    Growth kinetics 23
    Medium and nutrients 23
    pH 24
    Oxygen 26

CONTENTS

        Types of culture process *30*
        Summary of factors limiting scale-up *32*
   3 Monolayer culture *32*
        Introduction *32*
        Cell attachment *34*
        Scaling-up *35*
   4 Suspension culture *48*
        Adaptation to suspension culture *49*
        Static suspension culture *50*
        Small scale suspension culture *50*
        Scaling-up factors *53*
        Stirred bioreactors *54*
        Continuous-flow culture *56*
        Airlift fermenter *57*
   5 Immobilized cultures *58*
        Immurement cultures *59*
        Entrapment cultures *62*
        Porous carriers *62*
   References *66*

3 Cell line preservation and authentication
R. J. Hay, M. M. Cleland, S. Durkin, and Y. A. Reid

   1 Introduction *69*
   2 Cell line banking *70*
   3 Cell freezing and quantitation of recovery *71*
        Equipment *72*
        Preparation and freezing *73*
        Reconstitution and quantitating recovery *74*
   4 Cell line authentication *78*
        Species verification *78*
        Tests for microbial contamination *81*
        Testing for intraspecies cross-contamination *94*
        Miscellaneous characterizations and cell line availability *101*
   Acknowledgements *102*
   References *102*

4 Development of serum-free media
Soverin Karmiol

   1 Introduction *105*
   2 Role of serum and other undefined tissue extracts in cell culture systems *105*
   3 Response curves *106*
        Proliferating cultures *106*
        Non-proliferating cultures (hepatocytes) *111*
   4 Antimicrobials, phenol red, Hepes, and light *112*
        Phenol red *113*
        Gentamicin *114*

        Hepes   *115*
        Light   *116*
    5 Purity of components   *117*
    6 Fatty acids   *117*
       Acknowledgements   *120*
       References   *120*

5 Three-dimensional culture
L. A. Kunz-Schughart and W. Mueller-Kleiser

    1 Introduction   *123*
    2 Multicellular tumour spheroids (MCTS)   *124*
        MCTS monocultures   *124*
        MCTS co-cultures   *133*
    3 Experimental tissue modelling   *139*
        Current research on tissue modelling   *139*
        Tissue modelling of skin and mucosa   *140*
        Embryoid bodies   *143*
       References   *143*

6 Tissue engineering
Robert A. Brown and Rebecca A. Porter

    1 Introduction   *149*
    2 Design stages for tissue engineering   *150*
        Tissue engineered skin   *152*
        Tissue engineered urothelium   *152*
        Tissue engineered peripheral nerve implants   *153*
    3 Cell substrates and support materials   *154*
    4 Cell sources   *157*
    5 Orientation   *159*
        Mechanical cues   *161*
    6 Protocols   *163*
        Cell seeding of implantable materials   *163*
        Slow release systems for local control of TE constructs or repair sites   *165*
       Acknowledgements   *168*
       References   *168*

7 Cytotoxicity and viability assays
Anne P. Wilson

    1 Introduction   *175*
    2 Background   *176*
    3 Specific techniques   *177*
        Culture methods   *177*
        Duration of drug exposure and drug concentrations   *179*
        Recovery period   *182*

CONTENTS

    4 End-points *183*
        Cytotoxicity, viability, and survival *183*
        Cytotoxicity and viability *183*
        Survival (reproductive integrity) *191*

    5 Assay comparisons *192*

    6 Technical protocols *192*
        Drugs and drug solutions *192*
        Drug incubation *194*
        Assay by survival and proliferative capacity *194*
        Cytoxicity assays *201*

    7 Interpretation of results *210*
        Relationship between cell number and cytoxicity index *210*
        Dose-response curves *211*

    8 Pitfalls and troubleshooting *214*
        Large standard deviations *214*
        Variation between assays *214*
        Stimulation to above control levels *214*

    References *215*

8 Fluorescence *in situ* hybridization
W. Nicol Keith

    1 Introduction *221*

    2 Probes *222*

    3 Probe detection *228*

    4 A final word on the tricky bits *231*

    5 FISH resources *232*
        Solutions *232*
        Useful books *232*
        Useful Web sites *232*

    References *233*

9 Genetic modification
Majid Hafezparast

    1 Introduction *235*

    2 Transfection *235*
        Calcium phosphate–DNA co-precipitation *237*
        Lipid-mediated gene transfer (lipofection) *238*
        Electroporation *240*
        Staining of cells for expression of β-galactosidase *243*
        Rescue of episomal plasmids *244*

    3 Microcell-mediated chromosome transfer *245*
        Formation of micronuclei *246*
        Enucleation *247*
        Purification of microcells *248*
        Fusion of microcells to recipient cells *250*
        Selection of microcell hybrids *250*

4 Irradiation fusion gene transfer   *252*
   Use in mapping genes   *253*
   Use in positional cloning   *254*
   Irradiation doses   *255*

   References   *256*

10 Epithelial stem cell identification, isolation, and culture
David Hudson

1 Introduction   *259*
   Clinical application of cultured human stem cells   *259*
2 Basic principles for identification and purification of stem cells   *260*
3 Assessment of proliferative heterogeneity   *260*
4 Methods for the separation of different cell populations   *265*
   Isolation of cells by differential adhesion   *266*
   Separation of cultured and primary cells by flow and immunomagnetic sorting   *270*
5 Long-term maintenance of stem cells in culture   *273*
   Keratinocyte stem cells in culture   *273*
   Maintenance of non-epidermal epithelial cells in long-term culture   *273*
6 Stem cell characterization by immunocytochemistry   *274*
   Antibody markers of differentiated cell phenotypes   *274*
   Staining cell suspensions using cytospin preparations   *277*

   References   *278*

11 Senescence, apoptosis, and necrosis
Ian R. Kill and Richard G. A. Faragher

1 Introduction   *281*
   Cellular senescence   *281*
   Cell death   *282*
   Differentiation and de-differentiation   *282*
2 Simple measures of the population dynamics of primary cultures   *282*
   Measurement of the growth fraction   *283*
   Determination of the necrotic fraction of the population   *287*
   Determination of the senescent fraction of the population   *288*
   Determination of the apoptotic fraction of the population using TUNEL   *290*
3 Other techniques for analysing population dynamics   *291*
   Detection of apoptosis using DNA laddering   *292*
   Inhibition of apoptosis using peptide inhibitors of caspases   *293*
   Determination of the non-dividing fraction of a population by simplified haptotactic assays, 'Ponten Plates'   *295*
   Determination of telomerase activity and telomere length   *296*

   Acknowledgements   *301*

   References   *301*

A1 List of suppliers   *303*

Index   *311*

Protocol list

Monolayer culture
  Use of standard disposable roller cultures   36
  Growing cells in a bead bed reactor   40
  Microcarrier culture   44

Suspension culture
  Adaptation to suspension culture   49
  Spinner flask culture   52
  Growth of cells in continuous-flow culture   57

Immobilized cultures
  MiniPERM Bioreactor[a]   60
  Fixed porous bead (Siran) bed reactor   63

Cell freezing and quantitation of recovery
  Culture preparation and freezing procedure   73
  Recovery of cultures frozen in glass ampoules   74
  The dye exclusion test   75
  Estimation of colony-forming efficiency   76
  Measuring vitality of reconstituted cells   77

Cell line authentication
  Karyological procedure   80
  Microscopic examination   81
  Culture tests   82
  Testing a culture for the presence of mycoplasma   83
  Indirect test for mycoplasma   85
  Examining a culture for morphological evidence of viral contamination   87
  Preparation of double-stranded DNA templates for amplification   89
  PCR amplification reaction for HIV   90
  PCR amplification reaction for HBV   90
  Gel electrophoresis   91
  Electrophoretic transblotting   91
  Labelling oligonucleotides   92
  Pre-hybridization and hybridization   92
  Chemiluminescent detection   92
  Cell identification by Giemsa banding   94
  Preparation of genomic DNA from cell lines   98

PROTOCOL LIST

    Preparation of genomic DNA from cell lines using FTA™ paper   *99*
    PCR amplification and STR typing of cell lines   *100*

Antimicrobials, phenol red, Hepes, and light
    Procedure for medium component toxicity testing   *116*

Fatty acids
    Complexing fatty acids to albumin   *119*

Multicellular tumour spheroids (MCTS)
    Initiation and cultivation of tumour spheroids in spinner flasks   *128*
    Coating of 96-well plates or culture dishes with agarose   *130*
    Cultivation of liquid-overlay tumour spheroid cultures in 96-well plates   *131*
    Co-cultivation of MCTS and monocytes in liquid-overlay culture   *134*
    Co-cultivation of MCTS and fibroblast aggregates in liquid-overlay culture   *135*
    Co-cultivation of MCTS and endothelial cells   *137*
    Co-cultivation of MCTS and endothelial cells on basal membrane   *138*

Experimental tissue modelling
    Isolation and cultivation of different cell populations from oral mucosa   *141*

Tissue Engineering
    Human dermal fibroblast (hDF) seeding within collagen sponges   *163*
    Human dermal fibroblast (hDF) seeding within collagen lattices   *164*
    Urothelial cell seeding on collagen sponges (30)   *165*
    Phenytoin release from fibronectin mats   *165*
    Slow release growth factor depot formation   *167*

Technical protocols for cytotoxicity and viability assays
    Monolayer cloning   *195*
    Clonogenicity in soft agar using a double layer agar system   *196*
    Spheroid incubation   *198*
    Volume growth delay   *199*
    Clonogenic growth of disaggregated spheroids   *199*
    Outgrowth as cell monolayer from intact spheroids   *200*
    Preparation of microtitration plates   *201*
    Protein determination of cell monolayers using methylene blue (59, 60)   *202*
    Protein determination of cell monolayers using sulforhodamine B (SRB) (61, 62)   *202*
    Protein determination of cell monolayers using Kenacid blue (63, 101)   *203*
    Determination of amino acid incorporation into cell monolayers (39, 86)   *204*
    Incorporation of [^{3}H]nucleotides into DNA/RNA   *205*
    Total DNA synthesis measured by [^{125}I]Udr uptake (52)   *205*
    Measurement of viable cell number using neutral red   *206*
    Measurement of viable cell number using MTT (80)   *207*
    Measurement of viable fresh human tumour and normal cell cytotoxicity by differential staining cytotoxicity (DiSC) assay[a]   *207*
    Measurement of intracellular ATP using luminescence (83 and personal communication)   *209*
    Measurement of viable cell number using XTT (81, 82)   *210*

FISH Probes
    Probe labelling   *223*
    Probe precipitation   *223*

Chromosome preparation from lymphocytes 224
Chromosome preparation from cell lines 225
Fluorescence *in situ* hybridization: chromosome preparation 226
Target denaturation 226
Probe denaturation 227
Wash steps 227

FISH Probe detection

Avidin detection for biotinylated probes (FITC) 229
Digoxigenin antibody detection (FITC) 229
Probe detection using avidin and antibodies for two colours (FITC/Texas Red) 230
Two colour antibody detection 231

Transfection

Transfection by calcium phosphate–DNA co-precipitation 238
Optimization of conditions for lipofection 240
Determination of the best capacitance setting for electroporation of a particular mammalian cell type 242
Staining of cells for β-galactosidase expression 243
Plasmid rescue from cells transfected with EBV-based shuttle vectors 244

Microcell-mediated chromosome transfer

Preparation of microcells 248
Fusion of microcells with recipient cells and selection for microcell hybrid clones 252

Irradiation fusion gene transfer

Irradiation fusion gene transfer[a] 255

Assessment of proliferative heterogeneity

Determination of proliferative heterogeneity and colony-forming efficiency (CFE) 263
Cloning of epithelial colonies 264

Methods for the separation of different cell populations

Preparation of extracellular matrix coated dishes 267
Preparation of epithelial cell extracellular matrix coated dishes 268
Adhesion of rapidly attaching cells to extracellular matrix (ECM) proteins 268
Positive cell sorting with immunolabelled magnetic beads 271

Stem cell characterization by immunocytochemistry

Preparation of cells for use in immunocytochemistry: cytospin cell preparation system 277

Simple measures of the population dynamics of primary cultures

Cleaning of coverslips for cell culture 283
Labelling using 5-bromo-2'-deoxyuridine (BrdU) 284
Detection of total growth fraction using antisera to pKi-67 286
Determination of viable and necrotic fractions using calcein AM–ethidium bromide staining 288
Demonstration of senescent cells by senescence-associated β-galactosidase (SA-β) activity 289
Demonstration of apoptotic cells by TdT-mediated dUTP nick end-labelling (TUNEL) 290

PROTOCOL LIST

Other techniques for analysing population dynamics
- Analysis of DNA fragmentation *292*
- Inhibition of caspase activity by peptide inhibitors *294*
- Simplified haptotactic analysis *295*
- Demonstration of telomerase activity by TRAP *297*
- Determination of terminal restriction fragment (TRF) length by in-gel hybridization *299*

Abbreviations

ATCC	American Type Culture Collection
BrdU	5-bromo-2′-deoxyuridine
BSS	balanced salt solution
caspases	cysteinyl aspartate-specific proteinases
CFE	colony-forming efficiency
cpd	cumulative population doublings
CPE	cytopathogenic effect
DAPI	4-6-diaminophenylindole
DMEM	Dulbecco's modified Eagle's medium
DMF	dimethylformamide
DMSO	dimethyl sulfoxide
EBV	Epstein–Barr virus
ECM	extracellular matrix
EGF	epidermal growth factor
ES	embryonic stem
FACS	fluorescence activated cell sorting
FCS	fetal calf serum
FGF	fibroblast growth factor
FISH	fluorescence *in situ* hybridization
FITC	fluorescein isothiocyanate
HAT medium	medium containing hypoxanthine aminopterin and thymidine
HBV	hepatitis B virus
hDF	human dermal fibroblast
HIV	human immunodeficiency virus
ICE	interleukin 1β-converting enzyme
IFGT	irradiation fusion gene transfer
IL	interleukin
KGF	keratinocyte growth factor
LDH	lactate dehydrogenase
MCTS	multicellular tumour spheroids
MMCT	microcell-mediated chromosome transfer
MX medium	medium containing mycophenolic acid and xanthine
NaBt	sodium butyrate

ABBREVIATIONS

ORP	oxygen-reduction potential
PBS	phosphate-buffered saline
PCR	polymerase chain reaction
PD	population doublings
PDGF	platelet-derived growth factor
PDT	population doubling time
PEG	polyethylene glycol
PHA-P	phytohaemagglutinin-P
PHT	5,5-diphenylhydantoin
PMA	phorbol myristate acetate
PNA	peptide nucleic acid
SA-β	senescence-associated β-galactosidase
SDS	sodium dodecyl sulfate
STR	short tandem repeat
TE	tissue engineering
TGFβ	transforming growth factor β
TRAP	telomerase repeat amplification
TRF	terminal restriction fragment
TRITC	tetramethyl rhodamine isothiocyanate
TUNEL	TdT-mediated dUTP nick end-labelling
VNTR	variable number of tandem repeats
YAC	yeast artificial chromosome

Chapter 1
Introduction to basic principles

R. Ian Freshney
CRC Department of Medical Oncology, Garscube Estate, Switchback Road, Bearsden, Glasgow G61 1BD, U.K.

1 Background

Cell culture has become an indispensable technology in many branches of the life sciences. It provides the basis for studying the regulation of cell proliferation, differentiation and product formation in carefully controlled conditions, with processes and analytical tools which are scalable from the level of the single cell to in excess of 10 kg wet weight of cells. Cell culture has also provided the means to define almost the entire human genome, and to dissect the pathways of intracellular and intercellular signalling which ultimately regulate gene expression. From its ancestry in developmental biology and pathology, this discipline has now emerged as a tool for molecular geneticists, immunologists, surgeons, bioengineers, and manufacturers of pharmaceuticals, while still remaining a fundamental tool to the cell biologist, whose input is vital for the continuing development of the technology.

Cell culture has matured from a simple, microscope driven, observational science to a universally acknowledged technology with roots set as deep in industry as they are in academia. It stands among microelectronics, avionics, astrophysics and nuclear engineering, as one of the major bridges between fundamental research and industrial exploitation, and, in the current climate, perhaps the more commercial aspects will ensure its development for at least several more decades. The prospects for genetic therapy and tissue replacement are such that the questions are rapidly becoming ethical, as much as technical, as new opportunities arise for genetic manipulation, whole animal cloning and tissue transplantation.

Several textbooks are already available (1–4) to introduce the complete novice to the basic principles of preparation, sterilization, and cell propagation, so this book will concentrate on certain specialized aspects, many of which are essential for the complete understanding and correct application of the technology. This chapter will review some of the biology of cultured cells, their derivation and characterization, and give guidance in those areas which require critical observation and control.

2 Biology of cells in culture

2.1 Origin and characterization

The list of different cell types which can be grown in culture is extensive, includes representatives of most major cell types, and has significantly increased since the last edition of this book, due largely to the improved availability of selective media and specialized cell cultures through commercial sources such as Clonetics (see Chapter 4).

The use of markers that are cell type specific (see Chapter 3) has made it possible to determine the lineage from which many of these cultures were derived, although the position of the cells within the lineage is not always clear. During propagation, a precursor cell type will tend to predominate, rather than a differentiated cell. Consequently a cell line may appear to be heterogeneous, as some cultures, such as epidermal keratinocytes, can contain stem cells, precursor cells, and mature differentiated cells. There is constant renewal from the stem cells, proliferation and maturation in the precursor compartment, and terminal and irreversible differentiation in the mature compartment. Other cultures, such as dermal fibroblasts, contain a relatively uniform population of proliferating cells at low cell densities (about 10^4 cells/cm^2) and an equally uniform, more differentiated, non-proliferating population at high cell densities (10^5 cells/cm^2). This high density population of fibrocyte-like cells can re-enter the cell cycle if the cells are trypsinized or scraped (or 'wounded' by making a cut in the monolayer) to reduce the cell density or create a free edge. Most of the cells appear to be capable of proliferation, and there is little evidence of renewal from a stem cell compartment.

Culture heterogeneity also results from multiple lineages being present in the cell line. The only unifying factors are the selective conditions of the medium and substrate, and the predominance of the cell type (or types) which have the ability to survive and proliferate. This tends to select a common phenotype but, due to the interactive nature of growth control, may obscure the fact that the population contains several distinct phenotypes only detectable by cloning.

Nutritional factors like serum, Ca^{2+} ions (5), hormones (6, 7), cell and matrix interactions (8–11), and culture density (12), can affect differentiation and cell proliferation. Hence, it is not only essential to define the lineage of cells being used, but also to characterize and stabilize the stage of differentiation, by controlling cell density and the nutritional and hormonal environment to obtain a uniform population of cells, which will respond in a reproducible fashion to given signals.

As the dynamic properties of cell culture (proliferation, migration, nutrient utilization, product secretion) are sometimes difficult to control, and the complexity of cell interactions found *in vivo* can be difficult to recreate *in vitro*, there have been numerous attempts to either retain the structural integrity of the original tissue, using traditional organ culture, or to recreate it by combining propagated cells of different lineages in organotypic culture (see Chapter 5). Among the most successful examples of the latter are the so-called skin equivalent models, where epidermal keratinocytes are co-cultured with dermal fibroblasts and collagen in filter well inserts or some similar mechanical support. The result is a synthetic skin suitable for grafting and now being evaluated for tests of irritancy and inflammation.

2.2 Differentiation

As propagation of cell lines requires that the cell number increases, culture conditions have evolved to favour maximal cell proliferation. It is not surprising that these conditions are often not conducive to cell differentiation where cell growth is severely limited or completely abolished. Those conditions which favour cell proliferation are low cell density, low Ca^{2+} concentration (5) (100–600 mM), and the presence of growth factors such as epidermal growth factor (EGF), fibroblast growth factor (FGF), and platelet-derived growth factor (PDGF). High cell density ($> 1 \times 10^5$ cells/cm^2), high Ca^{2+} concentration (300–1500 mM), and the presence of differentiation inducers, hormones such as hydrocortisone (10), paracrine factors such as IL-6 and KGF (7, 9), and nerve growth factor (43), retinoids (13), and planar polar compounds, such as dimethyl sulfoxide (14), will favour cytostasis and differentiation.

The role of serum in differentiation is complex and depends on the cell type and medium used. While a low serum concentration promotes differentiation in oligodendrocytes (15), a high serum concentration causes squamous differentiation in bronchial epithelium (16). In the latter case, this is due to transforming growth factor β (TGFβ) released from platelets. Because of the undefined composition of serum and the ever-present risk of adventitious agents such as viruses, controlled studies on selective growth and differentiation are best conducted in serum-free media (see Chapter 4).

The establishment of the correct polarity and cell shape may also be important, particularly in epithelium (17). Many workers have shown that growing cells to high density on a floating collagen gel allows matrix interaction, access to medium on both sides and, particularly, to the basal surface where receptors and nutrient transporters are expressed. The plasticity of the substrate can facilitate a normal cell shape and normal polarity of the interactive ligands in the basement membrane (18).

Different conditions may be required for propagation and differentiation and hence an experimental protocol may require a growth phase for expansion and to allow for replicate samples, followed by a non-growth maturation phase to allow for increased expression of differentiated functions.

3 Choice of materials

3.1 Cell type

The cell type chosen will depend on the question being asked. For some processes, such as DNA synthesis, response to cytotoxins, or apoptosis, the cell type may not matter, provided the cells are competent. In other cases a specific process will require a particular cell type, for example surfactant synthesis in the lung will require a fresh isolate of type II pneumocytes or a cell line, such as NCI-H441, which still expresses surfactant proteins. A reasonable first step would be to determine from the literature whether a cell line exists with the required properties.

3.2 Source of tissue

3.2.1 Embryo or adult?

In general, cultures derived from embryonic tissues survive and proliferate better than those from the adult. This presumably reflects the lower level of specialization and higher proliferative potential in the embryo. Adult tissues usually have a lower growth fraction and a higher proportion of non-replicating specialized cells, often within a more structured and less readily disaggregated extracellular matrix. Initiation and propagation are more difficult, and the lifespan of the culture often shorter.

Embryonic or fetal tissue has many practical advantages, but it must be remembered that in some instances the cells will be different from adult cells and it cannot be assumed that they will mature into adult-type cells unless this can be confirmed by appropriate characterization.

Examples of widely used embryonic cell lines are the various 3T3 lines (primitive mouse mesodermal cells) and WI-38, MRC-5, and other human fetal lung fibroblasts. Mesodermally-derived cells (fibroblasts, endothelium, myoblasts) are on the whole easier to culture than epithelium, neurons, or endocrine tissue. This selectivity may reflect the extensive use of fibroblast cultures during the early development of culture media and the response of mesodermally-derived cells to mitogenic factors present in serum. Selective media have now been designed for epithelial and other cell types (see Chapter 4), and with some of these it has been shown that serum is inhibitory to growth and may promote differentiation (16). Primary culture of epithelial tissues, such as skin, lung, and mammary gland, is routine in some laboratories (19), and prepared cultures are available commercially (for example, Clonetics; Promocell).

3.2.2 Normal or neoplastic?

Normal tissue usually gives rise to cultures with a finite lifespan, while cultures from tumours can give rise to continuous cell lines. However, there are several examples of continuous cell lines (BHK-21 hamster kidney fibroblasts, MDCK dog kidney epithelium, 3T3 fibroblasts) which have been derived from normal tissues and which are non-tumorigenic (see Section 3.3.1).

Normal cells generally grow as an undifferentiated stem cell or precursor cell, and the onset of differentiation is accompanied by a cessation in cell proliferation which may be permanent. Some normal cells, such as fibrocytes or endothelium, are able to differentiate and still de-differentiate and resume proliferation and in turn re-differentiate, while others, such as squamous epithelium, skeletal muscle, and neurons, once committed to differentiate, appear to be incapable of resuming proliferation.

Cells cultured from neoplasms, such as B16 mouse melanoma, can express at least partial differentiation, while retaining the capacity to divide. Many studies of differentiation have taken advantage of this fact and used differentiated tumours such as hepatomas (36) and human and rodent neuroblastomas (20), although whether this differentiation is normal is not known.

Tumour cells can often be propagated in the syngeneic host, providing a cheap and simple method of producing large numbers of cells, albeit with lower purity. Where the natural host is not available, tumours can also be propagated in animals with a compromised immune system, with greater difficulty and cost, but similar advantages.

Many other differences between normal and neoplastic cells are similar to those between finite and continuous cell lines (see below) as immortalization is an important component of the process of transformation.

3.3 Subculture

Freshly isolated cultures are known as primary cultures until they are subcultured. They are usually heterogeneous and have a low growth fraction, but are more representative of the cell types in the tissue from which they were derived and in the expression of tissue-specific properties. Subculture allows the expansion of the culture (it is now known as a cell line), the possibility of cloning (see Chapter 8), characterization and preservation (Chapter 3), and greater uniformity, but may cause a loss of specialized cells and differentiated properties unless care is taken to select the correct lineage and preserve or re-induce differentiated properties. The greatest advantage of subculturing a primary culture into a cell line is the provision of large amounts of consistent material suitable for prolonged use.

3.3.1 Finite or continuous cell lines?

After several subcultures a cell line will either die out (finite cell line) or 'transform' to become a continuous cell line. It is not clear in all cases whether the stem line of a continuous culture pre-exists or arises during serial propagation. Because of the time taken for such cell lines to appear (often several months) and the differences in their properties, it has been assumed that a mutation occurs, but the pre-existence of immortalized cells, particularly in cultures from neoplasms, cannot be excluded.

Complementation analysis has shown that senescence is a dominant trait, and immortalization the result of mutations and/or deletions in genes such as p53 (21, 22) (see Chapter 11). The activity of telomerase is higher in immortal cell lines and may be sufficient to endow the cell line with an infinite lifespan (23).

The appearance of a continuous cell line is usually marked by an alteration in cytomorphology (giving cells that are smaller, less adherent, more rounded, and with a higher ratio of nucleus to cytoplasm), an increase in growth rate (population doubling time can decrease from 36–48 h to 12–36 h), a reduction in serum dependence, an increase in cloning efficiency, a reduction in anchorage dependence (as measured by an increased ability to proliferate in suspension as a liquid culture or cloned in agar), an increase in heteroploidy (chromosomal variation among cells) and aneuploidy (divergence from the donor, euploid, karyotype), and an increase in tumorigenicity. The resemblance between spontaneous *in vitro* transformation and malignant transformation is obvious but nevertheless the two are not necessarily identical although they have much in

Table 1. Commonly used cell lines

Designation	Species	Tissue	Cell type	Properties	Ref.
Normal					
MRC-5	Human	Embryo lung	Fibroblastic	Diploid, contact inhibited	31
3T3-L1	Mouse	Whole embryo	Fibroblastoid	Contact inhibited; can differentiate into adipose cells	32
BHK-21	Hamster	Kidney	Fibroblastic	Contact inhibited transformable with polyoma virus	33
MDCK	Dog	Kidney	Epithelial	Forms 'domes'	34
Transformed, solid tissue					
293	Human	Embryo kidney	Epithelial	Very sensitive to human adenoviruses	35
Hep-G2	Human	Hepatoma	Epithelial	Retains some drug metabolizing enzymes	36
HeLa-S$_3$	Human	Cervical carcinoma	Epithelial	Rapid growth; high plating efficiency; can be propagated in suspension	37, 38
Cos-7	Monkey	Kidney	Epithelioid	Good host for transfection	39
Caco-2	Human	Colo-rectal carcinoma	Epithelial	Forms tight monolayer with polarized transport	40
MCF-7	Human	Breast carcinoma	Epithelial	Oestrogen receptor positive	41
A549	Human	Lung carcinoma	Epithelial	Makes surfactant protein A	42
A431	Human	Squamous skin carcinoma	Epithelial	High levels of EGF receptor	42
PC-12	Rat	Pheochromo-cytoma	Neural	Neuronal differentiation with NGF	43
Leukaemic					
HL-60	Human	Myelocytic	Suspension	Differentiates in response to retinoids, NaBt, PMA	44
Jurkat E6-1	Human	T cell	Suspension	Produces IL-2, stimulated by phorbol esters	45
K562	Human	Erythrocytic	Suspension	Synthesizes haemoglobin when induced with sodium butyrate	46, 47

common. Normal cells can 'transform' to become continuous cell lines without becoming malignant, and malignant tumours can give rise to cultures which 'transform' and become more (or even less) tumorigenic, but acquire the other properties listed above. Transformation *in vitro* is primarily the acquisition of an infinite lifespan. Simultaneous or subsequent alterations in growth control can be under positive control by oncogenes or negative control by tumour suppressor genes (24).

The advantages of continuous cell lines are their faster growth rates to higher cell densities and resultant greater yield, their lower serum requirement and general ease of maintenance in commercially available media, and their ability to grow in suspension. Their disadvantages include greater chromosomal instability, divergence from the donor phenotype, and loss of tissue-specific markers. Examples of commonly used continuous cell lines are given in *Table 1*.

A number of techniques, including transfection or infection with viral genes (such as E6 and E7 from human papilloma virus, and SV40T) or viruses (such as Epstein–Barr virus, EBV), have been used to immortalize a wide range of cell types (25) (see Chapter 9). The retention of lineage-specific properties is variable.

3.3.2 Propagation in suspension

Most cultures are propagated as a monolayer attached to the substrate, but some, including transformed cells, haematopoietic cells, and cells from ascites, can be propagated in suspension. Suspension culture has advantages, including simpler propagation (subculture only requires dilution, no trypsinization), no requirement for increasing surface area with increasing bulk, ease of harvesting, and the possibility of achieving a 'steady state' or biostat culture if required (see Chapter 2).

3.4 Selection of medium

Regrettably, the choice of medium is still often empirical. What was used previously by others for the same cells, or what is currently being used in the laboratory for different cells, often dictates the choice of medium and serum. For continuous cell lines it may not matter as long as the conditions are consistent, but for specialized cell types, primary cultures, and growth in the absence of serum, the choice is more critical. This subject is dealt with in more detail in Chapter 4.

There are two major advantages of using more sophisticated media in the absence of serum: they may be selective for particular types of cell, and the isolation of purified products is easier in the absence of serum. Nevertheless, culture in the presence of serum is still easier and often no more expensive, though less controlled. Two major determinants regulate the use of serum-free media:

(a) Cost. Most people do not have the time, facilities, or inclination to make up their own media, and serum-free formulations, with their various additives, tend to be much more expensive than conventional media.

Table 2. Commonly used media

Medium	Properties	Ref.
RPMI 1640	No Ca^{2+}, low Mg^{2+}; designed for lymphoblastoid and useful for other adherent and non-adherent cell lines. Unsuitable for calcium phosphate transfection.	48
MEM/Hanks' salts	Classic broad specificity medium; low HCO_3^- for use in air.	49
MEM/Earles' salts	Classic broad specificity medium; high HCO_3^- for use in 5% CO_2.	49
F12	Large number of additional constituents (trace elements, copper, iron, additional vitamins, nucleosides, pyruvate, lipoic acid, but at low concentrations; suitable for cloning.	51
DMEM/F12	50:50 mixture of DMEM and F12; suitable for primary cultures and serum-free with appropriate supplementation.	52
L15	Bicarbonate-free; buffers in the absence of CO_2.	
MCDB 153	One of a series of serum-free media; suitable for growth of keratinocytes.	54

(b) Requirements for serum-free media are more cell type specific. Serum will cover many inadequacies revealed in its absence. Furthermore, because of their selectivity, a different medium may be required for each type of cell line. This problem may be particularly acute when culturing tumour cells where cell line variability may require modifications for cell lines from individual tumours.

In the final analysis the choice is often still empirical: read the literature and determine which medium has been used previously. If several media have been used, as is often the case, test them all, with others added if desired (see *Table 2*). Measure the growth (population doubling time (PDT) and saturation density), cloning efficiency (see Chapter 7), and expression of specific properties (differentiation, transfection efficiency, cell products, etc.). The choice of medium may not be the same in each case, for example differentiation of lung epithelium will proceed in serum, but propagation is better without. If possible, include one or more serum-free media in the panel to be tested, supplemented with growth factors, hormones, and trace elements as required (see Chapter 4).

Once a medium has been selected, try to keep this constant for as long as possible. Similarly, if serum is used, select a batch by testing samples from commercial suppliers and reserve enough to last six months to one year, before replacing it with another pre-tested batch. Testing procedures are as described above for media selection.

3.5 Gas phase

The composition of the gas phase is determined by:

- the type of medium (principally its sodium bicarbonate concentration)
- whether the culture vessel is open (Petri dishes, multiwell plates) or sealed (flasks, bottles)
- the amount of buffering required

Table 3. Relationship between CO_2 and bicarbonate

Open or closed vessel	Buffer capacity	[HCO_3^-]	Gas phase	Hepes
Closed	Low	4 mM	Air	–
Open or closed	Moderate	26 mM	5% CO_2	–
Open or closed	High	8 mM	2% CO_2	20 mM

Several variables are in play, but one major rule predominates and three basic conditions can be described. The rule is that the bicarbonate concentration and carbon dioxide tension must be in equilibrium. The three conditions are summarized in *Table 3*. It should be remembered that carbon dioxide/bicarbonate is essential to most cells, so a flask or dish cannot be vented without providing carbon dioxide in the atmosphere. Prepare medium to about pH 7.1–7.2 at room temperature, incubate a sample with the correct carbon dioxide tension for at least 0.5 h in a shallow dish, and check that the pH stabilizes at pH 7.4. Adjust with sterile 1 M HCl or 1 M NaOH if necessary.

Oxygen tension is usually maintained at atmospheric pressure, but variations have been described, such as elevated for organ culture (see Chapter 7) and reduced for cloning melanoma (26) and some haematopoietic cells (27).

3.6 Culture system

Originally, tissue culture was regarded as the culture of whole fragments of explanted tissue with the assumption that histological integrity was at least partially maintained. Now 'tissue culture' has become a generic term and includes organ culture, where a small fragment of tissue or whole embryonic organ is explanted to retain tissue architecture, and cell culture, where the tissue is dispersed mechanically or enzymatically, or the cells migrate from an explant, and the cells are propagated as a suspension or attached monolayer.

Cell cultures are usually devoid of structural organization, have lost their histotypic architecture and often the biochemical properties associated with it, and generally do not achieve a steady state unless special conditions are employed. They can, however, be propagated, expanded, and divided into identical replicates. They can be characterized and a defined cell population preserved by freezing, and they can be purified by growth in selective media (Chapter 4), physical cell separation, or cloning, to give a characterized cell strain with considerable uniformity.

Organ culture will preserve cell interaction and retain histological and biochemical differentiation for longer than cell culture. After the initial trauma of explantation and some central necrosis, organ cultures can remain in a steady state for a period of several days to years. However, they cannot be propagated easily, show greater experimental variation between replicates, and tend to be more difficult to use for quantitative determinations.

Purified cell lines can be maintained at high cell density to create histotypic cultures and different cell populations can be combined in organotypic culture (see Chapter 5), simulating some of the properties of organ culture.

3.6.1 Substrate

The nature of the substrate is determined largely by the type of cell and the use to which it will be put. Polystyrene which has been treated to make it wettable and give it a net negative charge is now used almost universally. In special cases (culture of neurons, muscle cells, capillary endothelial cells, and some epithelial cultures) the plastic is pre-coated with fibronectin, collagen, gelatin, or poly-L-lysine (which gives a net positive charge). Glass may also be used, but must be washed carefully with a non-toxic detergent.

3.6.2 Scale

Culture vessels vary in size from microtitration (30 mm^2, 100-200 µl) up through a range of dishes and flasks to 180 cm^2, and roller bottles and multisurface propagators (see Chapter 2) for large scale culture. The major determinants are the number of cells required (5×10^4 to 10^5/cm^2 maximum for most untransformed cells, 10^5 to 10^6/cm^2 for transformed), the number of replicates (96 or 144 in a microtitration plate), and the times of sampling: a 24-well plate is good for a large number of replicates for simultaneous sampling, but individual dishes, tubes, or bottles are preferable where sampling is carried out at different times. Petri dishes are cheaper than flasks and good for subsequent processing, e.g. staining or extractions. Flasks can be sealed, do not need a humid carbon dioxide incubator, and give better protection against contamination.

Volume is the main determinant for suspension cultures. Sparging (bubbling air through the culture) and agitation will become necessary as the depth increases (see Chapter 2).

Where a product is required rather than cells, there are advantages in perfusion systems which can be used to culture cells on membranes or hollow fibres. These supply nutrients across the membrane, and the product is collected either from the cell supernatant or the medium, depending on the molecular weight, which will determine the partitioning of the product on either side of the membrane.

Perfusion is also useful for time lapse studies, where cells are monitored on a microscope stage by a video camera, and for pharmacokinetic modelling where the duration and concentration of a test compound can be regulated precisely.

4 Procedures

4.1 Substrate

Most laboratories now utilize disposable plastics as substrates for tissue culture. They are optically clear, prepared for tissue culture use by modification of the plastic to make it wettable and suitable for cell attachment, and come sterilized for use. On the whole they are convenient and provide a reproducible source of vessels for both routine and experimental work.

INTRODUCTION TO BASIC PRINCIPLES

Some more fastidious cell types, such as bronchial epithelium, vascular endothelium, skeletal muscle, and neurons require the substrate to be coated with extracellular matrix materials such as fibronectin, collagen or laminin. Most matrix products are available individually (Becton Dickinson, Life Technologies, Sigma, Biofluids) or combined in Matrigel (Becton Dickinson), extracellular matrix produced by the Engelberth Holm Swarm sarcoma cell line. Alternatively, the substrate may be coated with extracellular matrix by growing cells on the plastic and then washing them off with 1% Triton X in ultrapure water (28).

Matrix coating can be carried out in three ways:

(a) By wetting the surface of the plastic with the matrix component(s), incubating for a short period (usually ~ 30 min), then removing the surplus and using the plastic with adsorbed matrix within seven to ten days (stored at 4°C if not used immediately).

(b) By wetting the plastic and removing the surplus matrix material and allowing the residue to dry.

(c) By adding collagen or Matrigel and allowing it to gel.

Wet or dry coating is used mainly for propagation, while gel coating is used to promote differentiation of cells growing on or in the gel.

4.2 Medium

Most of the commonly used media (see *Table 2*) are available commercially, pre-sterilized. For special formulations (see Chapter 4) or additions it may be necessary to prepare and sterilize some of the constituents. In general, stable solutions (water, salts, and media supplements such as tryptose or peptone) may be autoclaved at 121°C (100 kPa or 1 atmosphere above ambient) for 20 min, while labile solutions (media, trypsin and serum) must be filtered through a 0.2 μm porosity membrane filter (Millipore, Sartorius, Pall Gelman). Sterility testing (see Chapter 3) should be carried out on samples of each filtrate.

Where an automatic autoclave is used care must be taken to ensure that the timing of the run is determined by the temperature of the centre of the load and not just by drain or chamber temperature or pressure which will rise much faster than the load. The recorder probe should be placed in a package or bottle of fluid similar to the load and centrally located.

4.3 Cell culture

4.3.1 Primary cultures

The first step in preparing a primary culture is sterile dissection followed by mechanical or enzymatic disaggregation. The tissue may simply be chopped to around 1 mm^3 and the pieces attached to a dish by their own adhesiveness, by scratching the dish, or by using clotted plasma (29). In these cases cells will grow out from the fragment and may be used directly or subcultured. The fragment of tissue, or explant as it is called, may be transferred to a fresh dish or the outgrowth trypsinized to leave the explant and a new outgrowth generated.

When the cells from the outgrowth are trypsinized and reseeded into a fresh vessel they become a secondary culture, and the culture is now technically a cell line.

Primary cultures can also be generated by disaggregating tissue in enzymes such as trypsin (0.25% crude or 0.01–0.05% pure) or collagenase (200–2000 U/ml, crude) and the cell suspension allowed to settle on to, adhere, and spread out on the substrate (3, 4). This type of culture gives a higher yield of cells though it can be more selective, as only certain cells will survive dissociation. In practice, many successful primary cultures are generated using enzymes such as collagenase to reduce the tissue, particularly epithelium, to small clusters of cells which are then allowed to attach and grow out.

When primary cultures are initiated (see refs 3 and 4 for methods), all details of procedures should be carefully documented to form part of the provenance of any cell line that may arise and be found to be important. A sample of tissue, or DNA extracted from it, should be archived to be available for DNA fingerprinting or profiling for authentication of any cell lines that arise.

4.3.2 Subculture

A monolayer culture may be transferred to a second vessel and diluted by dissociating the cells of the monolayer in trypsin (suspension cultures need only be diluted). This is best done by rinsing the monolayer with PBS lacking Ca^{2+} and Mg^{2+} (PBSA) or PBSA containing 1 mM EDTA, and removing the rinse, adding cold trypsin (0.25% crude or 0.01–0.05% pure) for 30 sec, removing the trypsin, and incubating in the residue for 5–15 min, depending on the cell line. Cells are then resuspended in medium, counted and reseeded.

4.3.3 Growth curve

When cells are seeded into a flask they enter a lag period of 2–24 h, followed by a period of exponential growth (the 'log phase'), and finally enter a period of reduced or zero growth after they become confluent ('plateau phase') (see *Figure 1*). These phases are characteristic for each cell line and give rise to measurements which should be reproducible with each serial passage: the length of the lag period, the population doubling time (PDT) in mid-log phase, and the saturation density at plateau, given that the environmental conditions are kept constant. They should be determined when first handling a new cell line and at intervals of every few months thereafter. It is an important element of quality control to be able to demonstrate that the same seeding concentration will yield a reproducible number of cells at subculture, carried out after a consistent time interval, without necessarily performing a growth curve each time.

The determination of the growth cycle is important in designing routine subculture and experimental protocols. Cell behaviour and biochemistry changes significantly at each phase and it is therefore essential to control the stage of the growth cycle when drugs or reagents are added or cells harvested. The shape of the growth curve can also give information on the reproductive potential of the culture where differences in growth rate (PDT) (*Figure 1a*), adaptation or survival (*Figure 1b*), and density limitation of growth (level of saturation density in plateau) (*Figure 1c*), can be deduced from the shape of the curve. However it is generally recognized that the analysis of clonal growth is easier (30) (see Chapter 7), and less prone to ambiguity and misinterpretation (*Figure 1d*).

4.3.4 Feeding

Some rapidly growing cultures, such as transformed cell lines like HeLa, will require a medium change after three to four days in a seven day subculture cycle. This is usually indicated by an increase in acidity where the pH falls below pH 7.0. Medium can also deteriorate without a major pH change, as some constituents, like glutamine, are unstable, and others may be utilized without a major pH shift. It is, therefore, recommended that the medium is changed at least once per week.

4.3.5 Contamination

The problem of microbial contamination has been greatly reduced by the use of laminar airflow cabinets. The risk of contamination can also be reduced by use of antibiotics, but this should be reserved for high-risk procedures, such as primary culture, and cultures should be maintained in the absence of antibiotics so that chronic, cryptic contaminations are not harboured.

Check frequently for contamination by looking for a rapid change in pH (usually a fall, but some fungi can increase the pH), cloudiness in the medium, extracellular granularity under the microscope, or any unidentified material floating in the medium. If a contamination is detected, discard the flask unopened and autoclave. If in doubt, remove a sample and examine by phase microscopy, Gram's stain, or standard microbiological techniques (see Chapter 3).

Mycoplasma

Cultures can become contaminated by mycoplasma from media, sera, trypsin, imported cell lines, or the operator. Mycoplasma are not visible to the naked eye, and, while they can affect cell growth, their presence is often not obvious. It is important to test for mycoplasma at regular intervals (every one to three months) as they can seriously affect almost every aspect of cell metabolism, antigenicity, and growth characteristics. Several tests have been proposed, but the fluorescent DNA stain technique of Chen is the most widely used (see Chapter 3), although it is less sensitive than the PCR-based and culture methods.

4.3.6 Cross-contamination

This problem is dealt with in detail in Chapter 3. Its severity is often underrated, and consequently it still occurs with a higher frequency than many people admit. A significant number of cell lines, including Hep-2, KB, and Chang liver, are still in regular use without acknowledgement of their contamination with HeLa (55). Many other cell lines are cross-contaminated with cells other than

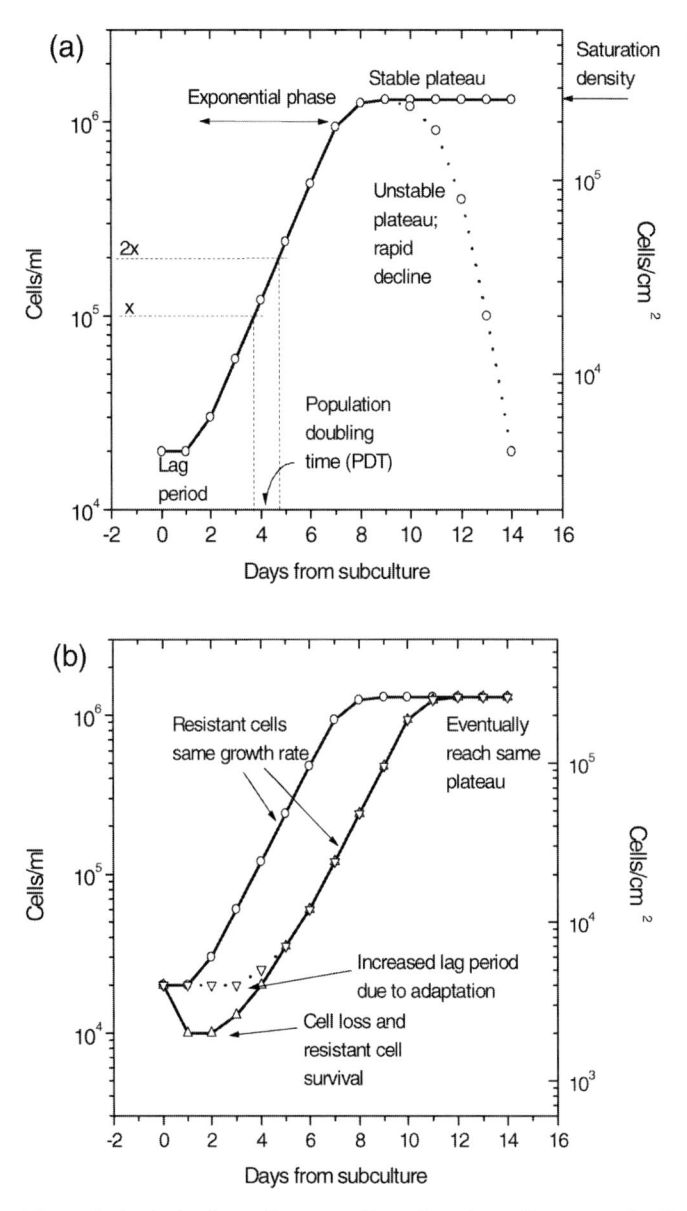

Figure 1. Analysis of growth curves. Examples of growth curves of cells following subculture are shown with cell concentration (cells/ml) on the left axis and cell density (cells/cm²) on the right axis. Both cell concentration and cell density are important for attached cells and unstirred suspensions, but only concentration is relevant to stirred suspensions. (a) Typical growth curve for a culture seeded at 2×10^4 cells/ml (5×10^3 cells/cm²) on day 0, showing a lag period of 24 h, followed by exponential growth ('log' phase) from day 1 to day 7, and entering a plateau phase on, or shortly after, day 8. The solid line represents a stable plateau, seen in most normal cells. The dotted line represents the rapid deterioration seen with many transformed cell lines, particularly suspension cultures (such as myeloma cultures), where the cells die soon after they have reached the plateau phase. (b) Cells plated as in (a), but in the right-hand curve there is evidence of reduced growth, as the curve displaced to the right. The dip at the beginning (solid line, upward pointing triangles) implies that many cells have not survived but that a resistant fraction has survived and has grown through with the same growth rate as the control. The dotted line (downward pointing triangles) would indicate that there was no cell loss, but that there was a prolonged period

of adaptation, followed by recovery at the same growth rate. (c) Cells plated as in (a), showing a reduced growth rate resulting from a longer population doubling time (PDT), with cells reaching same saturation density as control (squares), or plateauing at a lower saturation density (diamonds). Cells could also plateau at a lower saturation density while retaining the same PDT (not shown). (d) Cells plated as in (a), showing potential misinterpretations of data indicating reduced growth rates. The curve with upward facing triangles shows a reduced survival but the same growth rate. The curve denoted by squares shows a reduced PDT. Both curves converge at 8 days in this example, but could converge at any time with appropriate PDT and survival, indicating that single time points are unsuitable for correct analysis. In this example, the 8 day point could overestimate growth as the control has started to plateau. Although this occurs at 8 days in this example, it could occur earlier with a more rapidly growing cell line. Furthermore, all curves converge at the same saturation density if left long enough. Meaningful comparative observations can only be made while all cells are in exponential growth.

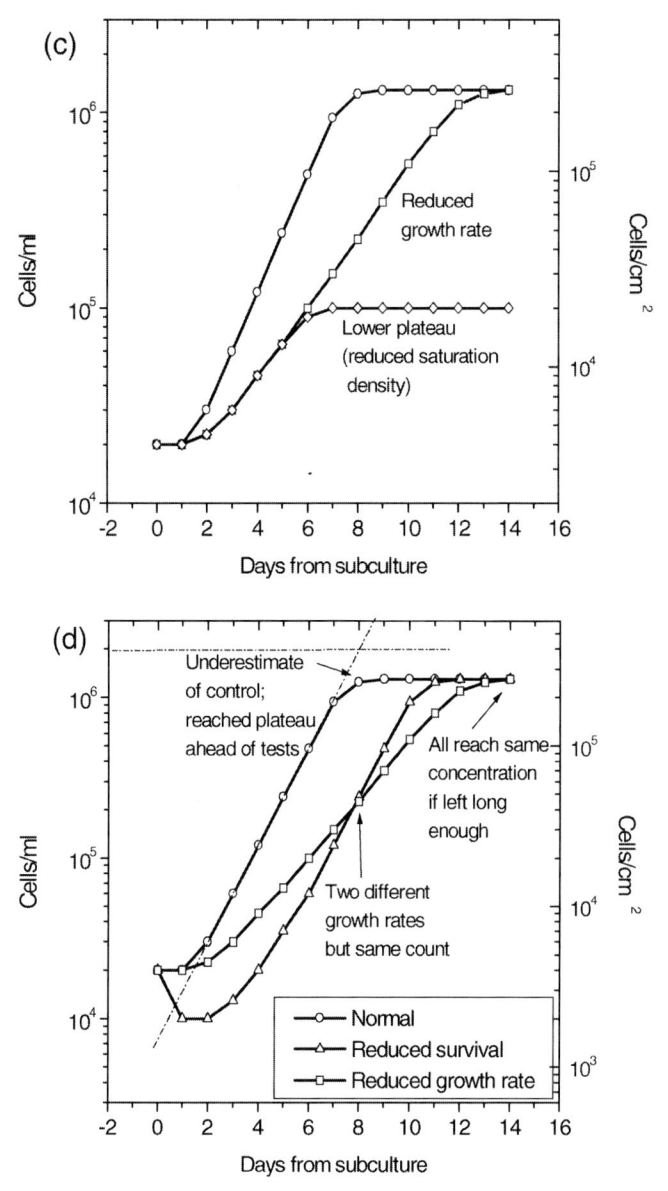

HeLa. Many more cross-contaminations are yet to be detected.
To avoid cross-contamination:

- do not share bottles of media or reagents among cell lines
- do not return a pipette, which has been in or near a flask or bottle containing cells, back to the medium bottle; use a fresh pipette
- do not share medium among operatives
- handle one cell line at a time

4.3.7 Instability and preservation

Early passage cell lines are unstable as they go through a period of adaptation to culture. However, between about the 5th and 35th generation (for human diploid fibroblasts; other cell types may be different) the culture is fairly consistent. As the culture will start to senesce as it gets older, finite cell lines should be used after the period of adaptation but before senescence. As continuous cell lines can be genetically unstable they should only be used continuously for approximately three months, before stock replacement. Some cell lines, such as 3T3 and other mouse lines which are immortal but not transformed, can transform spontaneously, and should not be propagated for prolonged periods.

Validated and authenticated frozen stocks of all cell lines should be maintained to protect against cell line instability, and to give insurance against contamination, incubator failure, or other accidental loss (see Chapter 3).

Animal cell culture is a precise discipline. Beware of those who say it is not, or it is 'magic', or due to 'green fingers'; they are not controlling all the variables. Consistency can be achieved, and the following chapters are intended to indicate how best to control conditions within the present limitations of our knowledge.

References

1. Doyle, A., Griffiths, J. B., and Newall, D. G. (ed.) (1990). *Cell and tissue culture: laboratory procedures.* Wiley, Chichester.
2. Davis, J. M. (ed.) (1994). *Basic cell culture: a practical approach.* IRL Press at Oxford University Press, Oxford.
3. Freshney, R. I. (2000). *Culture of animal cells: a manual of basic technique*, 4th edn. Wiley-Liss, New York.
4. Freshney, R. I. (1999). *Freshney's culture of animal cells, a multimedia guide.* Wiley-Liss, New York.
5. Boyce, S. T. and Ham, R. G. (1983). *J. Invest. Dermatol.*, **81** (Suppl 1), 33s.
6. Rooney, S. A., Young, S. L., and Mendelson, C. R. (1994). *FASEB J.*, **8**, 957.
7. McCormick, C. and Freshney, R. I. (2000). *Br. J. Cancer*, **82**, 881.
8. Speirs, V., Ray, K. P., and Freshney, R. I. (1991). *Br. J. Cancer*, **64**, 693.
9. Thomson, A. A., Foster, B. A., and Cunha, G. R. (1997). *Development*, **124**, 2431.
10. Yevdokimova, N. and Freshney, R. I. (1997). *Br. J. Cancer*, **76**, 261.
11. Berdichevsky, F., Gilbert, C., Shearer, M., and Taylor-Papadimitriou, J. (1992). *J. Cell Sci.*, **102**, 437.

12. Frame, M. C., Freshney, R. I., Vaughan, P. F., Graham, D. I., and Shaw, R. (1984). *Br. J. Cancer*, **49**, 269.
13. Sporn, M. B. and Roberts, A. B. (1983). *Cancer Res.*, **43**, 3034.
14. Friend, C., Scher, W., Holland, J. G., and Sato, T. (1971). *Proc. Natl. Acad. Sci. USA*, **68**, 378.
15. Raff, M. C., Miller, R. H., and Noble, M. (1983). *Nature*, **303**, 390.
16. Lechner, J. F., McClendon, I. A., Laveck, M. A., Shamsuddin, A. M., and Harris, C. C. (1983). *Cancer Res.*, **43**, 5915.
17. Chambard, M., Verrier, B., Gabrion, J., and Mauchamp, J. (1983). *J. Cell Biol.*, **96**, 1172.
18. Sattler, C. A., Michalopoulos, G., Sattler, G. L., and Pitot, H. C. (1978). *Cancer Res.*, **38**, 1539.
19. Freshney, R. I. (1992). *Culture of epithelial cells.* Wiley-Liss, New York.
20. Ghigo, D., Priotto, C., Migliorino, D., Geromin, D., Franchino, C., Todde, R., *et al.* (1998). *J. Cell. Physiol.*, **174**, 99.
21. Pereira-Smith, O. M. and Smith, J. (1988). *Proc. Natl. Acad. Sci. USA*, **85**, 6042.
22. Sasaki, M., Honda, T., Yamada, H., Wake, N., Barrett, J. C., and Oshimura, M. (1994). *Cancer Res.*, **54**, 6090.
23. Bodnar, A. G., Ouellette, M., Frolkis, M., Holt, S. E., Chiu, C.-P., Morin, G. B., *et al.* (1998). *Science*, **279**, 349.
24. Grander, D. (1998). *Med. Oncol.*, **15**, 20.
25. Freshney, R. I. and Freshney, M. G. (ed.) (1994). *Culture of immortalized cells.* Wiley-Liss, New York.
26. Courtenay, V. D., Selby, P. J., Smith, I. E., Mills, J., and Peckham, M. J. (1978). *Br. J. Cancer*, **38**, 77.
27. Freshney, M. G. (1994). In *Culture of hematopoietic cells* (ed. R. I. Freshney, I. B. Pragnell, and M. G. Freshney). Wiley-Liss, New York p. 265.
28. Vlodavsky, I., Lui, G. M., and Gospodarowicz, D. (1980). *Cell*, **19**, 607.
29. Nicosia, R. F. and Ottinetti, A. (1990). *In Vitro*, **26**, 119.
30. Ham, R. G. and McKeehan, W. L. (1978). *In Vitro*, **14**, 11.
31. Jacobs, J. P., Jones, C. M., and Baille, J. P. (1970). *Nature*, **227**, 168.
32. Green, H. and Kehinde, O. (1974). *Cell*, **1**, 113.
33. Macpherson, I. and Stoker, M. (1962). *Virology*, **16**, 147.
34. Gaush, C. R., Hard, W. L., and Smith, T. F. (1966). *Proc. Soc. Exp. Biol. Med.*, **122**, 931.
35. Graham, F. L., Smiley, J., Russell, W. C., and Nairn, R. (1977). *J. Gen. Virol.*, **36**, 59.
36. Knowles, B. B., Howe, C. C., and Aden, D. P. (1980). *Science*, **209**, 497.
37. Gey, G. O., Coffman, W. D., and Kubicek, M. T. (1952). *Cancer Res.*, **12**, 364.
38. Puck, T. T. and Marcus, P. I. (1955). *Proc. Natl. Acad. Sci. USA*, **41**, 432.
39. Gluzman, Y. (1981). *Cell*, **23**, 175.
40. Fogh, J., Wright, W. C., and Loveless, J. D. (1977). *J. Natl. Cancer Inst.*, **58**, 209.
41. Soule, H. D., Maloney, T. M., Wolman, S. R., Peterson, W. D., Brenz, R., McGrath, C. M., *et al.* (1990). *Cancer Res.*, **50**, 6075.
42. Giard, D. J., Aaronson, S. A., Todaro, G. J., Arnstein, P., Kersey, J. H., Dosik, K., *et al.* (1973). *J. Natl. Cancer Inst.*, **51**, 1417.
43. Greene, L. A. and Tischler, A. S. (1976). *Proc. Natl. Acad. Sci. USA*, **73**, 2424.
44. Gallagher, R., Collins, S., Trujillo, J., McCredie, K., Ahearn, M., Tsai, S., *et al.* (1979). *Blood*, **54**, 713.
45. Gillis, S. and Watson, J. (1980). *J. Exp. Med.*, **152**, 1709.
46. Andersson, L. C., Nilsson, K., and Gahmberg, C. G. (1979). *Int. J. Cancer*, **23**, 143.
47. Andersson, L. C., Jokinen, M., and Gahmberg, C. G. (1979). *Nature*, **278**, 364.
48. Moore, G. E., Gerner, R. E., and Franklin, H. A. (1967). *J. Am. Med. Assoc.*, **199**, 519.

49. Eagle, H. (1959). *Science*, **130**, 432.
50. Dulbecco, R. and Freeman, G. (1959). *Virology*, **8**, 396.
51. Ham, R. G. (1965). *Proc. Natl. Acad. Sci. USA*, **53**, 288.
52. Barnes, D. and Sato, G. (1980). *Anal. Biochem.*, **102**, 255.
53. Leibovitz, A. (1963). *Am. J. Hyg.*, **78**, 173.
54. Peehl, D. M. and Ham, R. G. (1980). *In Vitro*, **16**, 526.
55. Stacey, G. N., Masters, J. R. W., Hay, R. J., Drexler, H. G., MacLeod, R. A. F. and Freshney, R. I. (2000). Cell contamination leads to inaccurate data: we must take action now. *Nature*, 356.

Chapter 2
Scaling-up of animal cell cultures

Bryan Griffiths

5 Bourne Gardens, Porton, Salisbury SP4 0NU, U.K.

1 Introduction

Small scale culture of cells in flasks of up to 1 litre volume (175 cm^2 surface area) is the best means of establishing new cell lines in culture, for studying cell morphology and for comparing the effects of agents on growth and metabolism. However, there are many applications in which large numbers of cells are required, for example extraction of a cellular constituent (10^9 cells can provide 7 mg DNA); to produce viruses for vaccine production (typically 5 × 10^{10} cells per batch) or other cell products (interferon, plasminogen activator, interleukins, hormones, enzymes, erythropoietin, and antibodies); and to produce inocula for even larger cultures. Animal cell culture is a widely used production process in biotechnology, with systems in operation at scales over 10 000 litres. This has been achieved by graduating from multiples of small cultures, an approach which is tedious, labour-intensive, and expensive, to the use of large 'unit process' systems. Although unit processes are more cost-effective and efficient, achieving the necessary scale-up has required a series of modifications to overcome limiting factors such as oxygen limitation, shear damage, and metabolite toxicity. One of the aims of this chapter is to describe these limitations, indicate at what scale they are likely to occur, and suggest possible solutions. This theme has to be applied to cells that grow in suspension and also to those that will only grow when attached to a substrate (anchorage-dependent cells). Another aspect of scale-up that will be discussed is increasing unit cell density 50- to 100-fold by the use of cell immobilization and perfusion techniques.

Free suspension culture offers the easiest means of scale-up because a 1 litre vessel is conceptually very similar to a 1000 litre vessel. The changes concern the degree of environmental control and the means of maintaining the correct physiological conditions for cell growth, rather than significantly altering vessel design. Monolayer systems (for anchorage-dependent cells) are more difficult to scale-up in a single vessel, and consequently a wide range of diverse systems have evolved. The aim of increasing the surface area available to the cells and the total culture volume has been successfully achieved, and the most effective of these methods (microcarrier) will be described (Section 3.3.4).

2 General methods and culture parameters

Familiarity with certain biological concepts and methods is essential when understanding scale-up of a culture system. In small scale cultures there is leeway for some error. If the culture fails it is a nuisance, but not necessarily a disaster. Large scale culture failure is not only more serious in terms of cost, but also the system demands that conditions are more critically met. This section describes the factors that need to be considered as the culture size gets larger.

2.1 Cell quantification

Measuring total cell numbers (by haemocytometer counts of whole cells or stained nuclei) and total cell mass (by determining protein or dry weight) is easily achieved. It is far more difficult to get a reliable measure of cell viability because the methods employed either stress the cells or use a specific, and not necessarily typical, parameter of cell physiology. An additional difficulty is that in many culture systems the cells cannot be sampled (most anchorage-dependent cultures), or visually examined, and an indirect measurement has to be made.

2.1.1 Cell viability

The dye exclusion test is based on the concept that viable cells do not take up certain dyes, whereas dead cells are permeable to these dyes. Trypan blue (0.4%) is the most commonly used dye, but has the disadvantage of staining soluble protein. In the presence of serum, therefore, erythrocin B (0.4%) is often preferred. Cells are counted in the standard manner using a haemocytometer. Some caution should be used when interpreting results as the uptake of the dye is pH- and concentration-dependent, and there are situations in which misleading results can be obtained. Two relevant examples are membrane leakiness caused by recent trypsinization and freezing and thawing in the presence of dimethyl sulfoxide. A colorimetric method using the MTT assay (see Chapter 7) can be used both to measure viability after release of cytoplasmic contents into the medium from artificially lysed cells, and for microscopic visualization within the attached cell [1].

2.1.2 Indirect measurements

Indirect measurements of viability are based on metabolic activity. The most commonly used parameter is glucose utilization, but oxygen utilization, lactic or pyruvic acid production, or carbon dioxide production can also be used, as can the expression of a product, such as an enzyme. When cells are growing logarithmically, there is a very close correlation between nutrient utilization and cell numbers. However, during other growth phases, utilization rates, caused by maintenance rather than growth, can give misleading results. The measurements obtained can be expressed as a growth yield (Y) or specific utilization/respiration rate (Q):

$$\text{Growth yield (Y)} = \frac{\text{change in biomass concentration (dx)}}{\text{change in substrate concentration (ds)}} \qquad 1$$

Specific utilization/respiration rate $(Q_A) =$

$$\frac{\text{change in substrate concentration (ds)}}{\text{time (dt)} \times \text{cell mass/numbers (dx)}} \qquad 2$$

Typical values of growth yields for glucose (10^6 cells/g) are 385 (MRC-5), 620 (Vero), and 500 (BHK).

A method which is not so influenced by growth rate fluctuations is the lactate dehydrogenase (LDH) assay. LDH is measured in cell-free medium at 30°C by following the oxidation of NADH by the change in absorbance at 340 nm. The reaction is initiated by the addition of pyruvate (2). One unit of activity is defined as 1 μmol/min NADH consumed. LDH is released by dead/dying cells and is therefore a quantitative measurement of loss of cell viability. To measure viable cells a reverse assay can be performed by controlled lysis of the cells and measuring the increase in LDH.

2.2 Equipment and reagents

2.2.1 Culture vessel and growth surfaces

The standard non-disposable material for growth of animal cells is glass, although this is replaced by stainless steel in larger cultures. It is preferable to use borosilicate glass (e.g. Pyrex) because it is less alkaline than soda glass and withstands handling and autoclaving better. Cells usually attach readily to glass but, if necessary, attachment may be augmented by various surface treatments (see Section 3.2). In suspension culture, cell attachment has to be discouraged, and this is achieved by treatment of the culture vessel with a proprietary silicone preparation (siliconization). Examples are Dow Corning 1107 (which has to be baked on) or dimethyldichlorosilane (Repelcote, Hopkins and Williams) which requires thorough washing of the vessel in distilled water to remove the trichloroethane solvent.

Complex systems use a combination of stainless steel and silicone tubing to connect various components of the system. Silicone tubing is very permeable to gases, and loss of dissolved carbon dioxide can be a problem. It is also liable to rapid wear when used in a peristaltic pump. Thick-walled tubing with additional strengthening (sleeve) should be used. Custom-made connectors should be used to ensure good aseptic connection during process operation. (These are available from all fermenter supply companies; see Appendix.)

Safe removal of samples of the culture at frequent intervals is essential. An entry with a vaccine stopper through which a hypodermic syringe can be inserted provides a simple solution, but is only suitable for small cultures.

Repeated piercing of the vaccine stopper can lead to a loss of culture integrity. The use of specialized sampling devices, also available from fermenter supply companies, is recommended. These automatically enable the line to be cleared of static medium containing dead cells, thus avoiding the necessity of taking small initial samples which are then discarded, and increasing the chances of retaining sterility.

Air filters are required for the entry and exit of gases. Even if continuous gassing is not used, one filter entry is usually needed to equilibrate pressure and for forced input or withdrawal of medium. The filters should have a 0.22 μm rating and be non-wettable.

2.2.2 Non-nutritional medium supplements

Sodium carboxymethylcellulose (15–20 centrepoise, units of viscosity) is often added to media (at 0.1%) to help minimize mechanical damage to cells caused by shear forces generated by the stirrer impeller, forced aeration, or perfusion. This compound is more soluble than methylcellulose, and has a higher solubility at 4°C than at 37°C.

Pluronic F-68 (trade name for polyglycol) (BASF, Wyandot) is often added to media (at 0.1%) to reduce the amount of foaming that occurs in stirred and/or aerated cultures, especially when serum is present. It is also helpful in reducing cell attachment to glass by suppressing the action of serum in the attachment process. However, its most beneficial action is to protect cells from shear stress and bubble damage in stirred and sparged cultures, and it is especially effective in low serum or serum-free media.

2.3 Practical considerations

(a) Temperature of medium. Always pre-warm the medium to the operating temperature (usually 37°C) and stabilize the pH before adding cells. Shifts in pH during the initial stages of a culture create many problems, including a long lag phase and reduced yield.

(b) Growth phase of cells. Avoid using stationary phase cells as an inoculum since this will mean a long lag phase, or no growth at all. Ideally, cells in the late logarithmic phase should be used.

(c) Inoculation density. Always inoculate at a high enough cell density. There is no set rule as to the minimum inoculum level below which cells will not grow, as this varies between cell lines and depends on the complexity of the medium being used. As a guide, it may be between 5×10^4 and 2×10^5 cells/ml, or 5×10^3 and 2×10^4 cells/cm^2.

(d) Stirring rate. Find empirically the optimum stirring rate for a given culture vessel and cell line. This could vary between 100–500 r.p.m. for suspension cells, but is usually in the range 200–350 r.p.m., and between 20–100 r.p.m. for microcarrier cultures.

(e) Medium and surface area. The productivity of the system depends upon the quality and quantity of the medium and, for anchorage-dependent cells, the surface area for cell growth. A unit volume of medium is only capable of giving a finite yield of cells. Factors which affect the yield are: pH, oxygen limitation, accumulation of toxic products (e.g. NH_4), nutrient limitation (e.g. glutamine), spatial restrictions, and mechanical/shear stress. As soon as one of these factors comes into effect, the culture is finished and the remaining resources of the system are wasted. The aim is, therefore, to delay the onset of any one factor until the accumulated effect causes cessation of growth, at which point the system has been maximally utilized. Simple ways of achieving this are: a better buffering system (e.g. Hepes instead of bicarbonate), continuous gassing, generous headspace volume, enriched rather than basal media, with nutrient-sparing supplements such as non-essential amino acids or lactalbumin hydrolysate, perfusion loops through ultrafiltration membranes or dialysis tubing for detoxification (3) and oxygenation, and attention to culture and process design.

2.4 Growth kinetics

The standard format of a culture cycle beginning with a lag phase, proceeding through the logarithmic phase to a stationary phase and finally to the decline and death of cells, is well documented. Although cell growth usually implies increase in cell numbers, increase in cell mass can occur without any replication. The difference in mean cell mass between cell populations is considerable, as would be expected, but so is the variation within the same population.

Growth (increase in cell numbers or mass) can be defined in the following terms.

(a) Specific growth rate, μ (i.e. the rate of growth per unit amount (weight/numbers) of biomass):

$$\mu = (1/x)(dx/dt) h^{-1} \qquad 3$$

where dx is the increase in cell mass, dt is the time interval, and x is the cell mass. If the growth rate is constant (e.g. during logarithmic growth), then

$$\ln x = \ln x_o + \mu t \qquad 4$$

where x_o is the biomass at time t_o.

(b) Doubling time, td (i.e. the time for a population to double in number/mass):

$$t_d = \frac{\ln 2}{\mu} = \frac{0.693}{\mu} \qquad 5$$

(c) Degree of multiplication, n or number of doublings (i.e. the number of times the inoculum has replicated):

$$n = 3.32 \log(x/x_0) \qquad 6$$

2.5 Medium and nutrients

A given concentration of nutrients can only support a certain number of cells. Alternative nutrients can often be found by a cell when one becomes exhausted, but this is bad practice because the growth rate is always reduced (e.g. while alternative enzymes are being induced). If a minimal medium, such as Eagle's basal medium (BME) or minimum essential medium (MEM) is used with serum as the only supplement, then this problem is going to be met sooner than in cultures using complete media (e.g. 199), or in media supplemented with lactalbumin hydrolysate, peptone, or BSA (which provides many of the fatty acids). Nutrients likely to be exhausted first are glutamine, partly because it spontaneously cyclizes to pyrrolidone carboxylic acid and is enzymically converted (by serum and cellular enzymes) to glutamic acid, leucine, and isoleucine. Human diploid cells are almost unique in utilizing cystine heavily. A point to remember is that nutrients become growth-limiting before they become exhausted. As the concentration of amino acids falls, the cell finds it increasingly difficult to maintain sufficient intracellular pool levels. This is exaggerated in monolayer cultures because as cells become more tightly packed together, the surface area available for nutrient uptake becomes smaller (4).

Glucose is often another limiting factor as it is destructively utilized by cells and, rather than adding high concentrations at the beginning, it is more beneficial to supplement after two to three days. In order to maintain a culture, some additional feeding often has to be carried out either by complete, or partial, media changes or by perfusion.

Many cell types are either totally dependent upon certain additives or can only perform optimally when they are present. For many purposes it is highly desirable, or even essential, to reduce the serum level to 1% or below. In order to achieve this without a significant reduction in cell yield, various growth factors and hormones are added to the basal medium. The most common additives are insulin (5 mg/litre), transferrin (5–35 mg/litre), ethanolamine (20 µM), and selenium (5 µg/litre) (5).

Cell aggregation is often a problem in suspension cultures. Media lacking calcium and magnesium ions have been designed specifically for suspension cells because of the role of these ions in attachment (see Section 3.1). This problem has also been overcome by including very low levels of trypsin in the medium (2 µg/ml).

2.6 pH

Ideally pH should be near 7.4 at the initiation of a culture and not fall below a value of 7.0 during the culture, although many hybridoma lines appear to prefer a pH of 7.0 or lower. A pH below 6.8 is usually inhibitory to cell growth. Factors affecting the pH stability of the medium are buffer capacity and type, headspace, and glucose concentration.

The normal buffer system in tissue culture media is the carbon dioxide bicarbonate system analogous to that in blood. This is a weak buffer system, in that it has a pK_a well below the physiological optimum. It also requires the addition of carbon dioxide to the headspace above the medium to prevent the loss of carbon dioxide and an increase in hydroxyl ions. The buffering capacity of the medium is increased by the phosphates present in the balanced salt solution (BSS). Medium intended to equilibrate with 5% carbon dioxide usually contain Earle's BSS (25 mM NaHCO$_3$) but an alternative is Hanks' BSS (4 mM NaHCO$_3$) for equilibration with air. Improved buffering and pH stability in media is possible by using a zwitterionic buffer (6), such as Hepes (10–20 mM), either in addition to, or instead of, bicarbonate (include bicarbonate at 0.5 mM). Alternative buffer systems are provided by using specialist media such as Leibovitz's L-15 (7). This medium utilizes the buffering capacity of free amino acids, substitutes galactose and pyruvate for glucose, and omits sodium bicarbonate. It is suitable for open cultures.

The headspace volume in a closed culture is important because, in the initial stages of the culture, 5% carbon dioxide is needed to maintain a stable pH in the medium but, as the cells grow and generate carbon dioxide, it builds up in the headspace and this prevents it diffusing out of the medium. The result is an increase in weakly dissociated bicarbonate producing an excess of hydrogen ions in the medium and a fall in pH. Thus, a large headspace is required in closed cultures, typically tenfold greater than the medium volume (this volume is also needed to supply adequate oxygen). This generous headspace is not possible as cultures are scaled-up, and an open system with a continuous flow of air, supplied through one filter and extracted through another, is required.

The metabolism of glucose by cells results in the accumulation of pyruvic and lactic acids. Glucose is metabolized at a far greater rate than it is needed. Thus, glucose should ideally never be included in media at concentrations above 2 g/litre, and it is better to supplement during the culture than to increase the initial concentration. An alternative is to substitute glucose by galactose or fructose as this significantly reduces the formation of lactic acid, but it usually results in a slower growth rate. These precautions delay the onset of a non-physiological pH and are sufficient for small cultures. As scale-up increases, headspace volume and culture surface area in relation to the medium volume decrease. Also, many systems are developed in order to increase the surface area for cell attachment and cell density per unit volume. Thus pH problems occur far earlier in the culture cycle because carbon dioxide cannot escape as readily, and more cells means higher production of lactic acid and carbon dioxide. The answer is to carry out frequent medium changes or use perfusion, or have a pH control system.

The basis of a pH control system is an autoclavable pH probe (available from Pye Ingold, Russell). This feeds a signal to the pH controller which is converted to give a digital or analogue display of the pH. This is a pH monitor system. Control of pH requires the defining of high and low pH values beyond which the pH should not go. These two set points on the pH scale turn on a relay to activate a pump, or solenoid valve, which allows additions of acid or alkali to be made to the culture to bring it back to within the allowable units. It is rare to have the pH rise above the set point once the medium has equilibrated with the headspace, so the addition of acid can be disregarded. If an alkali is to be added, then sodium bicarbonate (5.5%) is recommended. Sodium hydroxide (0.2 M) can be used only with a fast stirring rate, which dilutes the alkali before localized concentrations can damage the cells, or if a perfusion loop is installed. Normally, liquid delivery pumps are supplied as part of the pH controller. Gas supply is controlled by a solenoid valve and 95% air directly. Above the set point carbon dioxide will be mixed with the air, but below this point only air will be delivered to the culture. This in itself is a controlling factor in that it helps remove carbon dioxide as well as meeting the oxygen requirements of the cultures. pH regulation is very readily adapted to computer control systems.

2.7 Oxygen

The scale-up of animal cell cultures is very dependent upon the ability to supply sufficient oxygen without causing cell damage. Oxygen is only sparingly soluble in culture media (7.6 µg/ml) and a survey of reported oxygen utilization rates by cells (8) reveals a mean value of 6 µg/10^6 cells/h. A typical culture of 2×10^6 cells/ml would, therefore, deplete the oxygen content of the medium (7.6 µg/ml) in under 1 h. It is necessary to supply oxygen to the medium throughout the life of the culture and the ability to do this adequately depends upon the oxygen transfer rate (OTR) of the system:

$$OTR = Kla\,(C^* - C) \qquad 7$$

where OTR is the amount of oxygen transferred per unit volume in unit time, Kl is the oxygen transfer coefficient, and a is the area of the interface across which oxygen transfer occurs (as this can only be measured in stationary and surface-aerated cultures, the value Kla is used; this is the mass transfer coefficient (vol/h). C^* is the concentration of dissolved oxygen when medium is saturated, and C is the actual concentration of oxygen at any given time.

The Kla (OTR/C when C = 0) is in units of h^{-1} hours^{-1} and is thus a measure of the time taken to oxygenate a given culture vessel completely under a particular set of conditions.

A culture can be aerated by one, or a combination, of the following methods: surface aeration, sparging, membrane diffusion, medium perfusion, increasing the partial pressure of oxygen, and increasing the atmospheric pressure (9).

2.7.1 Surface aeration: static cultures

In a closed system, such as a sealed flask, the important factors are the amount of oxygen in the system and the availability of this oxygen to the cells growing under 3–6 mm of medium. Normally a headspace/medium volume ratio of 10:1 is used in order to provide sufficient oxygen. Thus a 1 litre flask (e.g. a Roux bottle) with 900 ml of air and 100 ml of medium will initially contain 0.27 g oxygen (*Table 1*). This amount will support 10^8 cells for 450 h and is thus clearly adequate. The second factor is whether this oxygen can be made available to the cells. The transfer rate of oxygen from the gas phase into a liquid phase has been calculated at about 17 µg/cm^2/h (8). Again, this is well in excess of that required by cells in a 1 litre flask. However, if the surface is assumed to be

SCALING-UP OF ANIMAL CELL CULTURES

Table 1. Oxygen concentrations on the gas and liquid phases of a Roux bottle culture

Oxygen in 900 ml air:
900 × 0.21 × 32/22 400 = 0.27 g

Oxygen in 100 ml medium:
100 × 7.6 × 10^{-4} = 0.0076 g

Notes:
0.21 = proportion of oxygen in air;
32 = molecular weight of oxygen;
22 400 = gram molecular volume;
× 10^{-4} = solubility of oxygen in water at 37 °C when equilibrated with air.

saturated with dissolved oxygen and the concentration at the cell sheet is almost zero then, applying Fick's law of diffusion, the rate at which oxygen can diffuse to the cells is about 1.5 µg/cm^2/h. At this rate there is only sufficient oxygen to support about 50 × 10^6 cells in a 1 litre flask, a cell density which in practice many tissue culturists take as the norm. These calculations show the importance of maintaining a large headspace volume; otherwise, oxygen limitation could become one of the growth-limiting factors in static closed cultures. A fuller explanation of these calculations, including Fick's law of diffusion, can be found in a review by Spier and Griffiths (8).

2.7.2 Sparging

This is the bubbling of gas through the culture, and is a very efficient means of effecting oxygen transfer (as proven in bacterial fermentation). However, it may be damaging to animal cells due to the effect of the high surface energy of the bubble on the cell membrane. This damaging effect can be minimized by using large air bubbles (which have lower surface energies than small bubbles), by using a very low gassing rate (e.g. 5 ml/1 min), and by adding Pluronic F-68. A specialized form of sparging, the airlift fermenter (see Section 4.7) has also been used in large unit process monolayer cultures (e.g. multiple plate propagators). When sparging is used, efficiency of oxygenation is increased by using a culture vessel with a large height/diameter ratio. This creates a higher pressure at the base of the reactor, which increases oxygen solubility.

2.7.3 Membrane diffusion

Silicone tubing is very permeable to gases, and if long lengths of thin-walled tubing can be arranged in the culture vessel then sufficient diffusion of oxygen into the culture can be obtained. However, a lot of tubing is required (e.g. 30 m of 2.5 cm tubing for a 1000 litre culture). This method is expensive and inconvenient to use, and has the inherent problem that scale-up of the tubing required is mainly two-dimensional while that of the culture is three-dimensional. However, several commercial systems are available (e.g. Braun).

2.7.4 Medium perfusion

A closed-loop perfusion system continuously (or on demand) takes medium from the culture, passes it through an oxygenation chamber, and returns it to the culture. This method has many advantages if the medium can be conveniently separated from the cells for perfusion through the loop. The medium in the chamber can be vigorously sparged to ensure oxygen saturation and other additions, such as sodium hydroxide for pH control, which would damage the cells if put directly into the culture, can be made. This method is used in glass bead systems (Section 3.3.4) and has proved particularly effective in microcarrier systems (Section 3.3.4), where specially modified spin filters can be used.

2.7.5 Environmental supply

The dissolved oxygen concentration can be increased by increasing the headspace pO_2 (from atmospheric 21% to any value, using oxygen and nitrogen mixtures) and by raising the pressure of the culture by 100 kPa (about 1 atm) (which increases the solubility of oxygen and its diffusion rate). These methods should be employed only when the culture is well advanced, otherwise oxygen toxicity could occur. Finally, the geometry of the stirrer blade also affects the oxygen transfer rate.

2.7.6 Scale-up

Oxygen limitation is usually the first factor to be overcome in culture scale-up. This becomes a problem in conventional stirred cultures at volumes above 10 litres. However, with the current use of high density cultures maintained by perfusion, oxygen limitation can occur in a 2 litre culture. The relative effectiveness of some of the alternative oxygenation systems in large scale bioreactors is shown in *Table 2*, and the range of oxygenation procedures in various types of culture vessels has been reviewed (9, 10).

Table 2. Methods of oxygenating a 40 litre bioreactor (30 litre working volume with a 1.5:1 aspect ratio)

Oxygenating method	Oxygen delivery (mg/litre/h)	No. cells $\times 10^6$/ml supported
Air (10 ml/min at 40 r.p.m.)		
Surface aeration	0.5	0.08
Direct sparging	4.6	0.76
Spin filter sparging	3.0	0.40
Perfusion (1 vol/h)	12.6	2.10
Perfusion (1 vol/h) + spin filter sparging	15.9	2.65
Oxygen (10 ml/min at 80 r.p.m.)		
Spin filter sparging	51.0	8.50
+ perfusion (1 vol/h)	92.0	15.00

(assuming oxygen utilization rate of 2–6 µg/10^6 cells/h).

2.7.7 Redox potential

The oxidation-reduction potential (ORP), or redox potential, is a measure of the charge of the medium and is affected by the proportion of oxidative and reducing chemicals, the oxygen concentration, and the pH. When fresh medium is prepared and placed in the culture vessel it takes time for the redox potential to equilibrate, a phenomenon known as poising. The optimum level for many cell lines is +75 mV, which corresponds to a dissolved pO_2 of 8–10%. Some investigators find it beneficial to control the oxygen supply to the culture by means of redox, rather than an oxygen electrode. Alternatively, if the redox potential is monitored by means of a redox electrode and pH meter (with mV display), then an indication of how cell growth is progressing can be obtained (11). This is because the redox value falls during logarithmic growth and reaches a minimum value approximately 24 h before the onset of the stationary phase (*Figure 1*). This provides a useful guide to cell growth in cultures where cell sampling is not possible. It is also useful to be able to predict the end of the logarithmic growth phase so that medium changes, addition of virus, or product promoters can be given at the optimum time. The effect of redox potential on cell cultures has been reviewed by Griffiths (12).

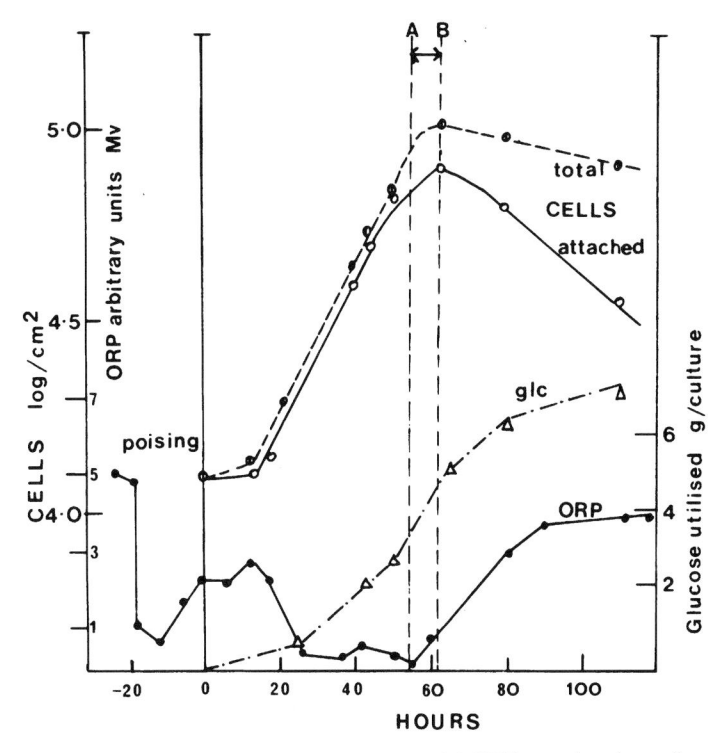

Figure 1. Changes in oxygen-reduction potential (ORP) correlated to cell growth and glucose (glc) utilization (11). The minimum ORP value (A) is reached 24 h before the end of logarithmic growth indicated by the maximum cell number (B). The upper curve shows the total number of cells, while the lower curve shows the number of attached cells. Further explanation is given in the text.

2.8 Types of culture process

2.8.1 Batch and continuous culture

In standard culture, known as *batch culture*, cells are inoculated into a fixed volume of medium and, as they grow, nutrients are consumed and metabolites accumulate. The environment is therefore continually changing, and this in turn enforces changes to cell metabolism, often referred to as physiological differentiation. Eventually cell multiplication ceases because of exhaustion of nutrient(s), accumulation of toxic waste products, or density-dependent limitation of growth in monolayer cultures. There are means of prolonging the life of a batch culture, and thus increasing the yield, by various substrate feed methods.

(a) Gradual addition of fresh medium, so increasing the volume of the culture (fed batch).

(b) Intermittently, by replacing a constant fraction of the culture with an equal volume of fresh medium (semi-continuous batch). All batch culture systems retain the accumulating waste products, to some degree, and have a fluctuating environment. All are suitable for both monolayer and suspension cells.

(c) Perfusion, by the continuous addition of medium to the culture and the withdrawal of an equal volume of used (cell-free) medium. Perfusion can be open, with complete removal of medium from the system, or closed, with recirculation of the medium, usually via a secondary vessel, back into the culture vessel. The secondary vessel is used to 'regenerate' the medium by gassing and pH correction.

(d) Continuous-flow culture, which gives true homeostatic conditions with no fluctuations of nutrients, metabolites, or cell numbers. It depends upon medium entering the culture with a corresponding withdrawal of medium plus cells. It is thus only suitable for suspension culture cells, or monolayer cells growing on microcarriers. Continuous-flow culture is described more fully in Section 4.6.

2.8.2 Comparison of batch, perfusion, and continuous-flow culture (*Figure 2*)

Continuous-flow culture is the only system in which the cellular content is homogeneous, and can be kept homogeneous for long periods of time (months). This can be vital for physiological studies, but may not be the most economical method for product generation. Production economics are calculated in terms of staff time, medium, equipment, and downstream processing costs. Also taken into account are the complexity and sophistication of the equipment and process, as this affects the calibre of the staff required and the reliability of the production process. Batch culture is more expensive on staff time and culture ingredients, because for every single harvest a sequence of inoculum build-up steps and then growth in the final vessel has to be carried out, and there is also downtime whilst the culture is prepared for its next run. Feeding routines for

SCALING-UP OF ANIMAL CELL CULTURES

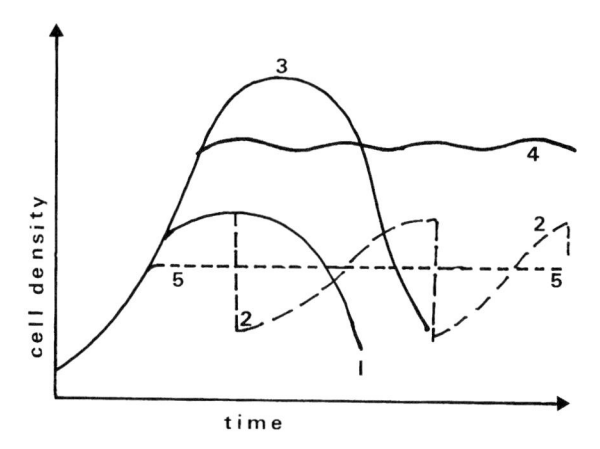

Figure 2. Comparison of culture processes. The numbers 1–5 are explained in the table below.

Culture type	Cell no. (millions)	Product yield per litre mg/week[a]	mg/month[a]
1 Batch	3	6	12
2 Semi-continuous batch	3	15	60
3 Fed batch	30	60	120
4 Continuous perfusion	10	70	280
5 Continuous flow	2	15	60

[a] Values allow for turn-around time of non-continuous cultures.

batch cultures can give repeated but smaller harvests, and the longer a culture can be maintained in a productive state then the more economical the whole process becomes.

Continuous-flow culture in the chemostat implies that cell yields are never maximal because a limiting growth factor is used to control the growth rate. If maximum yields are desired in this type of culture then the turbidostat option has to be used (see Section 4.6). Some applications, such as the production of a cytopathic virus, leave no choice other than batch culture. Maintenance of high yields, and therefore high product concentration, may be necessary to reduce downstream processing costs and these could outweigh medium expenses. For this purpose perfusion has to be used. Although for many processes this is more economical than batch culture, it does add to the complexity of the equipment and process, and increases the risk of a mechanical or electrical failure or microbial contamination prematurely ending the production run. There is no clear-cut answer to which type of culture process should be used—it depends upon the cell and product, the quantity of product, downstream processing problems, and product licensing regulations (batch definition of product, cell stability, and generation number). However, a relative ratio of unit costs for perfusion, continuous-flow, and batch culture in the production of monoclonal antibody is 1:2:3.5.

2.9 Summary of factors limiting scale-up

In the following sections a wide variety of alternative culture systems are described. The reason for this large number, apart from scale and different growth characteristics of cells, is the range of solutions which have been used to overcome limitations to scale-up. These will be briefly reviewed so that the underlying philosophy and evolution of the culture systems described in Sections 3–5 can be better understood (see refs 9 and 13).

As already explained, oxygen is the first limiting factor encountered in scale-up. To overcome this, reactors are available with membrane or hollow fibre devices to give bubble-free aeration, often by means of external medium loops. A related factor is mixing, and low-shear options such as airlift reactors and specially designed impellers have been developed. For anchorage-dependent cells the low unit surface area has limited effective scale-up, so a range of devices using plates, spirals, ceramics and, most effective of all, microcarriers, can be used.

The next problem to be overcome is nutrient limitation and toxic metabolite build-up. The most effective solution is to use perfusion by means of spin filters or hollow fibre loops. To maximize the benefit of perfusion, immobilization of suspension cells was needed to prevent cell wash-out, and this resulted in a collection of novel systems based on hollow fibres, membranes, encapsulation, and specialized matrices. Immobilization has many process advantages (see Section 5), and currently the effort is on scaling-up suitable entrapment matrices in fluidized and fixed beds.

3 Monolayer culture

3.1 Introduction

Tissue culture flasks and tubes giving surface areas of 5–200 cm^2 are familiar to all tissue culturists. The largest stationary flask routinely used in laboratories is the Roux bottle (or disposable plastic equivalent) which gives a surface area for cell attachment of 175–200 cm^2 (depending upon type), needs 100–150 ml medium, and utilizes 750–1000 cm^2 of storage space. This vessel will yield 2×10^7 diploid cells and up to 10^8 heteroploid cells. If one has to produce, for example, 10^{10} cells, then over 100 replicate cultures are needed (i.e. manipulations have to be repeated 100 times). In addition the cubic capacity of incubator space needed is over 100 litres. Clearly there comes a time in the scale-up of cell production when one has to use a more efficient culture system. Scale-up of anchorage-dependent cells reduces the number of cultures, is more efficient in the use of staff, and increases significantly the surface area/volume ratio. In order to do this a very wide and versatile range of tissue culture vessels and systems has been developed. Many of these are shown diagrammatically in *Figure 3*, and are described on the following pages. The methods with the most potential are those based on modifications to suspension culture systems because they allow a truly homogeneous unit process with enormous scale-up

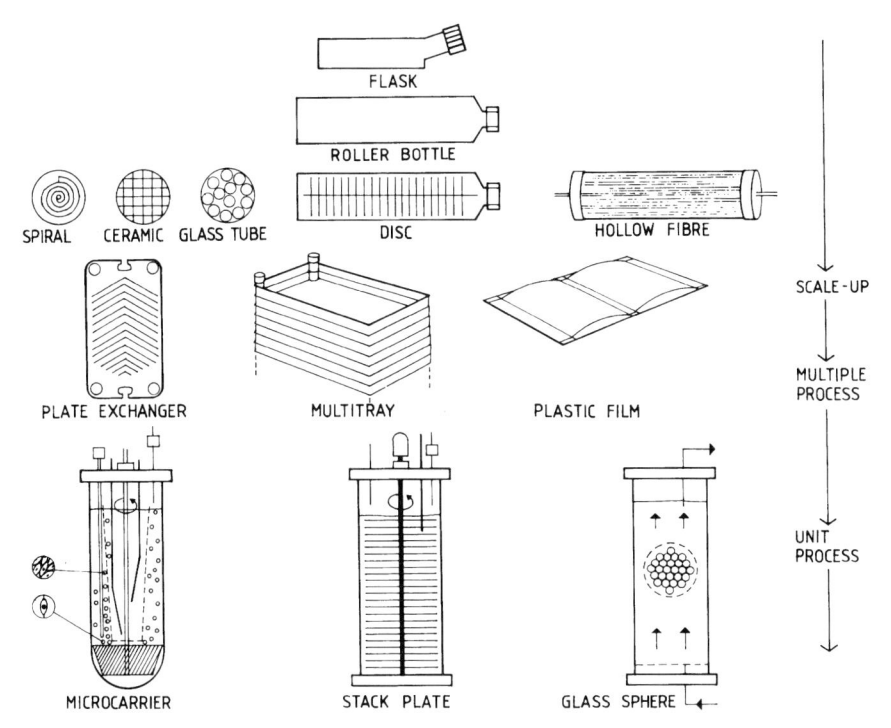

Figure 3. The scaling-up of culture systems for anchorage-dependent cells.

potential to be used. However, these systems should be attempted only if time and resources allow a lengthy development period.

Although suspension culture is the preferred method for increasing capacity, monolayer culture has the following advantages:

(a) It is very easy to change the medium completely and to wash the cell sheet before adding fresh medium. This is important in many applications when the growth is carried out in one set of conditions and product generation in another. A common requirement of a medium change is the transfer of cells from serum to serum-free conditions. The efficiency of medium changing in monolayer cultures is such that a total removal of the unwanted compound can be achieved.

(b) If artificially high cell densities are needed then these can be supported by using perfusion techniques. It is much easier to perfuse monolayer cultures because they are immobilized and a fine filter system (to withhold cells) is not required.

(c) Many cells will express a required product more efficiently when attached to a substrate.

(d) The same apparatus can be used with different media/cell ratios which, of course, can be easily changed during the course of an experiment.

(e) Monolayer cultures are more flexible because they can be used for all cell types. If a variety of cell types are to be used then a monolayer system might be a better investment.

It should be noted that the microcarrier system confers some of the advantages of a suspension culture system.

There are four main disadvantages of monolayer compared to suspension systems:

(a) They are difficult and expensive to scale-up.

(b) They require far more space.

(c) Cell growth cannot be monitored as effectively, because of the difficulty of sampling and counting an aliquot of cells.

(d) It is more difficult to measure and control parameters such as pH and oxygen, and to achieve homogeneity.

Microcarrier culture eliminates or at least reduces many of these disadvantages.

3.2 Cell attachment

Animal cell surfaces and the traditional glass and plastic culture surfaces are negatively charged, so for cell attachment to occur, crosslinking with glycoproteins and/or divalent cations (Ca^{2+}, Mg^{2+}) is required. The glycoprotein most studied in this respect is fibronectin, a compound of high molecular mass (220 000) synthesized by many cells and present in serum and other physiological fluids. Although cells can presumably attach by electrostatic forces alone, it has been found that the mechanism of attachment is similar, whatever the substrate charge (14). The important factor is the net negative charge, and surfaces such as glass and metal which have high surface energies are very suitable for cell attachment. Organic surfaces need to be wettable and negative, and this can be achieved by chemical treatment (e.g. oxidizing agents, strong acids) or physical treatment (e.g. high voltage discharge, UV light, high energy electron bombardment). One or more of these methods are used by manufacturers of tissue culture grade plastics. The result is to increase the net negative charge of the surface (for example by forming negative carboxyl groups) for electrostatic attachment.

Surfaces may also be coated to make them suitable for cell attachment. A tissue culture grade of collagen can be purchased which saves a tedious preparation procedure using rat tails. The usefulness of collagen as a growth surface is also demonstrated by the availability of collagen-coated microcarrier beads (Cytodex-3, Pharmacia).

3.2.1 Surfaces for cell attachment

i. Glass

Alum-borosilicate glass (e.g. Pyrex) is preferred because soda-lime glass releases alkali into the medium and needs to be detoxified (by boiling in weak acid) before use. After repeated use glassware can become less efficient for cell attachment, but efficiency can be regained by treatment with 1 mM magnesium acetate. After several hours soaking at room temperature the acetate is poured away and the glassware is rinsed with distilled water and autoclaved.

ii. Plastics

Polystyrene is the most used plastic for cell culture, but polyethylene, polycarbonate, Perspex, PVC, Teflon, cellophane, and cellulose acetate are all suitable when pre-treated correctly.

iii. Metals

Stainless steel and titanium are both suitable for cell growth because they are relatively chemically inert, but have a suitable high negative energy. There are many grades of stainless steel, and care has to be taken in choosing those which do not leak toxic metallic ions. The most common grade to use for culture applications is 316, but 321 and 304 may also be suitable. Stainless steel should be acid washed (10% nitric acid, 3.5% hydrofluoric acid, 86.5% water) to remove surface impurities and inclusions acquired during cutting.

3.3 Scaling-up

3.3.1 Step 1: roller bottle

The aims of scaling-up are to maximize the available surface area for growth and to minimize the volume of medium and headspace, while optimizing cell numbers and productivity. Stationary cultures have only one surface available for attachment and growth, and consequently they need a large medium volume. The medium volume can be reduced by rocking the culture or, more usually, by rolling a cylindrical vessel. The roller bottle (see *Protocol 1*) has nearly all its internal surface available for cell growth, although only 15-20% is covered by medium at any one time. Plastic disposable bottles are available in *c.* 750 cm^2 and *c.* 1500 cm^2 (1400-1750 cm^2) sizes. Rotation of the bottle subjects the cells to medium and air alternately, as compared with the near anaerobic conditions in a stationary culture. This method reduces the volume of medium required, but still requires a considerable headspace volume to maintain adequate oxygen and pH levels. The scale-up of a roller bottle requires that the diameter is kept as small as possible. The surface area can be doubled by doubling the diameter or the length. The first option increases the volume (medium and headspace) fourfold, the second option only twofold.

The only means of increasing the productivity of a roller bottle and decreasing its volume is by using a perfusion system, originally developed by Kruse *et al.* (15), and marketed by New Brunswick and Bellco (Autoharvester). This is an expensive option, as an intricate revolving connection has to be made for the supply lines to pass into the bottle. However, cell yields are considerably increased and extensive multilayering takes place.

Protocol 1
Use of standard disposable roller cultures

Equipment
- 1400 cm^2 (23 × 12 cm) plastic disposable bottle (e.g. Costar, Bibby Sterilin)
- Inverted microscope

Method
1. Add 300 ml of growth medium.
2. Add 1.5 × 10^7 cells.
3. Roll at 12 r.p.h. at 37°C for 2 h, to allow an even distribution of cells during the attachment phase.
4. Decrease the revolution rate to 5 r.p.h., and continue incubation.
5. Examine cells under an inverted microscope using an objective with a long working distance.
6. Harvest cells when visibly confluent (five to six days) by removing the medium, adding trypsin (0.25%), and rolling. Yields will be very similar to those obtained in flasks, assuming enough medium was added. This method has the advantage of allowing the medium volume/surface area ratio to be altered easily. Thus after a growth phase the medium volume can be reduced to get a higher product concentration.

3.3.2 Step 2: roller bottle modifications

The roller bottle system is still a multiple process, and thus inefficient in terms of staff resources and materials. To increase the surface area within the volume of a roller bottle, the following vessels have been developed (see also *Figure 3*).

i. SpiraCel

Bibby Sterilin have replaced their bulk cell culture vessel with a SpiraCel roller bottle. This is available with a spiral polystyrene cartridge in three sizes, 3000, 4500, and 6000 cm^2. It is crucial to get an even distribution of the cell inoculum throughout the spiral, otherwise very uneven growth and low yields are achieved. Cell growth can be visualized only on the outside of the spiral, and this can be misleading if the cell distribution is uneven.

ii. Glass tubes

A small scale example is the Bellco-Corbeil Culture System (Bellco). A roller bottle is packed with a parallel cluster of small glass tubes (separated by silicone spacer rings). Three versions are available giving surface areas of 5 × 10^3, 1 × 10^4, and 1.5 × 10^4 cm^2. Medium is perfused through the vessel from a reservoir. The method is ingenious in that it alternately rotates the bottle 360° clockwise and then 360° anticlockwise. This avoids the use of special caps for the supply of perfused medium.

An example of its use is the production of 3.2×10^9 Vero cells ($2.3 \times 10^5/cm^2$) over six days using 6.5 litres of medium (perfused at 50 ml/min) in the 10 000 cm² version.

iii. Increased surface area roller bottles

In place of the smooth surface in standard roller bottles the surface is 'corrugated', thus doubling the surface area within the same bottle dimensions; e.g. extended surface area roller bottle (ESRB) available from Bibby Sterilin (Corning), or the ImmobaSil surface which is a textured silicone rubber matrix surface (Ashby Scientific Ltd. or Integra Biosciences).

3.3.3 Step 3: large capacity stationary cultures

i. Cell factory (multitray unit)

The standard cell factory unit (Nunc) comprises ten chambers, each having a surface area of 600 cm², fixed together vertically and supplied with interconnecting channels. This enables all operations to be carried out once only for all chambers. It can thus be thought of as a flask with a 6000 cm² surface area using 2 litres of medium and taking up a total volume of 12 500 ml. In practice this unit is convenient to use and produces good results, similar to plastic flasks. It is made of tissue culture grade polystyrene and is disposable. One of the disadvantages of the system can be turned to good use. In practice it is difficult to wash out all the cells after harvesting with trypsin, etc. However, enough cells remain to inoculate a new culture when fresh medium is added. Given good aseptic technique, this disposable unit can be used repeatedly. The system is used commercially for interferon production (by linking together multiples of these units) (16). In addition, units are available giving 1200 (2 tray) and 24 000 cm² (40 tray).

ii. Costar CellCube

The CellCube (17) has parallel polystyrene trays in a modular closed-loop perfusion system with an oxygenator, pumps, and a system controller (pH, O_2, level control). The unit is compact, with the trays being only 1 mm apart; thus the smallest unit of 21 250 cm² is less than 5 litres total volume (1.25 litres medium). Additional units of 42 500 cm² (2.5 litres medium) and 85 000 cm² (5 litres medium) are available and four units can be run in parallel, giving 340 000 cm² growth area.

iii. Hollow fibre culture

Bundles of synthetic hollow fibres offer a matrix analogous to the vascular system, and allow cells to grow in tissue-like densities. Hollow fibres are usually used in ultrafiltration, selectively allowing passage of macromolecules through the spongy fibre wall while allowing a continuous flow of liquid through the lumen. When these fibres are enclosed in a cartridge and encapsulated at both ends, medium can be pumped in and will then perfuse through the fibre walls, which provide a large surface area for cell attachment and growth.

Culture chambers based on this principle are available from Amicon. The capillary fibres, which are made of acrylic polymer, are 350 μm in diameter with 75 μm walls. The pores through the internal lumen lining are available with molecular mass cut-offs between 10 000 and 100 000. It is difficult to calculate the total surface area available for growth but units are available in various sizes and these give a very high ratio of surface area to culture volume (in the region of 30 cm^2/ml). Up to 10^8 cells/ml have been maintained in this system. These cultures are mainly used for suspension cells (see Section 5.1.1) but are suitable for attached cells if the polysulfone type is used.

iv. Opticell culture system

This system (Cellex Biosciences Inc.) consists of a cylindrical ceramic cartridge (available in surface area between 0.4–12.5 m^2) with 1 mm^2 square channels running lengthways through the unit. A medium perfusion loop to a reservoir, in which environmental control is carried out, completes the system. It provides a large surface area/volume ratio (40:1) and its suitability for virus, cell surface antigen, and monoclonal antibody production is documented (18). Scale-up to 210 m^2 is possible with multiple cartridges arranged in parallel in a single controlled unit. Cartridges are available for both attached and suspension cells, which become entrapped in the rough porous ceramic texture.

v. Heli-Cel (Bibby Sterilin)

Twisted helical ribbons of polystyrene (3 mm × 5–10 mm × 100 μm) are used as packing material for the cultivation of anchorage-dependent cells. Medium is circulated through the bed by a pump, and the helical shape provides good hydrodynamic flow. The ribbons are transparent and therefore allow cell examination after removal from the bed.

3.3.4 Step 4: unit process systems

There are basically three systems which fit into fermentation (suspension culture) apparatus (see Figure 3):

- cells stationary, medium moves (e.g. glass bead reactor)
- heterogeneous mixing (e.g. stack plate reactor)
- homogeneous mixing (e.g. microcarrier)

i. Bead bed reactor

The use of a packed bed of 3–5 mm glass beads, through which medium is continually perfused, has been reported by a number of investigators since 1962. The potential of the system for scale-up was demonstrated by Whiteside and Spier (19) who used a 100 litre capacity system for the growth of BHK21 cells and systems can be commercially obtained (Meredos GmbH).

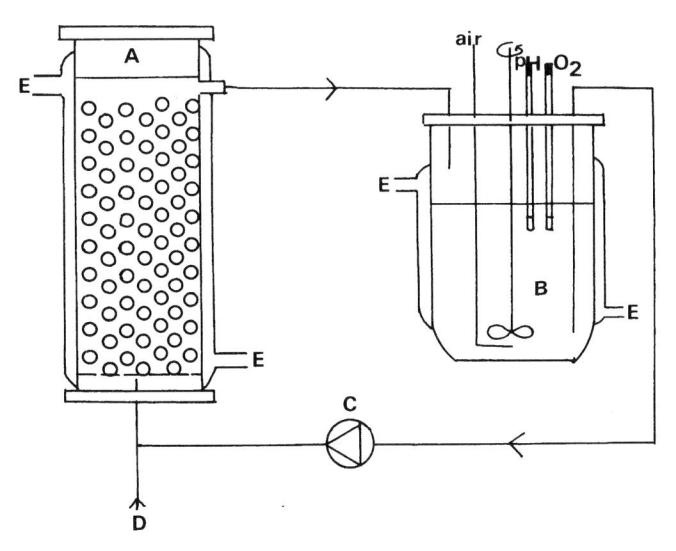

Figure 4. A glass bead bioreactor. (A) Glass bead bed; (B) reservoir; (C) pump; (D) inoculation and harvest line; (E) temperature-controlled water jackets.

Spheres of 3 mm diameter pack sufficiently tightly to prevent the packed bed from shifting, but allow sufficient flow of medium through the column so that fast flow rates, which would cause mechanical shear damage, are not needed. The physical properties of glass spheres are given in *Table 3*, and these data allow the necessary culture parameters to be calculated. A system which can be constructed in the laboratory is illustrated in *Figure 4* and demonstrated in *Protocol 2*. Medium is transferred by a peristaltic pump in this example, but an airlift driven system is also suitable and gives better oxygenation. Medium can be passed either up or down the column with no apparent difference in results.

Table 3. Properties of 3 mm and 5 mm glass beads

Glass sphere diameter	3 mm	5 mm
Weight/g	0.0375	0.0375
Surface area (cm^2/kg)	7546	4570
Volume (ml/kg)	990	625
Void fraction	0.46	0.40
Void volume (ml/kg)	455	250
Channel size (mm^2)	0.32	1.00
Vero cell yields:		
10^5/cm^2	0.78	2.5/3.1[a]
10^8/kg	4.00	8.0/9.9
10^6/ml	1.6	3.2/4.0

[a] Yield in 1 kg/10 kg bed reactor.

Protocol 2
Growing cells in a bead bed reactor

1. Prepare sufficient medium at the rate of 1-2 ml/10 cm^2 culture surface area.
2. Circulate medium between reservoir and vessel until the system is equilibrated with regard to temperature and pH (allow sufficient time for the glass beads to warm up to 37°C).
3. Add cell inoculum (1×10^4/cm^2) to a volume of medium equal to the void volume.
4. Drain the bead column into the reservoir, mix the prepared cell suspension thoroughly, and add to the bead column.
5. Clamp off all tubing and allow the cells to attach (3-8 h depending on cell type).
6. Once cells are attached start perfusing the medium, initially at a slow rate (1 linear cm/10 min). Visually inspect the effluent medium for cloudiness which would indicate that cells have not attached (stop perfusing if this is the case and resume after a further 8 h).
7. After 24 h this rate can be increased but should be kept below 5 cm/min, otherwise cells may detach (especially mitotic cells). The pH readout at the exit and a comparison of glucose concentrations in the input and exit sampling points will indicate whether the flow rate is sufficient.
8. Monitor cell growth by glucose determinations. Glucose utilization rates should be found empirically for each cell and medium, but a rough indication is $2-5 \times 10^8$ cells produced/g glucose.
9. When the culture is confluent, drain off the medium, wash the column with buffer, add the void volume of trypsin/versene or pronase, and allow to stand for 15-30 min. Cell harvesting can be accelerated by occasionally releasing the trypsin from the column into the harvest vessel and pushing it back into the culture again. Alternatively, feed gas bubbles through the column (only very slowly, otherwise the beads will shift and damage the cells).
10. The culture vessel should be washed immediately after use, otherwise it becomes very difficult to clean. Use a detergent such as Decon, and circulate continuously through the culture system, followed by tap-water, and then several changes of distilled water. Autoclave moist if control probes (oxygen, pH) are used. Use Pyrex glass beads in preference to the standard laboratory beads which are made of soda glass.

Although 3 mm beads are normally used in order to maximize the available surface area, the use of 5 mm spheres actually gives a higher total cell yield despite the reduced surface area per unit volume (9). A development of this system (porous beads) which significantly increases cell yields is described in Section 5.

ii. Heterogeneous reactor

Circular glass or stainless steel plates are fitted vertically, 5–7 mm apart, on a central shaft. This shaft may be stationary, with an airlift pump for mixing, or revolving around a vertical (6 r.p.h.) or horizontal (50–100 r.p.m.) axis. This multisurface propagator (20) was used at sizes ranging from 7.5–200 litres, giving a surface area of up to $2 \times 10^5/cm^2$. The author's experience is solely with the horizontal stirred plate type of vessel (*Figure 3*, stack plate), which is easier to use and has been successful for both heteroploid and human diploid cells. The main disadvantage with this type of culture is the high ratio of medium volume to surface area (1 ml to 1–2 cm²). This cannot be altered with the horizontal types, although it can be halved with the vertically revolving discs.

iii. Homogeneous systems (microcarrier)

When cells are grown on small spherical carriers they can be treated as a suspension culture, and advanced fermentation technology processes and apparatus can be utilized. The method was initiated by Van Wezel (21), who used dextran beads (Sephadex A-50). These were not entirely satisfactory, because the charge of the beads was unsuitable, and also possibly due to toxic effects. However, much developmental work has since resulted in many suitable microcarriers being commercially available (*Table 4*).

The author's experience has largely been with dextran-based microcarriers (Cytodex and Dormacell) and much of the following discussion is based on these products. The choice of this microcarrier was based on a preference for a dried product which could be accurately weighed and then prepared *in situ*, and the

Table 4. Comparison of microcarriers

Trade name	Manufacturer	Material	Specific gravity	Diameter (μm)	Area (cm²/g)
Biosilon	Nunc	Polystyrene	1.05	160–300	255
Bioglas	Solohill Eng.	Glass[a]	1.03	150–210	350
Bioplas	Solohill Eng.	Polystyrene[a]	1.04	150–210	350
Biospheres	Solohill Eng.	Collagen[a]	1.02	150–210	350
Cytodex-1	Pharmacia	DEAE Sephadex	1.03	160–230	6000
Cytodex-2	Pharmacia	DEAE Sephadex	1.04	115–200	5500
Cytodex-3	Pharmacia	Collagen	1.04	130–210	4600
Cytosphere	Lux	Polystyrene	1.04	160–230	250
Dormacell	Pfeifer & Langen	Dextran	1.05	140–240	7000
DE-53	Whatman	Cellulose	1.03	Fibres	4000
Gelibead	Hazelton Lab.	Gelatin	1.04	115–235	3800
Microdex	Dextran Prod.	DEAE dextran	1.03	150	250
Ventreglas	Ventrex	Glass	1.03	90–210	300
Ventregel	Ventrex	Gelatin	1.03	150–250	4300

[a] Biospheres (glass, plastic, collagen) available at specific gravity of 1.02 or 1.04 and diameters of 150–210 or 90–150 μm (manufactured by Solohill Eng. and distributed by Whatman and Cellon).

fact that with a density of only 1.03 g/ml this product could be used at concentrations of up to 15 g/litre (90 000 cm²/litre). However, this preference does not detract from the quality of other microcarriers, most of which have been used with equal success, and many of which offer particular advantages as discussed later.

Culture apparatus

A spinner vessel is not suitable unless the stirring system is modified. Paddles with large surface area are needed. These are commercially available (e.g. Bellco µ, and Cellon spinner vessels), but can be easily constructed out of a silicone rubber sheet and attached to the magnet with plastic ties. The advantage of constructing one within the laboratory is that the blades can be made to incline 20-30° from the vertical, thus giving much greater lift and mixing than vertical blades (*Figure 5*). Microcarrier cultures are stirred very slowly (maximum 75 r.p.m.) and it is essential to have a good quality magnetic stirrer that is capable of giving a smooth stirring action in the range of 20-100 r.p.m. Never use a stirrer mechanism that has moving surfaces in contact with each other in the medium, otherwise the microcarriers will be crushed. Thus, stirrers which revolve on the bottom of the vessel are unsuitable. As mixing, and thus mass transfer, is so poor in these cultures, the depth of medium should not exceed the diameter by more than a factor of two unless oxygenation systems or

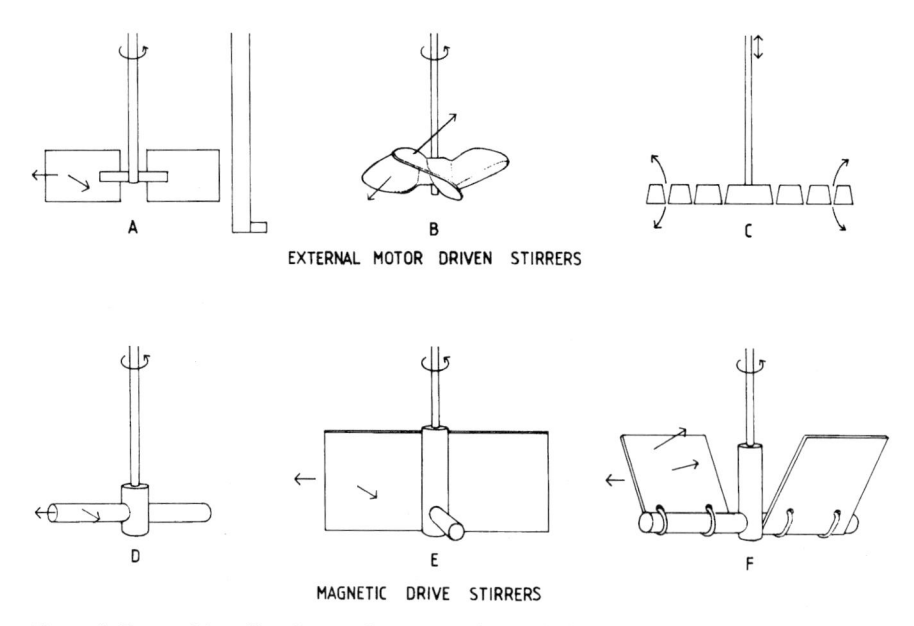

Figure 5. Types of impellers for growing suspension and microcarrier cells. (A) Flat disc turbine, d2:d1 = 0.33, radial flow, turbulent mixing; (B) marine impeller, d2:d1 = 0.33, axial flow, turbulent mixing, blade angle 25°; (C) Vibro-mixer; (D) stirrer bar, d2:d1 = 0.6, radial flow, laminar mixing; (E) vertical and (F) angled (25°) paddle, d1:d2 = 0.6–0.9, axial and radial flow, laminar and turbulent mixing. (d1 is the diameter of the vessel and d2 is the diameter of the impeller.)

SCALING-UP OF ANIMAL CELL CULTURES

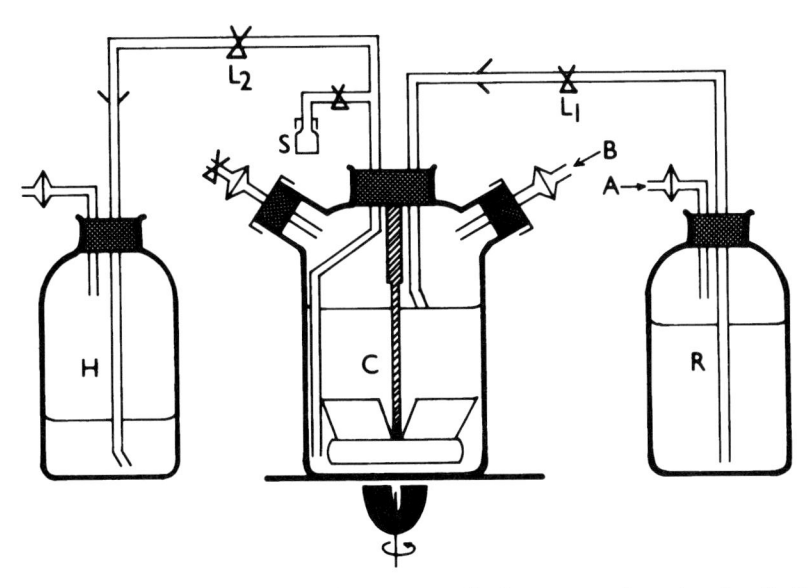

Figure 6. A simple microcarrier culture system allowing easy medium changing. To fill the culture (C) open clamp (L1) and push the medium from the reservoir (R) using air pressure (at A). To harvest, stop the stirring for 5 min, open L2, and push the medium from (C) to (H) using air pressure (at B). (S) is a sampling point.

regular medium changes are performed. In the basic culture systems, medium changes have to be carried out at frequent intervals. It is worthwhile, therefore, to make suitable connections to the culture vessel to enable this to be done conveniently *in situ*. This will pay dividends in reducing the chances of contaminating the culture. A simplified culture set-up is shown in *Figure 6*.

Initiation of the culture

Many of the factors discussed in Section 2 are critical when initiating a microcarrier culture. Microcarriers are spherical, and cells will always attach to an area of minimum curvature. Therefore a microcarrier surface can never be ideal, however suitable its chemical and physical properties.

Ensure the medium and beads are at a stable pH and temperature, and inoculate the cells (from a logarithmic, not a stationary culture) into a third of the final medium volume. This increases the chances of cells coming into contact with the microcarriers. Microcarrier concentrations of 2–3 g/litre should be used. Higher concentrations need environmental control or very frequent medium changes.

After the attachment period (3–8 h), slowly top up the culture to the working volume and increase the stirring rate to maintain completely homogeneous mixing. If these conditions are adhered to, and there are no changes in temperature and pH, then all cells that grow on plastic surfaces should readily initiate a microcarrier culture.

Maintenance of the culture

It is very easy to monitor the progress of cell growth in a microcarrier culture. Samples can be easily removed, cell counts (by nuclei counting) and glucose determinations carried out, and the cell morphology examined. As the cells grow, so the beads become heavier and need an increased stirring rate. After three days or so the culture will become acidic and need a medium change. Again, this is an extremely easy routine: turn off the stirrer, allow the beads to settle for 5 min, and decant off as much medium as desired. Top up gently with fresh medium (pre-warmed to 37 °C) and restart the stirring.

Harvesting

It is very difficult to harvest many cell types from microcarriers unless the cell density on the bead is very high, and cells do not multilayer on microcarriers to the same degree as in stationary cultures. Harvesting can be attempted by draining off the medium, washing the beads at least once in buffer, and adding the desired enzyme. Stir the culture fairly rapidly (75–125 r.p.m.) for 20–30 min. If the cells detach, a high proportion can be collected by allowing the beads to settle out for 2 min and then decanting off the supernatant. For a total harvest pour the mixture into a sterilized sintered glass funnel (porosity grade 1). The cells will pass through the filter, but the microcarriers will not. Alternatively, insert a similar filter (usually stainless steel) into the culture vessel.

If the microcarrier is dissolved then the cells can be released into suspension completely undamaged, and therefore of better quality than when the cells are trypsinized. Usually this is a far quicker method than trypsinization of cells. Gelatin beads are solubilized with trypsin and/or EDTA, collagenase acts on the collagen-coated beads, and dextranase can be used on the dextran microcarriers.

Protocol 3
Microcarrier culture

1. Add complete medium to spinner flask (200 ml in a 1 1itre flask), gas with 5% CO_2, and allow to equilibrate.

2. Decant PBS from sterilized stock solution of Cytodex-3 and replace with growth medium (1 g to 30–50 ml). Add Cytodex-3 to spinner vessel to give a final concentration of 2 g/litre (1–3 g/litre).

3. Put spinner on magnetic stirrer at 37 °C and allow temperature and physiological conditions to equilibrate (minimum 1 h).

4. Add cell inoculum obtained by trypsinization of late log phase cells (pre-stationary phase) at over five cells per bead (optimum to ensure cells on all beads is seven); inoculate at the same density per square centimetre as with other culture types, e.g. $5-10 \times 10^4/cm^2$. Cytodex-3 at 2 g/litre inoculated at six cells per bead gives: 8×10^6 microcarriers = 4.8×10^7 cells/litre

Protocol 3 continued

$$9500 \text{ cm}^2 = 5 \times 10^4 \text{ cells/cm}^2$$
$$200 \text{ ml} = 2.5 \times 10^5 \text{ cells/ml}$$

5. Place spinner flask on magnetic stirrer and stir at the minimum speed to ensure that all cells and carriers are in suspension (usually 20–30 r.p.m.). It is advantageous if cells and carriers are limited to the lower 60–70% of the culture for this purpose. Alternatively either:

    (a) Inoculate in 50% of the final medium volume and add the rest of the medium after 4–8 h.

    (b) Or stir intermittently (for 1 min every 20 min) for the first 4–8 h. However, only use this option for cells with very poor plating efficiency because it causes clumping and uneven distribution of cells per bead.

6. When the cells have attached (expect 90% attachment) the stirring speed can be increased to allow complete homogeneity (40–60 r.p.m.).

7. Monitor progress of culture by taking 1 ml samples at least daily; observe microscopically and carry out cell (nuclei) counts.

8. As the cell density increases there is often a tendency for microcarriers to begin clumping. This can be avoided by increasing the stirring speed to 75–90 r.p.m.

9. After three to four days the culture will become acid. Remove the spinner flask and re-gas the headspace and/or add sodium bicarbonate (5.5% stock solution). With some cells, or at microcarrier densities of 3 g/litre or more, partial medium changes should be carried out. Allow culture to settle (5–10 min), siphon off at least 50% (usually 70%) of the medium, and replace with pre-warmed fresh medium (serum can be reduced or omitted at this stage). Replace spinner flask on stirrer.

10. After four to five days cells reach a maximum cell density (confluency) at the same level as in static cultures, although multilayering is not so prevalent. Thus a cell yield of $1-2 \times 10^6$/ml ($2-4 \times 10^5$/cm^2) can be expected.

11. Cells can be harvested in the following way:

    (a) Allow culture to settle (10 min).

    (b) Decant off as much medium as possible (> 90%).

    (c) Add warm Ca^{2+}/Mg^{2+}-free PBS (or EDTA in PBS) and mix.

    (d) Allow culture to settle and decant off as much PBS as possible.

    (e) Add 0.2% trypsin (30 ml) and stir at 75–100 r.p.m. for 10–20 min at 37 °C.

    (f) Allow the beads to settle out (2 min). Either decant trypsin plus cells or pour mixture through a sterile, coarse-sinter glass filter or a specially designed filter such as the Cellector (Bellco).

    (g) Centrifuge cells (800 g for 5 min) and resuspend in fresh medium (with serum or trypsin inhibitor, e.g. soybean inhibitor at 0.5 mg/ml).

Note: as a guide to calculating settled volume and medium entrapment by microcarriers, 1 g of Cytodex, for example, has a volume of 15–18 ml.

Scaling-up of microcarrier culture

Scaling-up can be achieved by increasing the microcarrier concentration, or by increasing the culture size. In the first case nutrients and oxygen are very rapidly depleted, and the pH falls to non-physiological levels. Medium changes are not only tedious but provide rapidly changing environmental conditions. Perfusion, either to waste or by a closed-loop, must be used to achieve cultures with high microcarrier concentration. This can be brought about only by an efficient filtration system so that medium without cells and microcarriers can be withdrawn at a rapid rate. The only satisfactory means of doing this is with the type of filtration system illustrated in *Figure 7*. This is constructed of a stainless steel mesh with an absolute pore size in the 60–120 μm range. Attachment to the stirrer shaft means that a large surface area filter can be used and the revolving action discourages cell attachment and clogging.

However scaling-up is achieved, oxygen limitation is the chief factor to overcome. This is an especially difficult problem in microcarrier culture because stirring speeds are low (it was pointed out in Section 2.7 that the stirring speed has to be above 100 before it significantly affects the aeration rate). Sparging cannot be used as microcarriers get left above the medium level due to the foaming it causes. The perfusion filter previously described, however, does allow sparging into that part of the culture in which no beads are present. Unfortunately, this means that the most oxygenated medium in the culture is

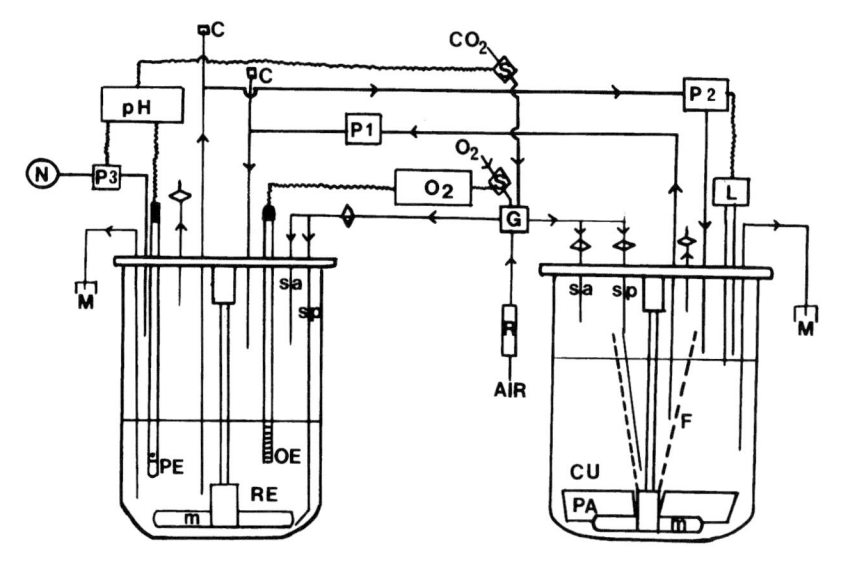

Figure 7. A closed-loop perfusion system with environmental control to allow high cell density microcarrier culture. (CU) Culture vessel; (RE) reservoir; (C) connector for medium changes and harvesting; (F) filter; (G) gas blender; (L) level controller; (M) sampling device; (m) magnet; (N) alkali (NaOH) reservoir; (OE) oxygen electrode; (PA) paddle; (PE) pH electrode; (P1–P3) pumps: 1, medium to reservoir (continuous), 2, medium to culture (controlled by L), 3, alkali to reservoir (controlled by PE); (R) rotameter measuring gas flow rate; (sa) gas supply for surface aeration; (sp) gas supply for sparging; (S) solenoid valve.

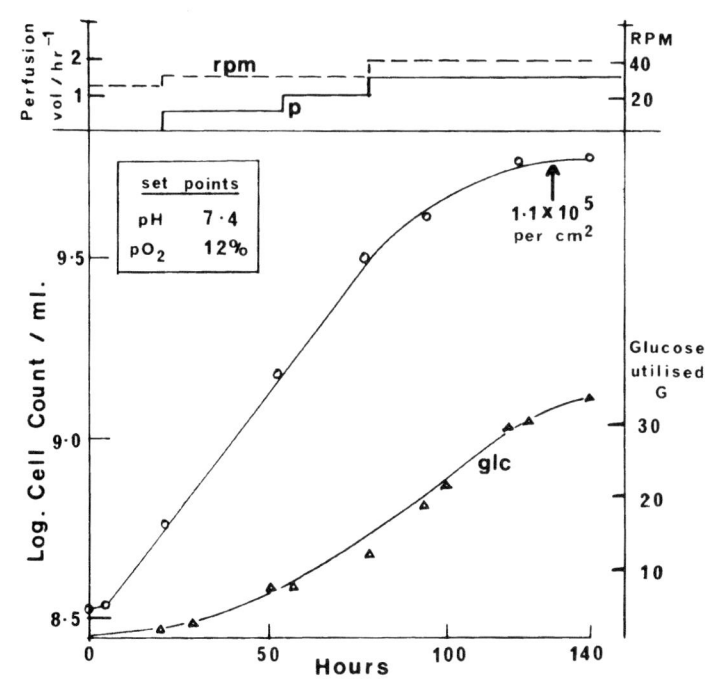

Figure 8. Growth of GPK epithelial cells on Cytodex-3 (10 g/litre) in the 10 litre culture system described in *Figure 7*. (o–o) Cell concentration; (∆–∆) glucose (glc) utilized; (p) perfusion rate.

removed by perfusion, but at a sparging rate of 10 cm³ of oxygen/litre considerable diffusion of oxygenated medium occurs into the main culture. However, modified spin filters are available which have a separate compartment for perfusion and sparging (22, 23) from Braun Biotech. Thus oxygen delivery to a microcarrier culture can utilize the following systems (see also *Table 2*):

(a) Surface aeration.
(b) Increasing the perfusion rate of fully oxygenated medium from the reservoir.
(c) Sparging into the filter compartment (23).

These three systems are used by the author in the system illustrated in *Figure 7* to run cultures at up to 15 g/litre Cytodex-3 in culture vessels between 2–20 litre working volume. A typical experiment is shown in *Figure 8*.

Summary of microcarrier culture

Microcarrier cultures are used commercially for vaccine and interferon production in fermenters up to 4000 litres. These processes use heteroploid or primary cells. One of the problems of using very large scale cultures is that the required seed inoculum gets progressively larger. Harvesting of cells from large unit scale cultures is not always very successful, although the availability of collagen and gelatin microcarriers eases this problem considerably. *In situ* trypsinization of cells is the usual method of choice, but it is difficult to remove a high percentage of the cells in a viable condition. Filters incorporated into the culture vessel provide the best solution and are most efficient with the 'hard' (glass, plastic) microcarriers which do not block the filter pores like the dextran and gelatin microcarriers. Some microcarriers can be washed and reused but this, of course, is not possible with the gelatin- or collagenase-treated beads. A development of microcarriers is the porous microcarrier system described in Section 5.3.

3.3.5 Summary: choice of equipment

A wide range of commercially available and laboratory-made equipment has been reviewed. This should enable a choice to be made, depending upon the amount and type of cells or product needed and the financial and staff resources available. To review the choices available, *Figure 3* and *Table 5* should be consulted.

4 Suspension culture

As indicated previously in this chapter, suspension culture is the preferred method for scaling-up cell cultures. Some cells, especially those of haemopoietic

Table 5. Comparison of monolayer culture systems, showing their relative productivity and scale-up potential

Culture	Unit surface area (cm^2)	Surface area: medium ratio[a]	Max. vol. (litre)
Multiple processes:			
Roux bottle	175–200	1–2	
Roller bottle	850–175	3	
–Corrugated	1700	6	
–Spiracell	6000	6	
–Corbeil	15 000	2.5	
Cell Factory	24 000	2.5	
Hollow fibres	18 000	72	
Opticell	120 000	30	
CellCube	340 000	17	
Unit processes:			
Heli-Cel	16 000/L	16	(1)[b]
Glass spheres	7500/L	18	200
Porous glass spheres	80 000/L	150	(10)[b]
Stack plates	1250/L	1.25	200
Microcarrier (5 g/litre)	35 000/L	35	4000
Porous microcarrier	300 000/L	300	(24)[b]

[a] Volume in bioreactor—does not allow for media changes or reservoir.
[b] Current experimental scales—scale-up potential.

derivation, grow best in suspension culture. Others, particularly transformed cells, can be adapted or selected, but some, for example human diploid cell strains (W1-38, MRC-5), will not survive in suspension at all. A further factor that dictates whether suspension systems can be used is that some cellular products are expressed only when the cell exists in a monolayer or if cell-to-cell contact is established (e.g. for the spread of intracellular viruses through a cell population).

4.1 Adaptation to suspension culture

Cell lines vary in the ease with which they can be persuaded to grow in suspension. For those that have the potential, there are two basic procedures that can be used to generate a suspension cell line from an anchorage-dependent one.

4.1.1 Selection

This method, as demonstrated by the derivation of the LS cell line from L-929 cells by Paul and Struthers (24), and the HeLa-S3 clone from HeLa cells, depends on the persistence of loosely attached variants within the population. A confluent monolayer culture is lightly tapped or gently swirled, the medium decanted from the culture, and cells in suspension recovered by centrifugation. Cells have to be collected from many cultures to provide a sufficient number to start a new culture at the required inoculum density (at least 2×10^5/ml). This procedure has to be repeated many times over a long period, because many of the cells that are collected are only in the mitotic phase, rather than potential suspension cells (cells round up and become very loosely attached to the substrate during mitosis). Eventually it is possible with some cell lines to derive a viable cell population that divides and grows in static suspension, just resting on the substrate rather than attaching and spreading out.

4.1.2 Adaptation

This method probably works the same way as the selection procedure, except that a selection pressure is exerted on the culture while it is maintained in suspension mechanically. However, with many cell lines the relatively large number that become anchorage-independent suggests it is more than just selection of a few variant cells. The method given in *Protocol 4*, used successfully by the author for many cell lines, is based on one published for BHK21 cells (25).

Protocol 4

Adaptation to suspension culture

1. Prepare the cell suspension by detaching monolayer cells with trypsin (0.1%) and versene (0.01%).
2. Make ready at least two, but preferably three or more, spinner vessels by adding the growth medium (calcium- and magnesium-free modification of any recognized culture medium).

Protocol 4 continued

3. Add cells at a concentration of at least 5×10^5/ml and commence stirring (the lowest rate which keeps the cells homogeneously mixed, e.g. 250 r.p.m.).
4. Sample and carry out viable cell counts every 24 h.
5. Every three days remove the medium from the culture, centrifuge (1000 g), and resuspend the cells in fresh medium at a density of at least 2.5×10^5/ml. Depending upon the degree of cell death, cultures will almost certainly have to be amalgamated to keep the cell density at the required level.
6. If, at a medium change, there is significant attachment of cells to the culture vessel, especially at the air–medium interface, then add trypsin (0.01%)/versene (0.01%) mixture to the empty vessel. Stir at 37°C for 30 min and recover the detached cells by centrifugation. If the cells in suspension are badly clumped, they too can be added to the trypsin/versene solution. This treatment is usually necessary at the first and second medium changes, but very rarely after this point. If cell clumping persists, then add 50 µg/ml trypsin or Dispase (Boehringer) to the growth medium.
7. Successful adaptation is recognized initially by an increase in cell numbers after a medium change, and subsequently by getting consistent cell yields in a given period of time (indicating a constant growth rate).

It is usually necessary to maintain the newly established cell strain in stirred suspension culture, because reversion to anchorage dependence can occur in static cultures. Sometimes cells will adhere to the substrate without becoming completely spread out and continue to grow and divide with a near-spherical morphology.

4.2 Static suspension culture

Many cell lines will grow as a suspension in a culture system used for monolayer cells (i.e. with no agitation or stirring). Cell lines that are capable of this form of growth include the many lymphoblast lines (e.g. MOLT, RAJI), hybridomas, and some non-haemopoietic lines, such as the LS cells described in the previous section. However, with the latter type of cells there is always the danger of reversion to a monolayer (e.g. a small proportion of the LS cell line always attaches and is discarded at each subcultivation). Static suspension culture is unsuitable for scale-up, for reasons already stated for monolayer culture.

4.3 Small scale suspension culture

For the purpose of this chapter, small scale means under 2 litres. This may seem an entirely arbitrary definition, but it is made on the basis that above this volume additional factors apply. The conventional laboratory suspension culture is the spinner flask, so-called because it contains a magnetic bar as the stirrer, and this is driven from below the vessel by a revolving magnet. Details of some of

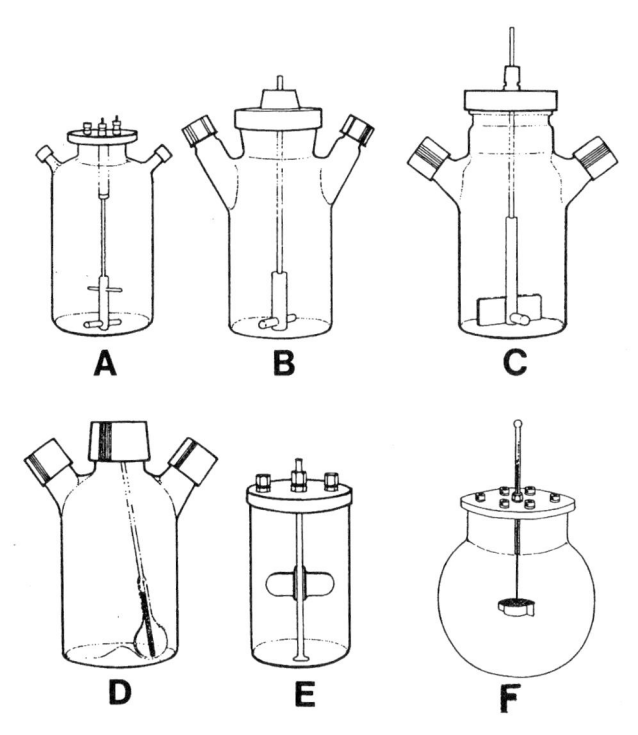

Figure 9. Commercially available spinner cultures. (A) LH Fermentation Biocul (1–20 litres); (B) Bellco and Wheaton Spinner Flasks (25 ml–2 litres); (C) Bellco and Cellon spinner (25 ml–2 litres); (D) Techne (25 ml–5 litres); (E) Techne Cytostat (1 litre); (F) Techne Br-06 Bioreactor (3 litres).

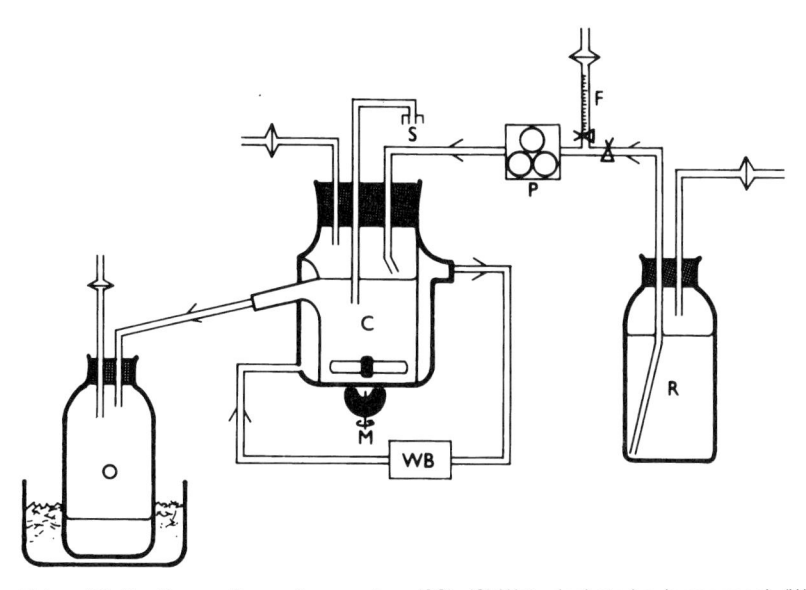

Figure 10. Continuous-flow culture system (28). (C) Water-jacketed culture vessel; (WB) water-bath and circulating system; (O) overflow vessel; (R) reservoir; (P) pump; (S) sampling device; (F) burette for measuring flow rate; (M) magnetic drive.

the readily available stirrer vessels, together with suppliers and available size range, are given in *Figure 9*.

The range of options are:

(a) Conventional vessels with spin bar (Bellco, Wheaton).
(b) Radial (pendulum) stirring system (*Figure 10*) for improved mixing at low speeds (under 100 r.p.m.) (Techne, Integra Biosciences CellSpin).
(c) Bellco dual overhead drive system (radial stirring and permits perfusion).
(d) Techne Br-06 floating impeller system allows working volumes of 500 ml to 3 litres to be used and increased during culture.

Protocol 5

Spinner flask culture

1. Add 200 ml medium to a 1 litre spinner flask, gas with 5% CO_2, and warm to 37°C.
2. Add $2-4 \times 10^7$ cells ($1-2 \times 10^5$/ml) harvested from a growing culture of cells (i.e. not a stationary or dying culture). Note well: this is a higher inoculum than used for stationary cultures.
3. Place spinner vessel on magnetic stirrer at 37°C and set at 100-200 r.p.m. (this is variable depending upon the vessel size, geometry, fluid volume, and cells. Set a speed which visually shows complete homogeneous mixing of the cells throughout the medium—150 r.p.m. is usually a safe choice).
4. Monitor the growth daily by removing a small sample through the side-arm in a Class II cabinet and carrying out a cell count and visual inspection for cell morphology and lack of microbial contamination.
5. Monitor the pH (colour change), especially in closed (non-gassed) systems, and should the culture become acidic (under pH 6.8), add sodium bicarbonate (5.5% stock solution) or re-gas with 5% CO_2. After three days a better option would be to allow the cells to settle out, remove 40-70% of the medium, add fresh pre-warmed medium to the culture, and continue stirring.
6. After four to five days the cells should reach their maximum density ($7-10 \times 10^5$/ml) and can be harvested. The timing will depend upon the kinetics of the cell (whether monoclonal antibody production (mAb) is growth or dying phase associated).

Note: if clumping or attachment to the culture vessel should occur the vessels can be siliconized with Dow Corning 1107 or Repelcote (dimethyldichlorosilane) (Hopkins and Williams), or medium with reduced Ca^{2+} and Mg^{2+} can be used.

4.3.1 SuperSpinner

The SuperSpinner (26) was developed to increase the productivity within a spinner flask without going to the complexity of the hollow fibre and membrane systems described below. It consists of a 1 litre Duran flask (Schott) equipped with a tumbling membrane stirrer moving a polypropylene hollow fibre through the medium to improve the oxygen supply. The small device is placed in a CO_2 incubator and a small membrane pump is connected via a sterilizing filter with the membrane (stirrer). The culture is placed on a magnetic stirrer and has a working volume from 300 ml to 1000 ml. Three different membrane lengths (1, 1.5, and 2 m) are available depending upon the degree of oxygenation the cell line needs. Cell concentrations in excess of 2×10^6/ml are achieved with mAb production of 160 mg/litre (specific productivity of mAb c. 50 μg/10^6 cells). Repeated batch culture (80% media replacement every two to three days) over 32 days produced 970 mg mAb (26). This relatively inexpensive modification of a spinner culture is available from Braun Biotech.

4.4 Scaling-up factors

In scale-up, both the physical and chemical requirements of cells have to be satisfied. The chemical factors require environmental monitoring and control to keep the cells in the proper physiological environment. These factors, which include oxygen, pH, medium components, and removal of waste products, are described in Section 2.

Physical parameters include the configuration of the bioreactor and the power supplied to it. The function of the stirrer impeller is to convert energy (measured as kW/m^3) into hydrodynamic motion in three dimensions (axial, radial, and tangential). The impeller has to circulate the whole liquid volume and to generate turbulence (i.e. it has to pump and to mix) in order to create a homogeneous blend, to keep cells in suspension, to optimize mass transfer rates between the different phases of the system (biological, liquid, and gaseous), and to facilitate heat transfer.

Good mixing becomes increasingly difficult with scaling-up, and the power needed to attain homogeneity can cause problems. The energy generated at the tip of the stirrer blade is a limiting factor as it gives rise to a damaging shear force. Shear forces are created by fluctuating liquid velocities in turbulent areas. The factors which affect this are: impeller shape (this dictates the primary induced flow direction) (*Figure 5*), the ratio of impeller to vessel diameter, and the impeller tip speed (a function of rotation rate and diameter). The greater the turbulence the more efficient the mixing, but a compromise has to be reached so that cells are not damaged. Large impellers running at low speeds give a low shear force and high pumping capacity, whereas smaller impellers need high stirring speeds and have high shear effects.

Magnetic bar stirring gives only radial mixing and no lift or turbulence. The marine impeller is more effective for cells than the flat-blade turbine impellers found in many bacterial systems, as it gives better mixing at low stirring speeds. If the cells are too fragile for stirring, or if sufficient mixing cannot be obtained without causing unacceptable shear rates, then an alternative system may have to be used. Pneumatic energy, for example mixing air bubbles (e.g. airlift fermenter) or hydraulic energy (e.g. medium perfusion), can be scaled-up without proportionally increasing the power. To improve the efficiency of mechanical stirring, the design of the stirrer paddle can be altered (e.g. as described for microcarrier culture), or multiple impellers can be used. Some examples are given in *Figure 5*.

A totally different stirring concept is the Vibro-mixer. This is a non-rotating agitator which produces a stirring effect by a vertical reciprocating motion with a path of 0.1-3 mm at a frequency of 50 Hz. The mixing disc is fixed horizontally to the agitator shaft, and conical-shaped holes in the disc cause a pumping action to occur as the shaft vibrates up and down. The shaft is driven by a motor which operates through an elastic diaphragm; this also provides a seal at the top of the culture. A fermentation system using this principle is available commercially (Vibro-Fermenter, Chemap). The advantages of this system are the greatly reduced shear forces, random mixing, reduced foaming, and reduced energy requirement, especially for scaling-up.

The significant effect on vessel design of moving from a magnetic stirrer to a direct drive system is the fact that the drive has to pass through the culture vessel. This means some complexity of design to ensure a perfect aseptic seal while transferring the drive, complete with lubricated bearings, through the bottom or top plate of the vessel. Culture vessels with magnetic coupling are becoming available at increasingly higher volumes, and overcome this problem of aseptic seals.

Scaling-up cannot be proportional; one cannot convert a 1 litre reactor into a 1000 litre reactor simply by increasing all dimensions by the same amount. The reasons for this are mathematical: doubling the diameter increases the volume threefold, and this affects the different physical parameters in different ways. One factor to be taken into consideration is the height/diameter ratio, as this is one of the most important fermenter design parameters. In sparged systems the taller the fermenter in relation to its diameter the better, as the air pressure will be higher at the bottom (increasing the oxygen solubility) and the residence time of bubbles longer. However, in non-sparged systems which are often used for animal cells and rely on surface aeration, the surface area:height ratio is more important and a 2:1 ratio should not be exceeded (preferably 1:1.5).

Mass transfer between the culture phases has been discussed in relation to oxygen (Section 2.7). It is this characteristic of scaling-up that demands the extra sophistication in culture design to maintain a physiologically correct environment. This sophistication includes impeller design, oxygen delivery systems, vessel geometry, perfusion loops, or a completely different concept in culture design to the stirred bioreactor (9, 10).

4.5 Stirred bioreactors

The move from externally driven magnetic spinner vessels to fermenters capable of scale-up from 1 litre to 1000 litres and beyond, has the following consequences at some stage:

(a) Change from glass to stainless steel vessels.

(b) Change from a mobile to a static system: connection to steam for *in situ* sterilization; requirement for water jacket or internal temperature control; need for a seed vessel, a medium holding vessel, and downstream processing capability.

(c) Greater sophistication in environmental control systems to meet the increasing mass transfer equipment.

In practice, the maximum size for a spinner vessel is 10 litres. Above this size there are difficulties in handling and autoclaving, as well as the difficulty of being able to agitate the culture adequately. Fermenters with motor driven stirrers are available from 500 ml, but these are chiefly for bacterial growth. It is a significant step to move above the 10–20 litre scale as the cost of the equipment is significant (e.g. $40–50 000 for a complete 35 litre system) and suitable laboratory facilities are required (steam, drainage, etc.). There is a wide range of vessels available from the various fermenter suppliers which include some of the following modifications:

- suitable impeller (e.g. marine)
- no baffles
- curved bottom for better mixing at low speeds
- water jacket rather than immersion heater type temperature control (to avoid localized heating at low stirring speeds)
- top-driven stirrer so that cells cannot become entangled between moving parts
- mirror internal finishes to reduce mechanical damage and cell attachment

As long as adequate mixing, and thus mass transfer of oxygen into the vessel, can be maintained without damaging the cells, there is no maximum to the scale-up potential. Namalva cells have been grown in 8000 litre vessels for the production of interferon. There are many heteroploid cell lines, such as Vero, HeLa, and hybridomas, which would grow in such systems. Whilst regulatory agencies require pharmaceutical products to be manufactured predominantly from normal diploid cells (which will not grow in suspension), the only motive for using this scale of culture was for veterinary products. However, the licensing of products such as tPA, EPO, interferon, and therapeutic monoclonal antibodies from heteroploid lines means that large scale bioreactors are in widespread use (but see Section 5). There are many applications for research products from various types of cells, and this is served by the 2–50 litre range of vessels. At present, the greatest incentive for large scale systems is to grow hybridoma cells (producing monoclonal antibody). In tissue culture the antibody yield is 50- to 100-fold lower than when the cells are passaged through the peritoneal cavity of mice, although the purity is greater, particularly if serum-free medium is used. The main need is to supply antibodies to meet the requirements for diagnostic and affinity chromatography purposes, but increasingly as new therapeutic and prophylactic drugs are being developed and licensed much larger quantities are required. Many of these cells are fragile and low yielding in culture, and the specialized techniques and apparatus described in Sections 4.7 and 5 are partly aimed towards this type of cell.

4.6 Continuous-flow culture

4.6.1 Introduction

At sub-maximal growth rates, the growth of a cell is determined by the concentration of a single growth-limiting nutrient. This is the basis of the chemostat, a fixed volume culture, in which medium is fed in at a constant rate, mixed with cells, and then leaves at the same rate. The culture begins as a batch culture while the inoculum grows to the maximum value that can be supported by the growth-limiting nutrient (assuming the dilution rate is less than the maximum growth rate). As the growth-limiting nutrient decreases in concentration, so the growth rate declines until it equals the dilution rate.

When this occurs, the culture is defined as being in a 'steady state' as both the cell numbers and nutrient concentrations remain constant.

When the culture is in a steady state, the cell growth rate (μ) is equal to the dilution rate (D). The dilution rate is the quotient of the medium flow rate per unit time and the culture volume (V):

$$\mu = D = f/V \text{ day} \qquad 8$$

As the growth rate is dependent on the medium flow rate, the mean generation (doubling) time can be calculated:

$$\mu = \ln 2/D \qquad 9$$

An alternative system to the chemostat is the turbidostat in which the cell density is held at a fixed value by altering the medium supply. The cell density (turbidity) is usually measured through a photoelectric cell. When the value is below the fixed point, medium supply is stopped to allow the cells to increase in number. Above the fixed point, medium is supplied to wash out the excess cells. This system really works well only when the cell growth rate is near maximum. However, this is its main advantage over the chemostat, which is least efficient or controllable when operating at the cell's maximum growth rate. Continuous flow culture of animal cells has been reviewed by Tovey (27).

4.6.2 Equipment

Complete chemostat systems can be purchased from dealers in fermentation equipment. However, systems can be easily constructed in the laboratory (*Figure 10*) (28). The culture vessel needs a side-arm overflow at the required liquid level, which should be approximately half the volume of the vessel. If a suitable 37°C cabinet is not available, then a water-jacketed vessel is needed. Apart from this, all other components are standard laboratory items. Vessel enclosures can be made from silicone (or white rubber) bungs wired onto the culture vessel. A good quality peristaltic pump, such as the Watson-Marlow range, is recommended.

4.6.3 Experimental procedures

Recommended cells are LS, HeLa-S3, or an established lymphoblastic cell line such as L1210 or a hybridoma. Growth-limiting factors can be chosen from the amino acids (e.g. cystine) or glucose (see also ref. 29).

Protocol 6
Growth of cells in continuous-flow culture

1. Inoculate the culture vessel at 10^5 cells/ml in preferred medium (e.g. Eagles MEM with 10% calf serum and the growth-limiting factor).
2. The chosen dilution rate is turned on after 24–48 h of growth. The maximum rate must not exceed the maximum growth of the cell line, which is usually within the range 14–27 h doubling time (although some cells can double their number in 9 h). Thus, the dilution rate will be in the range of 0.1/day (t_d = 166 h) to 1.2 (t_d = 14 h).
3. A steady state will become established within 100–200 h, although it may take up to 400 h, especially at the low dilution rates. This will be recognized by the fact that the daily cell counts will not vary by more than the expected counting error.
4. The culture can be maintained almost indefinitely, assuming it is kept sterile and no breakdown in components occurs. Sometimes the culture has to be terminated because of excessive attachment of cells at the interfaces. A duration of 1000 h is considered satisfactory.
5. Once the steady state has been achieved, the flow rate can be altered, and the response of the cell population in establishing new steady state conditions can be studied.
6. As well as carrying out routine cell counts, measurements can also be made to demonstrate the homogeneity of the cells and the medium. For instance, the size and chemical composition of the cells is remarkably consistent. Also, some of the medium components should be measured (e.g. glucose, an amino acid, lactic acid) to demonstrate again the consistency of the culture over a long period of time.

4.6.4 Uses

Continuous-flow culture provides a readily available continuous source of cells. Also, because optimal conditions or any desired physiological environment can be maintained, the culture is very suitable for product generation (as already shown for viruses and interferon) (27). For many purposes a two-stage chemostat is required so that optimal conditions can be met for cell growth (first stage) and product generation (second stage).

4.7 Airlift fermenter

The airlift fermenter relies on the bubble column principle both to agitate and to aerate a culture (30). Instead of mechanically stirring the cells, air bubbles are introduced into the bottom of the culture vessel. An inner (draft) tube is placed

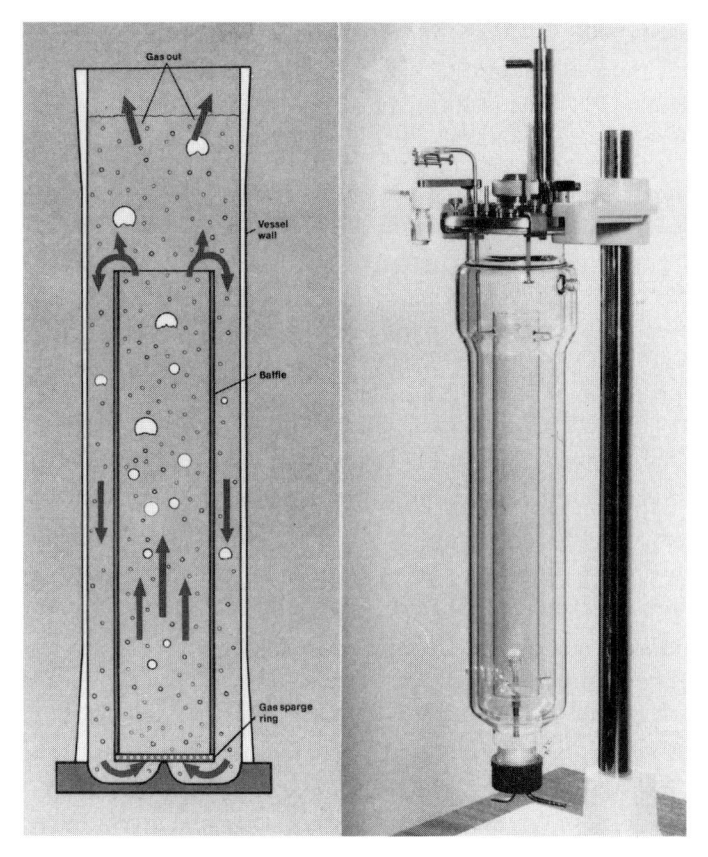

Figure 11. Principle of the airlift fermenter.

inside the vessel and mixing occurs because the air bubbles lift the medium (aerated medium has a lower density than non-aerated medium). The medium and cells which spill out from the top of the draft tube then circulate down the outside of the vessel. The amount of energy (compressed air) needed for the system is very low, shear forces are absent, and this method is thus ideal for fragile animal and plant cells. Also, as oxygen is continuously supplied to the culture, the large number of bubbles results in a high mass transfer rate. Culture units, as illustrated in *Figure 11*, are commercially available in sizes from 2–90 litres (Braun Biotech). The one disadvantage of the system is that scale-up is more or less linear (the 90 litre vessel requires nearly 4 m headroom). Whether it will be possible to use multiple draft tubes, and thus enable units with greatly increased diameters to be used, remains a developmental challenge. However, 2000 litre reactors are in operation for the production of monoclonal antibodies.

5 Immobilized cultures

Immobilized cultures are popular because they allow far higher unit cell densities to be achieved (50–200 $\times$ 10^6 cells/ml) and they also confer stability, and therefore longevity, to cultures. Cells *in vivo* are in a three-dimensional tissue matrix, therefore immobilization can mimic this physiological state. Higher cell density is achieved by facilitating perfusion of suspension cells and increasing the unit surface area for attached cells. In addition many immobilization materials protect cells from shear forces created by medium flow dynamics. As cells increase in unit density they become less dependent on the external supply of many growth factors provided by serum, so real cost savings can be achieved.

The emphasis has been on developing systems for suspension cells because commercial production of monoclonal antibody has been such a dominant factor. Basically two approaches have been used: immurement (confining cells within a medium-permeable barrier), and entrapment (enmeshing cells within an open matrix through which medium can flow unhindered) (9).

Immobilization techniques that can be scaled-up, such as the porous carriers described in Section 5.3, will become a dominant production technology once the beads and process parameters have been optimized. They are currently excellent laboratory systems for the manufacture of cell products and, with gelatin carriers, production of cells.

5.1 Immurement cultures

5.1.1 Hollow fibres

These have been discussed in Section 3.3.3, and are very effective for suspension cells at scales up to about 1 litre (1–2 × 10^8 cells/ml). Simple systems can be set up in the laboratory by purchasing individual hollow fibre cartridges (e.g. Amicon, Microgon). All that is needed is a medium reservoir and a pump to circulate the medium through the intracapillary section of the cartridge, and a harvest line from the extracapillary compartment in which the cells and product reside. To get better performance there are a number of 'turn-key' systems that can be purchased, from relatively simple units, e.g. Kinetek, Cellco, Cell-Pharm, Amicon (Mini Flo-Path), Asahi Medical Co (Cultureflo), to extremely complex and sophisticated units (e.g. Cellex Acusyst) capable of producing up to 40 g of monoclonal antibody per month.

The Acusyst (Endotronics, now Cellex Biosciences Inc.) system made the breakthrough by using pressure differentials to simulate *in vivo* arterial flow (31). The unit has a dual medium circuit, one passing through the lumen, the other through the extracapillary (cell) space. By cyclically alternating the pressure between the two circuits, medium is made to pass either into, or out of, the lumen. This allows a flushing of medium through the cell compartment and overcomes the gradient problem, and also gives the possibility of concentrating the product for harvesting. There are a range of Acusyst units, the Acusyst MiniMax and Maximiser being the laboratory models using respectively 2–3 and 3–8 litres of medium per day, and capable of running continuously for 50–150 days (four to five months being a typical run length). There is an initial growth phase of 10–25 days depending upon inoculum size, medium composition, and cell line, and when a density of 1–2 × 10^{10} cells is reached (1.1 m^2

cartridge, e.g. Acusyst Maximiser) production starts and daily harvesting of mAb begins. Cells cannot be sampled during the culture so reliance is placed upon a number of process measurements made on a regular basis—on-line pH and oxygen; off-line glucose, lactate, and if necessary ammonia and LDH (for loss of viability) at least weekly. The cultures are purchased as a turn-key unit with all accessories and full operating instructions so it is inappropriate to produce a protocol. An objective description of the operation of the Acusyst-Jr is published by Hanak and Davis (32).

The Tecnomouse (33) (Integra Biosciences) is another hollow fibre bioreactor containing up to five flat culture cassettes containing hollow fibres surrounded by a silicone membrane that gives uniform oxygenation and nutrient supply of the culture and ensures homogeneity within the culture. The system comprises a control unit, a gas and medium supply unit, and the five culture cassettes.

The culture is initiated with cells from 3×225 cm^2 flasks (5×10^7) and grows to 5×10^8 cells. The continuous medium supply is controllable and programmable with typical flow rates (perfusion or recirculation) of 30 ml/h increasing to 70 ml/h in 5 ml steps. After six days harvesting can be started and repeated every two to three days for 30–70 days. Each cassette gives 7 ml culture with 2.5–5.5 mg/ml mAb (equivalent to an average of 10 mg/day). Thus about 10 litres will yield 400 mg mAb/month and 1.5 g/month using the five cassettes (data based on manufacturer's information).

5.1.2 Membrane culture systems (miniPERM)

There have been a number of membrane-based culture units developed (9), many based on dialysis tubing, and even available as large fermenters (Bioengineering AG Membrane Laboratory Fermenter with a Cuphron dialysis membrane of 10 000 Dalton molecular weight cut-off forming an inner chamber). One of the more successful and currently available systems is described here—the miniPERM Bioreactor (Heraeus Instruments).

Protocol 7
MiniPERM Bioreactor[a]

1. Add 35 ml of cell suspension at 1–2×10^6 cells/ml in a syringe to the production module through the Luer lock.
2. Add 300–400 ml culture medium to the nutrient module.
3. Place the bioreactor on the bottle turning device in the gassed incubator and set the speed to 10 r.p.m.
4. Sample daily by removing aliquots of culture with a syringe via the Luer lock and carry out cell counts (and optionally glucose measurements).
5. At a cell density of 6×10^6/ml replace the medium in the nutrient module every four days, at higher densities every day. Serum-free medium can be used at this stage.

SCALING-UP OF ANIMAL CELL CULTURES

Protocol 7 continued

6. Typically a cell density of 15–30 × 10⁶ cells/ml is maintained after day 6 with a daily yield of 0.8 mg mAb/ml (range for various hybridomas 0.5–12 mg/day).
7. Keep the culture going until viability is lost (after 40 days).

[a] Based on Falkenberg et al. (34).

The miniPERM Bioreactor (34) consists of two components, the production module (40 ml) containing the cells and the nutrient module (600 ml of medium). The modules are separated by a semi-permeable dialysis membrane (MWCO 12.5 kDa) which retains the cells and mAb in the production module but allows metabolic waste products to diffuse out to the nutrient module. There is a permeable silicone rubber membrane for oxygenation and gas exchange in the production module. The whole unit rotates (up to 40 r.p.m.) within a CO_2 incubator. It can be purchased as a complete disposable ready to use unit or the nutrient module (polycarbonate) can be autoclaved and reused at least ten times.

5.1.3 Encapsulation

The entrapment of cells in semi-solid matrices, or spheres, has many applications, but the basic function is to stabilize the cell and thus protect it from sub-optimal conditions. Cells can be immobilized by adsorption, covalent bonding, crosslinking, or entrapment in a polymeric matrix. Materials that can be used are gelatin, polylysine, alginate, and agarose; the choice largely depends upon the problem being addressed (35, 36). The matrix allows free diffusion of nutrients and generated product between the enclosed microenvironment and the external medium. Alginate is a polysaccharide and is crosslinked with Ca^{2+} ions. The rate of crosslinking is dependent on the concentration of Ca^{2+} (e.g. 40 min with 10 mM $CaCl_2$). A recommended technique is to suspend the cells in isotonic NaCl buffered with Tris (1 mM) and 4% sodium alginate, and to add this mixture dropwise into a stirred solution of isotonic NaCl, 1 mM Tris, 10 mM Ca^{2+} at pH 7.4. The resulting spheres are 2–3 mm in diameter. The entrapped cells can be harvested by dissolving the polymer in 0.1 M EDTA or 35 mM sodium citrate. Disadvantages of alginate are that calcium must be present and phosphate absent and that large molecules, such as monoclonal antibodies, cannot diffuse out. For these reasons, agarose in a suspension of paraffin oil provides a more suitable alternative. 5% agarose in PBS free of Ca^{2+} and Mg^{2+} is melted at 70 °C, cooled to 40 °C, and mixed with cells suspended in their normal growth medium. This mixture is added to an equal volume of paraffin oil and emulsified with a Vibro-mixer. The emulsion is cooled in an ice-bath, growth medium added, and, after centrifugation, the oil is removed. The spheres (80–200 μm) are washed in medium, centrifuged and, after removing the remaining oil, transferred to the culture vessel.

A custom-made unit can be purchased (Bellco Bioreactor). This 3 litre fluidized bed culture (bubble column) actually encapsulates the cells in hydrogel beads within the culture vessel and comes with a range of modular units to control all process parameters.

5.2 Entrapment cultures

5.2.1 Opticell

This system has already been described (Section 3.3.3). The special ceramic cartridges for suspension cells (S Core) entrap the cells within the porous ceramic walls of the unit. They are available in sizes from 0.42 m^2 to 210 m^2 (multiple cartridges), which will support 5×10^{10} cells with a feed/harvest rate of 500 litres/day and give a yield of about 50 g of monoclonal antibody per day.

5.2.2 Fibres

A simple laboratory method is to enmesh cells in cellulose fibres (DEAE, TLC, QAE, TEAE; all from Sigma). The fibres are autoclaved at 30 mg/ml in PBS, washed twice in sterile PBS, and added to the medium at a final concentration of 3 mg/ml in a spinner stirred bioreactor. This method has even been found suitable for human diploid cells (37).

5.3 Porous carriers (38, 39)

Microcarriers and glass spheres are restricted to attached cells and, because a sphere has a low surface area/volume ratio, restricted in their cell density potential. Their advantages are summarized in *Table 6*. A change from a solid to porous sphere of open, interconnecting pores (*Figure 12*), increases their potential enormously. There are various types of porous (micro)carrier commercially available (*Table 4*). A characteristic of these porous carriers is their equal suitability

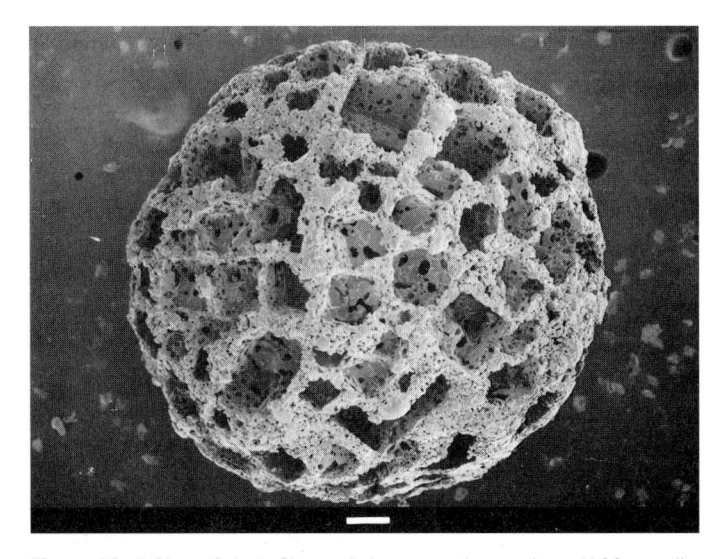

Figure 12. A Siran (Schott Glaswerke) porous glass sphere (120 μm diameter).

SCALING-UP OF ANIMAL CELL CULTURES

Table 6. Porous carriers—advantages compared to solid carriers

(a)	Unit cell density 20- to 50-fold higher.
(b)	Support both attached and suspension cells.
(c)	Immobilization in 3D configuration easily achieved.
(d)	Short diffusion paths into a sphere.
(e)	Suitable for stirred, fluidized, or fixed bed reactors.
(f)	Good scale-up potential by comparison with analogous systems (e.g. microcarrier at 4000 litre).
(g)	Cells protected from shear.
(h)	Capable of long-term continuous culture.

for suspension cells (by entrapment) and anchorage-dependent cells (huge surface area).

The problem with many immobilization materials is that diffusion paths become too long, preventing scale-up. A sphere is ideal in that cells and nutrients have only to penetrate 30% of the diameter to occupy 70% of the total volume. This facilitates scale-up as each sphere, whether in a stirred, fluidized, or fixed bed culture, can be considered an individual mini-bioreactor.

5.3.1 Fixed bed (Porosphere)

The apparatus and experimental procedures described above (Section 3.3.2.i) for solid spheres is equally suited to porous Siran spheres (Schott) of 4–6 mm diameter (39). The only differences are that the bed should be packed with oven-dried spheres and the void volume of cells plus medium inoculated (2×10^6/ml) directly to the bed (dry beads permit better penetration of cells). A larger medium volume is required, and faster perfusion rates (5–20 linear cm/min) should be used (cells are protected from medium shear within the matrix), e.g. 1 litre packed bed needs at least a 15 litre reservoir. After the initial 72 h period, 10 litres of fresh medium is added daily.

Protocol 8
Fixed porous bead (Siran) bed reactor

This protocol is based on a 1 litre fixed bed (water jacket temperature control) of 5 mm Siran beads (Schott) perfused from a 15 litre (11 litre working volume) Applikon BV stirred tank with pH and oxygen control as a medium reservoir.

1. Preliminary preparation:
   (a) Beads. Boil in distilled water for 2 h, rinse, and oven dry (4 h at 100°C). Rinse again and heat dry, then heat sterilize (180°C for 2 h) in a sealed beaker. Add aseptically to autoclaved culture vessel.
   (b) Reservoir vessel. Calibrate probes, assemble, and autoclave.
2. Suspend inoculum (5×10^9 cells per litre bed volume) in 800 ml pre-warmed medium (it is recommended that the medium is circulated for several hours

Protocol 8 continued

   through the bed to both equilibrate the system, check the probes, and condition the beads before inoculation).

3. Add cell suspension to the inoculation vessel, connect to the bottom connector of the glass bead vessel, and slowly add the inoculum by air pressure (by hand pump) being careful not to introduce air bubbles.

4. Drain and refill the bed twice.

5. Start perfusing the medium at 40 ml/min or a linear flow velocity of 2 cm/min.

6. Monitor the reservoir daily for glucose concentration and when the level falls below 2 mg/ml change the medium for a fresh supply (the harvest). Alternatively the system can be run in a continuous mode where the medium feed rate is adjusted to maintain a glucose concentration of 2 mg/ml.

7. Run the culture until the cells lose viability or the bed becomes blocked with cell debris, usually after 50–60 days. This can be delayed further if occasionally the perfusion rate is significantly increased for a short time, or the culture is set up for alternate up and down-flow.

8. At the end of the culture, should a cell count be needed, remove a measured aliquot of beads (e.g. 20 cc), add 100 ml of 0.1% citric acid and 0.2% Triton X-100, place on an orbital shaker (150 rev/min) at 37°C for 2 h. Collect supernatant, measure the volume, wash beads with PBS, add crystal violet to the supernatant (0.1% final concentration), and count on a haemocytometer.

Typical results in the author's laboratory (38–40) with *Protocol 8* are 2.75×10^{10} viable cells/litre giving an average yield of 166 mg/litre/day (compared to stirred reactor and airlift cultures of the same hybridoma of 25.5 and 18.5 mg/litre respectively). This method is a low investment introduction to high productivity production of mAb which is simple to use and reliable with low maintenance, at least for the first 50 days of culture.

5.3.2 Fluidized beds

Porous microcarrier technology, currently the most successful scale-up method for high density perfused cultures, was pioneered by the Verax Corporation (now Cellex Biosciences Inc.). Turn-key units were available from 16 ml to 24 litres fluidized beds (41). The smallest system in the range, Verax System One, is a benchtop continuous perfusion fluidized bioreactor suitable for process assessment and development, and also for laboratory scale production of mAbs. Cells are immobilized in porous collagen microspheres, weighted to give a specific gravity of 1.6, which allows high recycle flow rates (typically 75 cm/min) to give efficient fluidization. The microspheres have a sponge-like structure with a pore size of 20–40 μm and internal pore volume of 85% allowing immobilization of cells to densities of $1-4 \times 10^8$/ml. They are fluidized in the form of a slurry.

The Verax system comprises a bioreactor (fluidization tube), a control system

Table 7. Commercially available porous (micro)carriers

Trade name	Supplier	Material	Diam (μm)	Culture mode[a]
Cultispher G	HyClone	Gelatin	170–270 300–500	F
Cytocell	Pharmacia	Cellulose	180–210	S
Cellsnow	Kirin Ltd.	Cellulose	800–1000	S
Cytoline 1, 2	Pharmacia	Polyethylene	1200–1500	F
ImmobaSil	Ashby Scientific	Silicone rubber	1000	S
Microsphere	Cellex Biosciences	Collagen	500–600	F
Siran	Schott Glasswerke	Glass	600–1000 4000–6000	F X
Microcarrier	Asahi Chem. Ind.	Cellulose	300	S

[a] S, stirred; F, fluidized; X, fixed bed.

(for pH, oxygen, medium flow rates), gas and heat exchanger and medium supply, and harvest vessels. The system is run continuously for long periods (typically over 100 days). In the author's laboratory it produced 15×10^{10} cells/litre and 540 mg mAb/litre/day (compared to 166 in the fixed bed described above, 25.5 in a stirred reactor, and 18.5 mg in an airlift fermenter). Protocols for its operation come with the equipment and versions have been published (42). In summary, it is probably the most productive system available, giving the cells a very high specific production rate, but does require some skill to operate to its maximum potential.

An alternative commercial system is the Cytopilot (Pharmacia Vogelbusch) which is a fluidized system using polyethylene carriers (Cytoline) and supports 12×10^7 cells/ml carrier (43). It is available as the Cytopilot-Mini (400 ml bed) for laboratory scale operation as well as sizes up to 25 litres. The unit has a magnetic stirrer that drives the medium up through a distribution plate into the upper chamber in which the microcarriers are lifted by hydrodynamic pressure. The degree of fluidization is controlled by the stirrer speed and a clear boundary layer is kept at the top of the culture so that clear medium can flow through the internal central circulation tube (loop) back to the stirrer. The culture is oxygenated by means of a mini-sparger delivering microbubbles and a medium feed rate of up to 25 bed matrix volumes per day giving high productivity. The unit can also be used as a packed bed bioreactor if the circulation system is reversed.

There is a range of porous microcarriers available (*Table 7*) for fluidized systems. Alternatively, some of the carriers are designed for stirred cultures.

5.3.3 Stirred cultures

The Cultispher-G (gelatin), Cellsnow (cellulose), and ImmobaSil (silicone rubber) (44) microcarriers are the most suited to stirred bioreactors, and can be used in an identical manner to solid microcarriers, i.e. 2 g/litre in shake flasks, spinner flasks, or stirred fermenters. The silicone rubber of the ImmobaSil microcarriers facilitates oxygen diffusion (44) and this offers a great advantage over other formulations. About a 40-fold higher concentration of attached cells, and even greater densities of suspension cells, can be achieved over solid microcarriers. The cells can be released from the microcarriers by collagenase treatment.

References

1. Mosmann, T. (1983). *J. Immunol. Methods*, **65**, 55.
2. Racher, A., Looby, D., and Griffiths, J. B. (1990). *Cytotechnology*, **3**, 301.
3. Ronning, O. (1993). In *Cell and tissue culture: laboratory procedures* (ed. A. Doyle, J. B. Griffiths, and D. Newell), Module 8C:2.1. John Wiley & Sons, Chichester.
4. Griffiths, J. B. (1972). *J. Cell Sci.*, **10**, 512.
5. Murakami, H. (1989). In *Advances in biotechnological processes*, Vol. 11, p. 107. Alan R. Liss, New York.
6. Good, N. E. (1963). *Biochemistry*, **5**, 467.
7. Leibovitz, A. (1963). *Am. J. Hyg.*, **78**, 173.
8. Spier, R. E. and Griffiths, J. B. (1984). *Dev. Biol. Stand.*, **55**, 81.
9. Griffiths, J. B. (1988). In *Animal cell biotechnology* (ed. R. E. Spier and J. B. Griffiths), Vol. 3, p. 179. Academic Press, London.
10. Prokop, A. and Rosenberg, M. Z. (1989). *Adv. Biochem. Eng.*, **39**, 29
11. Toth, G. M. (1977). In *Cell culture and its applications* (ed. R. T. Acton and J. D. Lynn), p. 617. Academic Press, New York.
12. Griffiths, J. B. (1984). *Dev. Biol. Stand.*, **55**, 113.
13. Griffiths, J. B. (1990). *Cytotechnology*, **3**, 106.
14. Maroudas, N. G. (1975). *J. Theor. Biol.*, **49**, 417.
15. Kruse, P. J., Keen, L. N., and Whittle, W. L. (1970). *In Vitro*, **6**, 75.
16. Skoda, R., Pakos, V., Hormann, A., Spath, O., and Johansson, A. (1979). *Dev. Biol. Stand.*, **42**, 121.
17. Beeksma, I. A. and Kompier, R. (1995). In *Animal cell technology; developments towards the 21st Century* (ed. E. C. Beuvery, J. B. Griffiths, and W. P. Zeeijlemaker), p. 661. Kluwer Academic Publishers, Dordrecht.
18. Berg, G. J. and Bodeker, G. D. (1988). In *Animal cell biotechnology* (ed. R. E. Spier and J. B. Griffiths), Vol. 3, p. 322. Academic Press, London.
19. Whiteside, J. P. and Spier, R. E. (1981). *Biotech. Bioeng.*, **23**, 551.
20. Weiss, R. E. and Schleiter, J. B. (1968). *Biotech. Bioeng.*, **10**, 601.
21. van Wezel, A. L. (1967). *Nature*, **216**, 64.
22. Spier, R. E. and Whiteside, J. P. (1984). *Dev. Biol. Stand.*, **55**, 151.
23. Griffiths, J. B., Cameron, D. R., and Looby, D. (1987). *Dev. Biol. Stand.*, **66**, 331.
24. Paul, J. and Struthers, M. G. (1963). *Biochem. Biophys. Res. Commun.*, **11**, 135.
25. Capstick, P. B., Garland, A. J., Masters, R. C., and Chapman, W. G. (1966). *Exp. Cell Res.*, **44**, 119.
26. Heidemann, R., Riese, U., Lutkeymeyer, D., Buntemeyer, H., and Lehmann, J. (1994). *Cytotechnology*, **14**, 1.
27. Tovey, M. G. (1980). *Adv. Cancer Res.*, **33**, 1.
28. Pirt, S. J. and Callow, D. S. (1964). *Exp. Cell Res.*, **33**, 413.
29. Hayter, P. (1993). In *Cell and tissue culture: laboratory procedures* (ed. A. Doyle, J. B. Griffiths, and D. Newell), Module 28B:4.1. John Wiley & Sons, Chichester.
30. Merchuk, J. C. and Siegal, M. H. (1988). *J. Chem. Technol. Biotechnol.*, **41**, 105.
31. Tyo, M. A., Bulbulian, B. J., Menken, B. Z., and Murphy, T. J. (1988). In *Animal cell biotechnology* (ed. R. E. Spier and J. B. Griffiths), Vol. 3, p. 357. Academic Press, London.

32. Hanak, J. A. J. and Davis, J. M. (1993). In *Cell and tissue culture: laboratory procedures* (ed. A. Doyle, J. B. Griffiths, and D. G. Newell), p. 28D:3.1. John Wiley & Sons, Chichester.
33. Singh, R. P., Fassnacht, D., Perani, A., Simpson, N. H., Goldenzon, C., Portner, R., *et al.* (1998). In *New developments and new applications in animal cell technology* (ed. O.-W. Merten, P. Perrin, and J. B. Griffiths), p. 235. Kluwer Academic Publishers, Dordrecht.
34. Falkenberg, F. W., Weichart, H., Krane, M., Bartels, I., Palme, M., Nagels, H.-O., *et al.* (1995). *J. Immunol. Methods*, **179**, 13.
35. Nilsson, K. and Mosbach, K. (1980). *FEBS Lett.*, **118**, 145.
36. Nilsson, K., Scheirer, M., Merten, O. W., Ostberg, L., Liehl, E., and Katinger, H. W. D. (1983). *Nature*, **302**, 629.
37. Litwin, J. (1985). *Dev. Biol. Stand.*, **60**, 237.
38. Griffiths, J. B. (1990). In *Animal cell biotechnology* (ed. R. E. Spier and J. B. Griffiths), Vol. 4, p. 147. Academic Press, London.
39. Looby, D. and Griffiths, J. B. (1989). *Cytotechnology*, **2**, 339.
40. Looby, D. and Griffiths, J. B. (1998). In *Laboratory procedures in biotechnology* (ed. A. Doyle and J. B. Griffiths), p. 268. John Wiley & Sons, Chichester.
41. Runstadler, P. W. Jr., Tung, A. S., Hayman, E. G., Ray, N. G., Sample, J. V. G., and DeLucia, D. E. (1989). In *Large scale mammalian cell culture technology* (ed. A. S. Lubiniecki), p. 363. Marcel Dekker, New York.
42. Looby, D. (1993). In *Cell and tissue culture: laboratory procedures* (ed. A. Doyle, J. B. Griffiths, and D. G. Newell), p. 28D:1.1. John Wiley & Sons, Chichester.
43. Valle, M. A., Kaufman, J., Bentley, W. E., and Shiloach, J. (1998). In *New developments and new applications in animal cell technology* (ed. O.-W. Merten, P. Perrin, and J. B. Griffiths), p. 381. Kluwer Academic Publishers, Dordrecht.
44. Looby, D., Racher, A. J., Fuller, J. P., and Griffiths, J. B. (1995). In *Animal cell technology: developments towards the 21st Century* (ed. E. C. Beuvery, J. B. Griffiths, and W. P. Zeeijlemaker), p. 783. Kluwer Academic Publishers, Dordrecht.

Chapter 3
Cell line preservation and authentication

R. J. Hay, M. Miranda-Cleland, S. Durkin, and
Y. A. Reid

American Type Culture Collection, 10801 University Boulevard, Manassas,
VA 20110-2209, USA.

1 Introduction

Literally thousands of different cell lines have been derived from human and other metazoan tissues. Many of these originate from normal tissues and exhibit a definable, limited doubling potential. Other cell lines may be propagated continuously, either having gone through an engineered or spontaneous genetic change from the normal primary population, or having been developed initially from tumour tissue. Both finite lines of sufficient doubling potential and continuous lines can be expanded to produce a large number of aliquots, frozen, and authenticated for widespread use in research. Note: use of the terms cell line and cell strain is as recommended by The Tissue Culture Association committee on nomenclature (1).

The advantages of working with a well-defined cell line free from contaminating organisms would appear obvious. Unfortunately, however, the potential pitfalls associated with the use of cell lines obtained and processed casually require repeated emphasis. Numerous occasions where cell lines exchanged among co-operating laboratories have been contaminated with cells of other species have been detailed and documented elsewhere (2, 3). For example, lines supposed to be human have been found to be monkey, mouse, or mongoose while others thought to be monkey or mink were identified as rat and dog (2). Similarly, the problem of intraspecies cross-contamination among cultured human cell lines has been recognized for over thirty years and detailed reviews are available on the subject (4). The loss of time and research funds as a result of these problems is incalculable.

While bacterial and fungal contaminations represent an added concern, in most instances they are overt and easily detected, and are therefore of less serious consequence than the more insidious contaminations by mycoplasma. That the presence of these micro-organisms in cultured cell lines often negates research findings has been stated repeatedly (5, 6). However, the prevalence of contaminated cultures in the research community suggests that the problem cannot be overemphasized.

In this chapter the procedures used to preserve and authenticate cell lines and hybridomas at the American Type Culture Collection (ATCC) are outlined. They have evolved over the past 37 years as awareness of the difficulties of microbial and cellular cross-contaminations became apparent and as the need for authenticated cell lines increased.

2 Cell line banking

Cell lines pertinent for accessioning are selected by ATCC scientists and advisors during regular reviews of the literature. The originators themselves also frequently offer the lines directly for consideration. Detailed information with regard to specific groups of cell lines accessioned is available elsewhere (7-9). Generally, starter cultures or ampoules are obtained from the donor, and progeny are propagated according to instructions to yield the first 'token' freeze. Cultures derived from such token material are then subjected to critical characterizations as described below. If these tests suggest that further efforts are warranted, the material is expanded to produce seed and distribution stocks. Note especially that the major authentication efforts are applied to cell populations in the initial seed stock of ampoules. The distribution stock consists of ampoules that are distributed on request to investigators. The reference seed stock, on the other hand, is retained to generate further distribution stocks as the initial stock becomes depleted.

Although this procedure has been developed to suit the needs of a large central repository, it is also applicable in smaller laboratories. Even where the number of cell lines and users may be limited, it is important to separate 'seed stock' from 'working or distribution stock'. Otherwise the frequent replacement of cultured material, recommended to prevent phenotypic drift or senescence, may deplete valuable seed stock which may be difficult and expensive to replace.

These various steps in the overall accessioning scheme are summarized in *Figure 1*. It is important to recognize that the characterized seed stock serves as a frozen 'reservoir' for production of distribution stocks over the years. Because seed stock ampoules are used to generate new distribution material, one can assure recipients that all the cultures obtained closely resemble those received ten or more years previously. This is a critical consideration for design of cell banking procedures. Problems associated with genetic instability, cell line selection, senescence, or transformation may be minimized or avoided entirely by strict adherence to this principle.

It is prudent to handle all primary tissues or cell lines not specifically shown to be free of adventitious agents as biohazards in a Class II vertical laminar flow hood (BS5726; NIH Spec. 03-112). This precaution protects both the cell culture technician and the laboratory from infection or contamination. Some recommend furthermore that all human tumour lines be treated with similar caution due to the known presence of oncogenes.

Accessioning Scheme

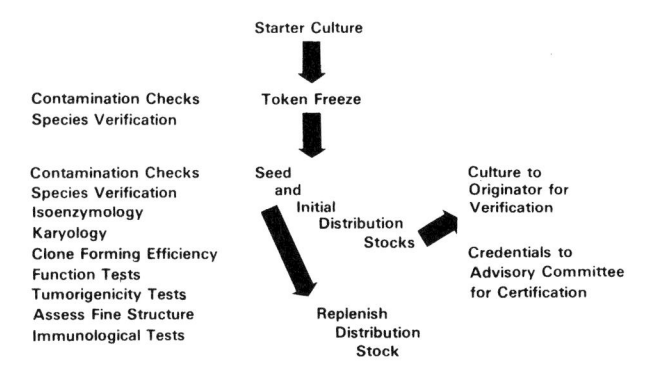

```
                        Starter Culture
                              ↓
Contamination Checks    Token Freeze
Species Verification          ↓
Contamination Checks    Seed              Culture to
Species Verification    and               Originator for
Isoenzymology           Initial           Verification
Karyology               Distribution
Clone Forming Efficiency        Stocks
Function Tests                            Credentials to
Tumorigenicity Tests                      Advisory Committee
Assess Fine Structure   Replenish         for Certification
Immunological Tests     Distribution
                        Stock
```

Figure 1. Accessioning scheme for new cell lines. Scheme illustrating a recommended plan for the accession of cell lines to be banked for general distribution. At least the first two of the various characterizations indicated on the left-hand side should be performed before release of any cell line.

3 Cell freezing and quantitation of recovery

Cellular damage induced by freezing and thawing is generally believed to be caused by intracellular ice crystals and osmotic effects. The addition of a cryoprotective agent, such as dimethyl sulfoxide (DMSO) or glycerol, and the selection of suitable freezing and thawing rates minimizes cellular injury.

While short-term storage of cell lines using mechanical freezers ($-75\,°C$) is possible, storage in liquid nitrogen ($-196\,°C$) or its vapour (to $-135\,°C$) is much preferred. The use of liquid nitrogen refrigerators is advantageous not only because of the lower temperatures and, consequently, almost infinite storage times possible, but also due to the total absence of risk of mechanical failures and the prolonged holding times now available. Certainly for all but the smallest cell line banking activity, storage in a liquid nitrogen refrigerator is essential.

Two considerations on safety in processing cell lines require special emphasis. First, the cell culture technician may be endangered due to the possibility that liquid nitrogen can penetrate ampoules via hairline leaks during storage. On warming, rapid evaporation of the nitrogen within the confines of such an ampoule can cause a sharp explosion with shattered glass or plastic flying at high force almost instantaneously in all directions. Fortunately the frequency of such traumatic accidents declines dramatically as the operator gains experience in ampoule sealing and testing (Sections 3.1.1 and 3.3). However, even with highly accomplished laboratory workers, the remote possibility of explosion still exists. Thus, a protective face mask should be worn whenever ampoules are removed from liquid nitrogen storage and until they have been safely opened in the laminar flow hood.

Secondly, DMSO can solubilize organic substances and, by virtue of its penetrability through rubber and skin, carry these to the circulation. Thus, special precautions should be exercised when using DMSO to avoid contamination with hazardous chemicals and minimize skin contact.

3.1 Equipment

3.1.1 Ampoules, marking, and sealing devices

A decision on whether to utilize glass or plastic ampoules will depend on the scale of operation and the extent of anticipated distribution. Glass ampoules can be sterilized, loaded with the appropriate cell suspension, and permanently sealed in comparatively large quantities. They are recommended for large lots of cells (20 ampoules or more) being prepared for long-term use or general distribution. Smaller numbers (1–20) of plastic ampoules are easier than glass to handle, mark, and seal. Problems with the seal may occur in some cases, however, especially if frequent handling or manipulations for shipment are necessary.

The marking of ampoules requires special consideration in that legibility can easily be obscured as the ampoules are frozen, snapped on and off storage canes, transferred between freezers or shipping containers, and so forth. The use of paper labels, ballpoint pen, or standard laboratory markers all are problematical and this is especially true with glass ampoules. These should be labelled in advance with ceramic ink that can be heat annealed to the glass surface. The markings can be applied by hand with a straight pen or, if large lots are being processed, through use of a mechanical labeller (e.g. from Markem Co., Keene, New Hampshire).

Glass ampoules can best be sealed by pulling on the neck of the ampoule as it is rotated in the highest heat zone of the flame from a gas–oxygen torch. This pull-seal technique is preferred since it reduces the risk of permitting pin-hole leaks in the sealed tip. Torches for manual sealing can be obtained from scientific supply houses. For large lots, a torch can be attached to a semi-automatic sealing device available as Bench sealer model 161 from Morgan Sheet Metal Co., Sarasota, FL, USA.

3.1.2 Slow freezing apparatus

The optimum freezing rate for cell lines (for most cells about $-1\,°C/min$) can be achieved through use of apparatus varying in complexity from a tailor-made styrofoam box to a completely programmable freezing unit. The former should have a wall thickness of about 2 cm to approximate the $-1\,°C$ cooling rate when placed in a mechanical freezer at $-70\,°C$. Alternatively, manufacturers of liquid nitrogen refrigerators supply adapted refrigerator neck plugs, at modest cost, which can be adjusted for slow freezing of numbers of ampoules. For those who produce larger quantities of ampoules and require more precise control of the freezing rate, a controlled rate freezer (e.g. Cryo-Med; Planer; Union Carbide) should be considered.

3.1.3 Liquid nitrogen refrigerators

Choice of an appropriate refrigerator will require considerations of economy, both in terms of liquid nitrogen consumption and initial outlay, storage capacity required, and desired ease of entry and retrieval. Freezers with narrower neck tube openings are generally more economical.

Ampoules of cells may be stored immersed in liquid nitrogen or in the vapour phase. The latter has the advantage that ampoules with pin-hole leaks will not be exposed to liquid so the danger of explosion is eliminated. The slightly higher temperature probably offers no disadvantage except perhaps with seed stocks retained for extremely long periods.

Useful accessories for refrigerators include roller bases for ease of movement, alarm systems to warn of dangerously low levels of liquid nitrogen, and racking systems, with larger refrigerators, for ready storage and recovery. Even when an automatic alarm system is used this should be backed-up with a regular manual check using a dip-stick. Electronic systems can fail and it has been known for both the automatic fill system and its back-up alarm to fail.

3.2 Preparation and freezing

Cultures in the late logarithmic or just pre-confluent phase of growth should be selected to give the highest possible initial viability. Treat the cultures with trypsin if necessary to produce a uniform single cell suspension as if for routine subcultivation and proceed as in *Protocol 1*.

Protocol 1
Culture preparation and freezing procedure

Equipment and reagents

- Ampoules
- Liquid nitrogen
- Cryoprotective agent: DMSO or glycerol
- Aqueous methylene blue (0.05%)

Method

1. Just before use, prepare the freeze medium by simple admixture of fresh tissue culture medium and the required cryoprotective agent, generally DMSO or glycerol. The concentrations of choice vary slightly with the cell line and cryoprotective agent used, ranging from 5–10% (v/v). Add serum last to avoid protein precipitation by high DMSO concentrations.

2. Collect the cells in pellet form by centrifugation at about 200 g for 10 min and resuspend in an appropriate volume of freeze medium at room temperature. Concentrations of 10^6 to 10^7 cells/ml are generally satisfactory and practical. For some applications it may be desirable to increase or decrease this by as much as 1 log.

3. If the total volume is large, as for a production-level freeze, maintain the cell suspension with gentle agitation in an appropriate stirring vessel. Set this up in

> **Protocol 1** continued
>
> advance complete with a Cornwall automatic syringe. If the cell suspension is in a small volume (10–20 ml), dispense by means of a syringe fitted with an 18 gauge cannula. Mix repeatedly during this process.
>
> 4. Maintain the pH when necessary by gassing with an appropriate mixture of water-saturated air/CO_2.
>
> 5. Dispense the cell suspension in 1 ml aliquots to the ampoule taking care to proceed rapidly and with uniformity. Cells sediment quickly under unit gravity and inaccuracies can result if the process is interrupted.
>
> 6. Seal each ampoule, place on an ampoule cane or in a suitable rack, and immerse totally upright in aqueous methylene blue (0.05%) at 4°C. After 30–45 min remove, wash in cold tap-water, and discard any ampoule containing the blue dye.
>
> 7. Dry the ampoules thoroughly and begin slow freezing. The optimum rate of cooling is usually −1°C/min but varies among different cell lines and can best be determined empirically before any large scale freeze is attempted.
>
> 8. When ampoules have reached −70°C or lower, immerse them in liquid nitrogen and then transfer them immediately and rapidly to the liquid nitrogen refrigerator.
>
> 9. Record the location and specifics of the cell line in question. Two separate cross-index files providing both subject and location cards can be used for this purpose. Alternatively, computerized systems with the essential back-up may be devised.

3.3 Reconstitution and quantitating recovery

Rapid thawing of the cell suspension is essential for optimal recovery.

A variety of methods may be used to quantitate cell recovery after freezing. Of course this is the first characterization step to be performed after preservation.

Protocol 2

Recovery of cultures frozen in glass ampoules

Equipment and reagents

- Full face mask
- Water-bath
- 70% ethanol
- Complete growth medium

Method

1. With face and neck protected by a full face mask, retrieve the selected ampoule rapidly and plunge directly into a warm water-bath (37°C). If ampoules have been stored submerged in liquid nitrogen each ampoule must be thawed separately by depositing directly into about 10 cm of water at 37°C in a 1.5–2 litre bucket with a lid, snapping the lid closed immediately after the ampoule is inserted.

CELL LINE PRESERVATION AND AUTHENTICATION

Protocol 2 continued

2. Agitate the ampoule contents until the suspension has thawed completely (20–60 sec).
3. Immerse the ampoule in 70% ethanol at room temperature.
4. Score standard ampoules using a small file that has been dipped previously in ethanol. Pre-scored ampoules require no such treatment.
5. Use a sterile towel to pick up the ampoule; break sharply at the neck at the pre-scored point. **CONTINUE WEARING FACE MASK AT LEAST THROUGH THIS STAGE.**
6. Transfer the contents, by means of a sterile Pasteur pipette, to a centrifuge tube or culture vessel. Add 10 ml of complete growth medium. In some cases, slow addition over a 1–2 min interval at this point may be beneficial.
7. Centrifuge at 200 g for 10 min. Remove the supernatant, add a fresh aliquot of medium, and mix the suspension. This centrifugation step may be omitted in some cases since the residual concentration of cryoprotective agent is low and the stress of centrifugation can be harmful.
8. Initiate the culture by standard procedure. If the centrifugation step was omitted, change the medium after 24 h.

3.3.1 Dye exclusion for quantitating cell viability

A very approximate estimate of the viability of cells in a suspension may be obtained by the dye exclusion test. A solution of the dye in saline is added to the suspension and the percentage of cells which do not take up the stain is determined by direct count using a haemocytometer. Trypan blue or erythrocin B are the stains most commonly used for this procedure. The former reportedly has a higher affinity for protein in solution than for non-viable cells. This may reduce the accuracy of the estimate if the suspension contains much more than 1% serum. Furthermore, because solutions of erythrocin B are clear, microbial growth or precipitates are immediately apparent. This is not true for stock solutions of trypan blue. *Protocol 3* describes the recommended procedure.

Protocol 3
The dye exclusion test

Reagents
- Stock dye solution: 100 mg erythrocin B/100 ml of an isotonic phosphate-buffered saline (PBS) adjusted to pH 7.2–7.4 with 1 M NaOH

Method
1. Obtain a uniform suspension of cells in growth by any standard procedure. If the estimate is for freeze characterization, reconstitution of contents from about 5% of the ampoules prepared is recommended.

Protocol 3 continued

2. Dilute an accurately measured aliquot of the cell suspension using the stock dye solution. For ease and accuracy of count a final density of about 0.3–2×10^6 cells/ml is satisfactory.
3. Score the number of stained versus total number of cells keeping in mind that, in general, a larger sampling (300 cells or more) will give a more accurate quantitative estimate. Ideally the count should be performed within 1–5 min after mixing of the dye and cell suspension.

3.3.2 Colony-forming efficiency

The dye exclusion test for cell viability generally overestimates recovery. For lines consisting of adherent cells, the colony-forming ability of cells from the reconstituted population represents a more accurate overall estimate of survival. Of course the choice of growth medium used, the substrate on which the colonies develop, and the incubation time all may have an effect on the end-result. Thus for comparisons among different freezes, conditions for selected lines must be standardized. A representative outline as used at the ATCC is provided in *Protocol 4*.

Protocol 4
Estimation of colony-forming efficiency

Equipment and reagents

- T-75 flasks
- Dissecting microscope or automated colony counter
- Complete growth medium
- 10% solution of formaldehyde or other suitable fixative
- 1% aqueous toluidine blue or other simple stain

Method

1. Pool the reconstituted contents of ampoules from about 5% of a freeze lot and serially dilute the suspension to provide inocula of 100–10^4 cells/culture depending upon the cell line, plate size, and expected recoveries. For example, cells of a line with high plating efficiency could be added at 10 cells, 100 cells, 1000 cells, and 10 000 cells to four separate plate sets, assuming 9 cm plates were used.
2. Make the serial dilutions by tenfold reductions in cell number by transferring 1 ml of suspension to 9 ml of the selected, complete growth medium. Discard the 1 ml pipette and use a fresh 1 ml pipette to mix and make each subsequent, lower density transfer for dilution.
3. Inoculate 1 ml of each dilution, beginning with the lowest cell concentration, to each of three T-75 flasks (or other suitable culture vessel) containing 8 ml of growth medium and incubate at 37 °C.

CELL LINE PRESERVATION AND AUTHENTICATION

Protocol 4 continued

Note: if media with high bicarbonate are used with low cell inocula, even sealed plastic flasks will have to be equilibrated *and incubated* with the appropriate air/CO$_2$ mixture to maintain pH.

4. Renew the fluid on test cultures on the fourth or fifth day and thereafter every third or fourth day.
5. After 12–14 days total, remove the fluid, and fix the culture with a 10% solution of formaldehyde or other suitable fixative.
6. Remove the fixative, rinse the flask interior gently with several changes of tap-water, and stain with 1% aqueous toluidine blue or other simple stain for 1–5 min.
7. Remove the staining solution, rinse out residual fluid with tap-water, and count colonies consisting of 32 or more cells through use of a dissecting microscope or automated colony counter.
8. Calculate the per cent colony-forming efficiency:

$$\frac{\text{colonies formed}}{\text{number of cells inoculated}} \times 100$$

3.3.3 Proliferation in mass culture

The vitality of reconstituted cells from either adherent or non-adherent lines can also be documented by simply quantitating proliferation during the initial one to two weeks after recovery from liquid nitrogen, as in *Protocol 5*.

Protocol 5

Measuring vitality of reconstituted cells

Equipment and reagents

- Culture vessels (e.g. T-25 flasks)
- Electronic cell counter or haemocytometer
- Culture fluid

Method

1. Using the cell suspension pooled from 5% of a freeze lot, inoculate three sets of culture vessels (e.g. T-25 flasks) at different densities. Typically one might choose the inoculum expected to yield a confluent or maximum density culture in seven days for one set; twice that for a second set; and three to five times that (depending on viable count) for the third set.
2. Renew the culture fluid (or add fresh medium for a suspension culture) after four days.
3. Harvest the three sets of cultures by standard trypsinization if an adherent culture

> **Protocol 5** continued
>
> or by direct sampling if a suspension culture, and determine the cell yield by either electronic cell count or use of a haemocytometer.
>
> 4. Calculate the fold increase (number of cells recovered/number of cells inoculated) and compare this with what one would expect under similar conditions with cells taken during logarithmic or pre-confluent growth phase.

4 Cell line authentication

In addition to recoverability from liquid nitrogen, the absolute minimum recommended authentication steps include verification of species and demonstration that the cell line is free of bacterial, fungal or mycoplasmal contamination.

4.1 Species verification

Species of origin can be determined for cell lines using isoenzyme profiles or by cytogenetics (10). These two methods are used at ATCC.

4.1.1 Isoenzyme profiles

Isozyme analyses performed on homogenates of cell lines from over 25 species have demonstrated the utility of this biochemical characteristic for species verification (11). By determining the mobilities of multiple isozyme systems one can identify the species of origin of cell lines with a high degree of certainty. The procedures are relatively straightforward, provide consistent results, and do not require expensive equipment.

A kit (AuthentiKit™, Innovative Chemistry, Inc.) has been developed for this purpose. Pre-cast 1% agarose gels on a polystyrene film backing, buffers, enzyme substrates with stabilizers, and appropriate control extracts are available. Specially constructed electrophoretic chambers are utilized after coupling to power supplies. Over 15 different enzyme systems can be evaluated.

The advantages of the kit include the convenience of ready-made gels and reagents plus the significantly lower times required for electrophoretic separations (15–45 min). After drying, the gels can be retained to document cell line characteristics (*Figure 2*).

4.1.2 Cytogenetics

Karyologic techniques have been used to monitor for interspecies contamination among cell lines. In many instances the chromosomal constitutions are so dramatically different that even cursory microscopic observations are adequate. In others, as for example comparisons among cell lines from closely related primates, careful evaluation of banded preparations (10) is required (see Section 4.3.1).

The standard method described in *Protocol 6* involves the swelling of metaphase-arrested cells by brief exposure to hypotonic saline (*Table 1*). The cells

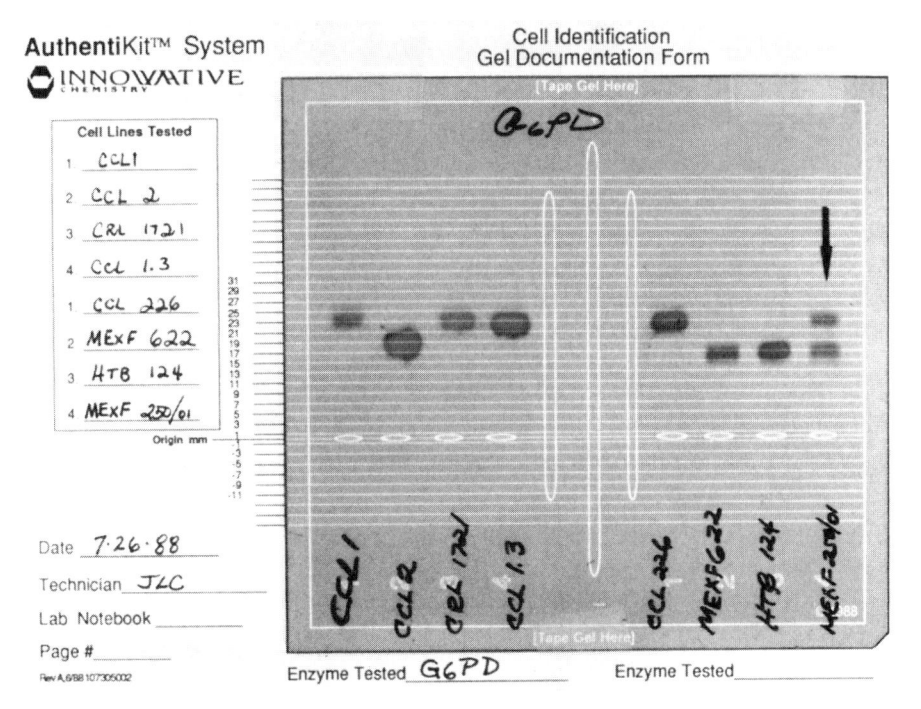

Figure 2. Isoenzyme gel. An agarose gel processed for glucose-6-phosphate dehydrogenase using the AuthentiKit™. The gel was loaded with extracts from murine and human cells plus one murine/human mixture. Lanes 2 in the left section (LHS) and 2 and 3 in the right section (RHS) are human. Lanes 1, 3, and 4 (LHS) and 1 (RHS) are from extracts of murine lines. Lane 4 (RHS and *arrow*) is from the mixed line.

Table 1. Stock solutions for routine karyology

Colcemid stock:	1 mg/50 ml double glass distilled water (DGDW); store frozen in small aliquots
KCl:	0.075 M in DGDW
Fixative:	3 parts anhydrous methyl alcohol, 1 part glacial acetic acid (combine just before use)
Acetic orcein stain:	2 g natural orcein dissolved in 100 ml of 45% acetic acid
Giemsa stain:	10% in 0.01 M phosphate buffer at pH 7 (solution available commercially)

are then fixed, applied to slides to optimize spreading, stained, and mounted for microscopic observation.

The frequency of introduction of artefacts through this method will vary depending upon the cell line and the degree of experience. Rupturing of cells will occur, for example, and apparent losses or gains in chromosomes will result. However, by counting the chromosomes in 50–100 well-spread metaphases and recording the modal number the cytogeneticist can obtain a reliable estimate for a specific line.

Protocol 6
Karyological procedure

Equipment and reagents
- 15 ml centrifuge tube
- Slides
- Microscope
- Colcemid
- Trypsin
- Hypotonic KCl solution
- Fixative
- Acetic acid orcein or Giemsa

Method
1. To a culture (T-75) in the exponential growth phase add colcemid to give a final concentration of 0.1–0.4 µg/ml.
2. Incubate for 1–6 h, selecting the length of this period according to the cycling time of the cell population under study. Diploid human cells with relatively long doubling times would generally require longer incubation than a rapidly-proliferating Chinese hamster ovary cell line.
3. Gently decant the supernatant and treat the adherent layer with trypsin as for a standard subcultivation. Omit this step if working with a suspension culture. Place the suspension in a 15 ml centrifuge tube. Add serum to a final concentration of approximately 10% of the cell suspension to neutralize further trypsin action.
4. Collect the suspended cells by centrifugation, discard the supernatant, and resuspend the pellet in the hypotonic KCl solution at 37 °C.
5. After 10–15 min incubation at 37 °C, sediment the cells by centrifugation at 100 g for 10 min and decant the supernatant. Resuspend the cells in a small amount of hypotonic KCl (approx. 1–2 times the volume of the cell pellet).
6. Slowly add 5 ml of freshly-made fixative while agitating the tube manually by pipetting or using a vortex mixer.
7. After 20 min or more repeat the centrifugation step, decant the supernatant, add 5 ml of fresh fixative, mix by agitation, and let stand at room temperature for 10–15 min.
8. Repeat step 7, centrifuge at 100 g for 2 min, remove the supernatant, and resuspend the cells in a small amount of fresh fixative (approx. 10–15 times the volume of cell pellet).
9. To prepare slides, add the suspension dropwise onto clean, cold (4 °C) slides held at about a 45° angle, blow gently to spread the cell suspension over the slide, and allow to dry completely in air at room temperature.
10. Examine the preparation for general arrangement using phase-contrast optics. Metaphases should be spread evenly over the slide surface without overlapping. The densities can be adjusted either by altering the number of drops of suspension applied or by changing the concentration of cells in the suspension.
11. Stain with either acetic acid orcein for 3–5 min or Giemsa for 10–15 min, rinse in tap-water, and air dry. Mount in Permount, if so desired.

The karyotype is constructed by cutting chromosomes from a photomicrograph either manually or using a software package and arranging them according to arm length, position of centromere, presence of secondary constrictions, and so forth. Consult the *Atlas of mammalian chromosomes* (12) for examples of conventionally stained preparations from over 550 species. For a more critical karyotypic analysis, chromosome banding techniques are required (see Section 4.3.1).

4.2 Tests for microbial contamination

The tests included here are suitable for detection of most micro-organisms that would be expected to survive as contaminants in cell lines or culture fluids. Techniques for detecting protozoan contamination are not presented as these organisms are rarely found in continuous lines and descriptions of the methodology are available elsewhere (13).

Commercial dry powders are entirely satisfactory for preparation of test media for bacteria, fungi, and mycoplasma provided that positive controls are included at least with the initial trials on each lot obtained.

4.2.1 Detection of bacteria and fungi

To examine cell cultures or suspect media for bacterial or fungal contaminants proceed as in *Protocol 7*.

Protocol 7

Microscopic examination

Equipment
- Inverted microscope equipped with phase-contrast optics

Method

1. Using an inverted microscope equipped with phase-contrast optics, examine cell culture vessels individually. Scrutiny should be especially rigorous in cases in which large scale production is involved. Check each culture first using low power.

2. After moving the cultures to a suitable isolated area, remove aliquots of fluid from cultures that are suspect and retain these for further examination. Alternatively, autoclave and discard all such cultures.

3. Prepare wet mounts using drops of the test fluids and observe under high power.

4. Prepare smears, heat-fix, and stain by any conventional method (e.g. Wright's stain), and examine under oil immersion.

5. Consult refs 14 and 15 for photomicrographs of representative contaminants and further details.

Microscopic examination is only sufficient for detection of gross contaminations and even some of these cannot be readily detected by simple observation. Therefore, an extensive series of culture tests is also required to provide reasonable assurance that a cell line stock or medium is free of fungi and bacteria. To perform these on stocks of frozen cells, follow *Protocol 8*.

Protocol 8
Culture tests

Equipment and reagents
- Syringe with a 18 gauge cannula
- Antibiotic-free medium
- See *Table 2*

Method

1. Pool and mix the contents of about 5% of the ampoules from each freeze lot prepared using a syringe with a 18 gauge cannula. It is generally recommended that antibiotics be omitted from media used to cultivate and preserve stock cell populations. If antibiotics are used, the pooled suspension should be centrifuged at 2000 g for 20 min and the pellet should be resuspended in antibiotic-free medium. A series of three such washes with antibiotic-free medium prior to testing will reduce the concentration of antibiotics which would obscure contamination in some cases.

2. From each pool, inoculate each of the test media listed in *Table 2* with a minimum of 0.3 ml of the test cell suspension and incubate under the conditions indicated. Include positive and negative controls comprising a suitable range of bacteria and fungi which might be anticipated. A recommended grouping consists of *Pseudomonas aeruginosa, Micrococcus salivarius, Escherichia coli, Bacteroides distasonis, Penicillium notatum, Aspergillus niger*, and *Candida albicans*.

3. Observe as suggested for 14–21 days before concluding that the test is negative. Contamination is indicated if colonies appear on solid media or if any of the liquid media become turbid.

4.2.2 Mycoplasma detection

Contamination of cell cultures by mycoplasma can be a much more insidious problem than that created by growth of bacteria or fungi. While the presence of some mycoplasma species may be apparent due to the degenerative effects induced, others metabolize and proliferate actively in the culture without producing any overt morphological change in the contaminated cell line. Thus cell culture studies relating to metabolism, surface receptors, virus–host interactions, and so forth are certainly suspect, if not negated in interpretation entirely, when conducted with cell lines harbouring mycoplasma.

Nine general methods are available for detection of mycoplasma contamination (5, 6). The direct culture test, a sensitive PCR analytical kit, and the 'in-

CELL LINE PRESERVATION AND AUTHENTICATION

Table 2. Suggested regimen for detecting bacterial or fungal contamination in cell cultures[a]

Test medium	Temperature (°C)	Aerobic state	Observation time (days)
Blood agar with fresh defibrinated rabbit blood (5%)	37 37	Aerobic Anaerobic	14
Thioglycollate broth	37 26	Aerobic	14
Trypticase soy broth	37 26	Aerobic	14
Brain heart infusion broth	37 26	Aerobic	14
Sabouraud broth	37 26	Aerobic	21
YM broth	37 26	Aerobic	21
Nutrient broth with 2% yeast extract	37 26	Aerobic	21

[a] For further detail see ref. 15.

direct' test employing a bisbenzimidazole fluorochrome stain (Hoechst 33258) for DNA are used routinely at the ATCC to check incoming cell lines and all cell distribution stocks for mycoplasma (16, 17).

The serum, yeast extract, and other ingredients used for mycoplasma isolation and propagation should be pre-tested for absence of toxicity and for growth-promoting properties before use (5). Positive controls consisting of *M. arginini*, *M. orale*, and *A. laidlawii* are recommended since they are among the most prominent species isolated from cultured cells (6, 16).

Protocol 9
Testing a culture for the presence of mycoplasma

Equipment and reagents
- Rubber policeman
- Inverted microscope
- Mycoplasma broth and mycoplasma agar

Method

1. To test a culture, be sure that the growth medium is free of antibiotics, especially gentamycin, tylocine, or other anti-mycoplasmal inhibitors.
2. Remove all but about 3 ml of fluid from a confluent or dense culture which has not been fed for at least three days, and scrape off parts of the monolayer with a rubber policeman. Suspension cultures near saturation may be sampled directly.
3. Inoculate 1.0 ml of the cell suspension to a tube of mycoplasma broth and 0.1 ml onto a plate of mycoplasma agar.

Protocol 9 continued

4. Incubate the broth aerobically at 37°C and the agar at 37°C in humidified 95% nitrogen/5% CO_2. A change in the pH of the broth or development of turbidity warns of mycoplasma contamination.
5. At weekly intervals for the following 14 days transfer 0.1 ml of the broth to a fresh plate of mycoplasma agar, and incubate at 37°C anaerobically.
6. Using an inverted microscope examine all agar plates at × 100 and × 300 weekly for a minimum of three weeks. A photomicrograph of mycoplasma colonies on agar is shown in *Figure 3*.

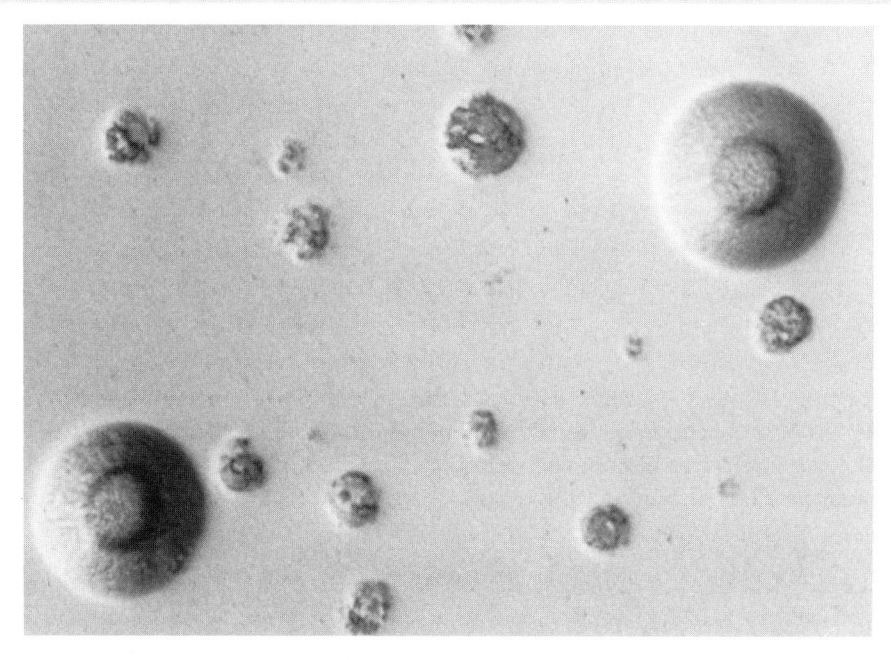

Figure 3. Mycoplasma colonies. Microphotograph showing colonies of mycoplasma in agar plate. The smaller material between colonies is from the inoculum of cultured cells and debris.

Since many strains of mycoplasma, especially *M. hyorhinis*, are difficult or impossible to cultivate in artificial media, an 'indirect' test method should also be included.

Positive and negative controls should be included with each test series. For this indirect test the Vero or 3T6 cells inoculated with medium only serve as negative controls. *M. hyorhinis*, *M. arginini*, *M. orale*, and *A. laidlawii* are suitable controls with the order of preference as listed. The nucleic acid of the organisms is visible as particulate or fibrillar matter over the cytoplasm with the cultured cell nuclei more prominent (*Figure 4*). With heavy infections, intercellular spaces also show staining.

Protocol 10

Indirect test for mycoplasma

Equipment and reagents

- 60 mm Petri plates
- 10.5 × 22 mm glass coverslips (No. 1)
- Fluorescence microscope
- Eagle's minimum essential medium supplemented with 10% fetal bovine serum
- Microscope slides
- Fixative
- Stain
- Mounting fluid

Method

1. Inoculate Petri plates (60 mm) containing 10.5 × 22 mm glass coverslips (No. 1) with 10^5 indicator cells in 4 ml of Eagle's minimum essential medium supplemented with 10% fetal bovine serum. The Vero or 3T6 lines are commonly used. These cells help to amplify low level infections which may otherwise be undetectable.

2. After incubating these indicator cultures for 16–24 h at 37°C in an air (95%)/CO_2 (5%) atmosphere, add 0.5 ml of the test cell suspension obtained as outlined for the direct test.

3. Return the plates to the incubator for an additional six days.

4. Remove the culture plates from the incubator, aspirate the fluid, and immediately add fixative. It is important not to let the culture dry before fixation as this may introduce artefacts.

5. After 5 min replace the 'spent' fixative with a fresh volume of the same solution such that the specimen-coverslip is well immersed.

6. Aspirate, allow the specimen to air dry, and prepare a fresh working stain solution from the 1000 × stain stock.

7. Add 5 ml of the staining solution to each dish. Cover and let stand for 30 min at room temperature.

8. Aspirate and wash the culture three times with 3–5 ml of distilled water, removing the final wash completely to allow the coverslip to dry.

9. Remove the coverslip and place with cells up on a drop of mounting fluid on a 1 × 3 microscope slide.

10. Add another drop of mounting fluid on top of the specimen-coverslip and place a larger clean coverslip above this, using a standard mounting technique to avoid trapping bubbles.

11. Examine each specimen at × 500 using a fluorescence microscope fitted with number 50 barrier (LP440 nm) and BG12 exciter (330/380 nm) filters.

Note: confluent cells do not spread sufficiently to stain adequately with Hoechst 33258. Ensure that cultures are subconfluent at the time of staining.

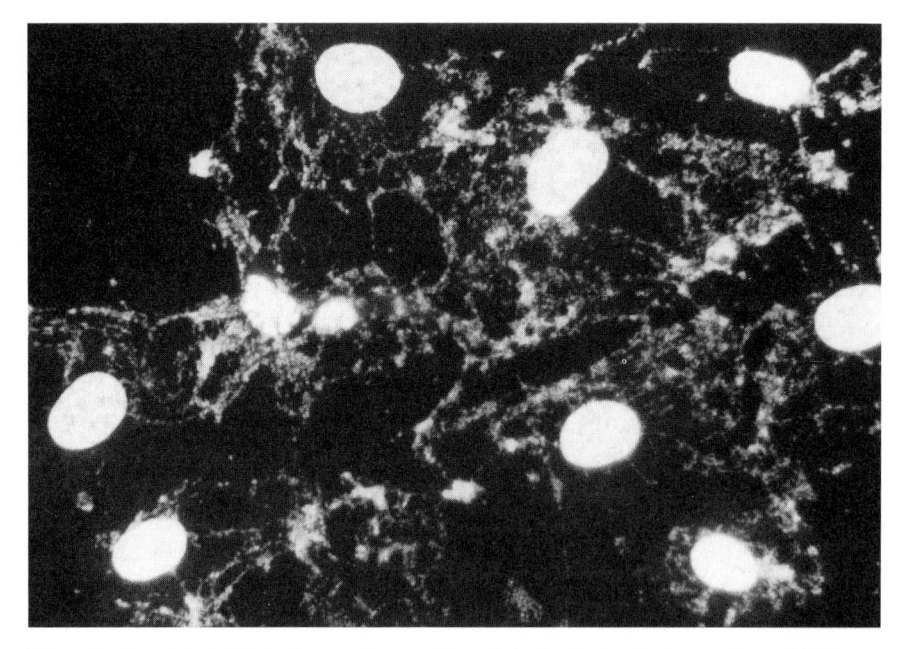

Figure 4. Hoechst staining of mycoplasma. Microphotograph showing mycoplasmal DNA demonstrated by Hoechst staining. The larger fluorescing bodies are nuclei of the substrate cell line Vero. (Photograph courtesy of M. L. Macy.)

Preparation of positive controls, i.e. deliberately infecting cultures, creates a potential hazard to other stocks, so infected cultures should be prepared at a time when, or in a place where, other cultures will not be at risk. Fix and store positive controls at −4°C over desiccant until required.

In all quality control work with mycoplasma or unknown cultures it is prudent to work in a vertical laminar flow hood, preferably isolated from other standard cell culture activity. One should be aware continually of the danger of contaminating clean cultures by aerosols from mycoplasma-containing cultures manipulated in the same area or manually during processing of multiple culture flasks. Appropriate disinfection between uses of a hood or work area is strongly recommended.

Various polymerase chain reaction (PCR) tests for mycoplasma have been developed, including those used at ATCC. Eight commonly encountered mycoplasma contaminants can be detected: *M. arginini*, *M. fermentans*, *M. hominis*, *M. hyorhinis*, *M. orale*, *M. pirum*, *M. salivarium*, and *Acholeplasma laidlawii*; which account for more than 95% of mycoplasma contaminants in cell cultures. Mycoplasma DNA in plants and insects can also be detected. The amplification product can be used to identify the species of mycoplasma by size determination and restriction enzyme digestion.

Further detail on these methods, the reagents required, and other aspects of mycoplasma infection of cultured cells are available in refs 5, 6, 16, and 17.

4.2.3 Testing for the presence of viruses

Of the various tests applied for detection of adventitious agents associated with cultured cells, those for endogenous and contaminant viruses are the most problematical. Development of an overt and characteristic cytopathogenic effect (CPE) will certainly provide an early indication of viral contamination (*Protocol 11*).

Protocol 11
Examining a culture for morphological evidence of viral contamination

Equipment and reagents
- Inverted microscope equipped with phase-contrast optics
- Fixative and stain

Method

1. Hold each flask or bottle so that light is transmitted through the monolayer and look for plaques, foci, or areas that lack uniformity. Cultures from frozen stocks should be set up from the pooled contents of about 5% of the ampoules from each lot.

2. Using an inverted microscope equipped with phase-contrast optics, examine cell culture vessels individually, paying special attention to any uneven areas in gross morphology observed previously.

3. Prepare coverslip cultures if higher power or additional study is required. The coverslips with monolayers can be fixed and stained by standard histological procedure, and the morphology can be compared with that of appropriate positive and negative controls.

The absence of CPE, however, definitely does not indicate that the culture is virus-free. In fact, persistent or latent infections may exist in cell lines and remain undetected until the appropriate biological, immunological, cytological, ultrastructural, and/or biochemical tests are applied. Additional host systems or manipulations, for example, treatment with halogenated nucleosides, may be required for virus activation and isolation.

Molecular techniques for detection and identification of specific viral infections may be diagnostic. In this regard, a non-radioactive method has been adapted at ATCC to screen for human immunodeficiency (HIV) and hepatitis viruses (e.g. HBV) in human fetal cells and tissues.

As cited by David Kellogg and co-workers (18) SK38 and SK39 primers have been successfully used to detect HIV-1 in infants born to seropositive mothers.

Table 3. Reagents for detection of HIV

Primers:	SK38 5'-ATAATCCACCTATCCCAGTAGGAGAAAT-3'
	SK39 5'-TTTGGTCCTTGTCTTATGTCCAGAATGC-3'
DNA probes:	SK-19 ATCCTGGGATTAAATAAAATAGTAAGAATGTATAGCCCTAC
Positive control:	ATCC cell line CRL-8993 (8E5)
Negative control:	ATCC cell line CCl-171 (MRC-5)

Table 4. Reagents for detection of HBV

Primers:	MD06 (5')-CTTGGATCCTATGGGAGTGG
	MD03 (5')-CTCAAGCTTCATCATCCATATA
DNA probe:	MD09 (5')-GGCCTCAGTCCGTTTCTCTTGGCTCAGTTTACTAGTGCCATTTGTTC
Positive control:	ATCC cell line HB-8064 (Hep-3B)
Negative control:	ATCC cell line CCL-2

Table 5. Solutions for HIV/HBV assays

$1 \times$ TBE buffer:	0.05 M Tris, 0.05 M boric acid, 1 mM EDTA pH 8.0
$20 \times$ SSC buffer:	3 M NaCl, 0.3 M Na$_3$ citrate·2H$_2$O pH 7.0
Lysis buffer:	0.15 M NaCl, 0.1 M EDTA, 0.02 M Tris, 1% (w/v) SDS pH 8.0
	20 mg/ml proteinase K stock in 0.15 M NaCl
	10 mg/ml RNase stock in 10 mM Tris pH 7.5
Maleic acid:	0.1 M maleic acid, 0.15 M NaCl pH 7.5
Standard hybridization buffer (1 litre):	0.1% N' lauroylsarcosine, 0.02% SDS, 1% blocking reagent, 125 ml of $20 \times$ SSC
Washing buffer:	0.1 M maleic acid, 0.15 M NaCl, 0.3% (w/v) Tween 20 pH 7.5

These oligonucleotides primers (SK38, SK39) amplify a 115 bp fragment of the HIV gag gene (Table 3).

MD06 and MD03 oligonucleotides are unique for HBV. The first oligonucleotide (MD06) spans position map genome 636–648, and the second position (MD03) 736–746. During the PCR amplification those two nucleotides amplify the synthesis of the 110 bp fragments (19) (Table 4).

The amplicons are separated by a 10% polyacrylamide gel, then transferred to a nylon membrane by electrophoretic transblotting, and detected by hybridization with digoxigenin labelled oligonucleotides and chemiluminescence (Table 5).

Equipment, reagents, and supplies for these methods are commercially available.

Protocol 12
Preparation of double-stranded DNA templates for amplification

Equipment and reagents
- Bacteriological loop
- Spectrophotometer
- Liquid nitrogen
- Microcentrifuge tube/pestle
- Polypropylene 15 ml conical centrifuge tube
- PBS (Mg^{2+}, Ca^{2+} free)
- Lysis buffer
- DNase-free RNase
- Proteinase K
- Phenol/chloroform/isoamyl alcohol (25:24:1)
- 3 M sodium acetate pH 7.2
- Ethanol

A. Cells
1. Pellet the cells and wash twice with 1 × PBS (Mg^{2+}, Ca^{2+} free).
2. Resuspend cells in lysis buffer (0.3 ml/3 × 10^6 cells). Add DNase-free RNase (1 mg/ml final concentration). Incubate with shaking at 37°C for 1 h.
3. Add proteinase K (0.1 mg/ml final concentration). Incubate with shaking at 55°C overnight.
4. Incubate for 8 min at 95°C to inactivate the proteinase K.
5. Extract with phenol/chloroform/isoamyl alcohol (25:24:1).
6. Separate the phases by centrifugation. Transfer the aqueous phase to a new tube.
7. Add 0.1 vol. of 3 M sodium acetate pH 7.2 to the solution of DNA, plus 2 vol. of ice-cold absolute ethanol. Invert tube gently a few times to precipitate the DNA.
8. Collect DNA on a bacteriological loop and transfer to a tube containing 70% cold ethanol. Wash well to remove all residual salt and phenol.
9. Resuspend DNA in TE buffer and incubate the tube at 4°C until the DNA has completely dissolved. Determine the concentration spectrophotometrically.

B. Tissue
1. Under liquid nitrogen grind the tissue using microcentrifuge tube/pestle. Transfer the powder to a polypropylene 15 ml conical centrifuge tube.
2. Add lysis buffer (1.2 ml/100 mg of tissue) plus proteinase K (final concentration 0.1 mg/ml).
3. Follow part A, steps 4-9.

Protocol 13
PCR amplification reaction for HIV

Equipment and reagents
- Thermal cycler
- 10 × PCR buffer without MgCl$_2$
- 1.25 mM dNTP mix: dATP, dGTP, dCTP, dTTP
- Primers
- *Taq* polymerase
- 25 mM MgCl$_2$

Method
1. Mix in order:
   - 10.0 µl of 10 × PCR buffer without MgCl$_2$
   - 16.0 µl of 1.25 mM dNTP mix
   - 50 pmole of each primer
   - 2.5 U of *Taq* polymerase
   - 10.0 µl of 25 mM MgCl$_2$
   - Sterile ddH$_2$O to yield a final volume of 100 µl
   - 13.0 ng DNA
2. Add 100 µl of mineral oil.
3. Amplify DNA for 30 cycles. Denature at 95 °C for 30 sec, anneal at 60 °C for 1 min, extend at 72 °C for 1 min. Finally extend at 72 °C for 10 min.

Protocol 14
PCR amplification reaction for HBV

Equipment and reagents
- See *Protocol 13*

Method
1. Mix in order:
   - 10.0 µl of 10 × PCR buffer without MgCl$_2$
   - 16.0 µl of 1.25 mM dNTP mix
   - 100 pmole of each primer
   - 2.5 U of *Taq* polymerase
   - 8.0 µl of 25 mM MgCl$_2$
   - 10 ng of DNA
   - Sterile ddH$_2$O to yield a final volume of 100 µl
2. Add 100 µl mineral oil.
3. Amplify for 30 cycles. Denature at 95 °C for 30 sec, anneal at 55 °C for 30 sec, and extend at 72 °C for 60 sec.

Protocol 15

Gel electrophoresis

Equipment and reagents

- 10% polyacrylamide vertical mini gel
- Microcentrifuge tube
- Loading dye
- TBE
- 0.5 ng/ml ethidium bromide

Method

1. Remove the mineral oil, transfer PCR product to a microcentrifuge tube (10.0 μl for HIV or 15.0 μl for HBV).
2. Add 2.0 μl of loading dye. Load the mix onto a 10% polyacrylamide vertical mini gel submerged in 0.5 × TBE. Run at 100 V until the dye migrates to the bottom of the gel.
3. Stain in 0.5 ng/ml ethidium bromide for about 15 min. Visualize the DNA fragments under UV illumination. Photograph stained gel.

Protocol 16

Electrophoretic transblotting

Equipment and reagents

- Electrophoretic transblotting apparatus
- 4 cm magnetic stir bar
- Bio-ICE cooling unit
- Nylon membrane
- Whatman 3MM filter paper
- TBE
- 0.4 M NaOH
- SSC

Method

1. Cut a piece of nylon membrane and six pieces of Whatman 3MM filter paper to the same size as the gel and one piece of filter paper slightly larger than the membrane.
2. Soak the membrane in 0.5 × TBE buffer for at least 30 min. Soak the fibre pads (see apparatus instructions) in 0.5 × TBE buffer.
3. Wet the six pre-cut pieces of 3MM paper in 0.5 × TBE buffer. Assemble as a sandwich, 3MM paper, gel, and membrane in the cassette (see apparatus instructions).
4. Place a 4 cm magnetic stir bar and Bio-ICE cooling unit in the tank buffer.
5. Place the gel holder cassette in the buffer tank, with the grey side of the cassette facing the black side of the tank. Fill the tank with 0.5 × TBE buffer at 4°C. Stir. Transfer for 1 h at 100 V, changing the Bio-ICE cooling unit every 20 min.
6. Denature the DNA by soaking in 0.4 M NaOH for 10 min. Rinse the membrane well with 2 × SSC. Fix the DNA by baking the membrane for 30 min at 80°C.

Protocol 17
Labelling oligonucleotides

Reagents

- Genius 5 Oligonucleotide 3' End Labeling Kit (Boehringer Mannheim)
- Genius 2 DNA Labeling Kit (Boehringer Mannheim)

Method

1. Probe labelling. Label the probe with digoxigenin using the Genius 5 Oligonucleotide 3' End Labeling Kit. Follow the kit instructions.
2. DNA marker labelling. Label the marker with digoxigenin-11-dUTP using the Genius 2 DNA Labeling Kit. Follow the kit instructions.

Protocol 18
Pre-hybridization and hybridization

Equipment and reagents

- Shaking water-bath
- Pre-hybridization buffer
- Labelled probe
- SSC, SDS

Method

1. Heat the pre-hybridization buffer at 65 °C.
2. Place the membrane in a plastic container and add 15 ml of pre-hybridization buffer, cover tightly. Incubate in a 65 °C shaking water-bath for 4 h.
3. Discard the pre-hybridization buffer, place the membrane in a clean plastic container, and add the hybridization buffer containing labelled probe (20 ng /ml for HIV, 18 ng/ml HBV). Cover tightly, incubate in a 65 °C shaking water-bath, overnight.
4. Wash the membrane twice in 2 × SSC, 0.1% SDS at room temperature for 5 min.
5. Wash the membrane twice in 0.5 × SSC, 0.1% SDS at 65 °C for 10 min.

Protocol 19
Chemiluminescent detection

Equipment and reagents

- X-ray film
- Washing buffer
- 1% blocking solution
- Antibody solution
- Detection buffer
- CSPD (disodium 3-(4-methoxyspiro {1,2-dioxetane-3,2'-(5'-chloro) tricyclo[3.3.1.1^{3,7}]decan}-4-yl)phenyl phosphate) (Boehringer Mannheim)

CELL LINE PRESERVATION AND AUTHENTICATION

Protocol 19 continued

Method

1. Soak the membrane in washing buffer for 1 min. Incubate in 1% blocking solution for 30 min with gentle shaking.

2. Incubate the membrane in antibody solution for 30 min with gentle shaking (dilute the anti-Dig-alkaline phosphatase 1:10 000).

3. Wash the membrane twice in washing buffer for 15 min. Soak the membrane in detection buffer for 15 min.

4. Spread at least 2 ml of CSPD on the membrane and incubate with gentle shaking. Wrap the membrane with plastic wrap. Remove the bubbles by rolling with a pipette. Expose the membrane to X-ray film.

HBV amplicons and controls revealed by chemiluminescence are provided as *Figure 5*.

It should be emphasized that, in spite of these screens, latent viruses and viruses which do not produce overt CPE or haemadsorption will escape detection. Some of these could be potentially dangerous for the cell culture technician. For additional detail and methodology consult refs 20–23.

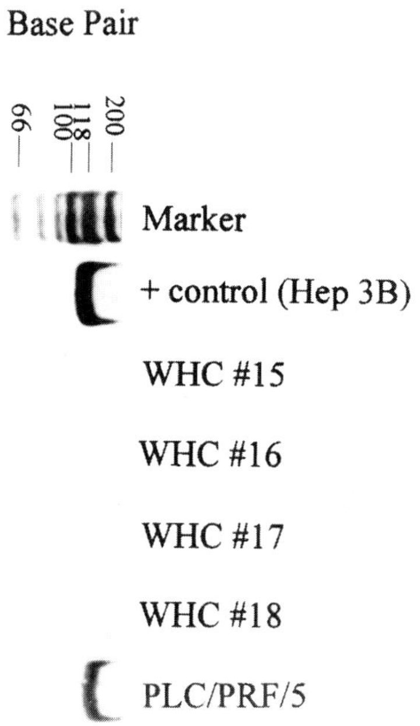

Figure 5. Photo of PCR for HBV. Screen for human hepatitis B sequences in fetal cells. Positive controls show amplicons from cell lines Hep-3B and PLC/PRF5 known to be infected with hepatitis B (ATCC HB-8064 and CRL-8024). Negative control and test cultures are indicated by letters WHC.

4.3 Testing for intraspecies cross-contamination

With the dramatic increase in numbers of cell lines being developed, the risk of intraspecies cross-contamination rises proportionately. The problem is especially acute in laboratories where work is in progress with the many different cell lines of human and murine origin (hybridomas) available today. Tests for unique karyology, polymorphic isoenzymes and DNA, and surface marker antigens are all important tools to detect cellular cross-contamination within a given species. Methods for chromosome banding and PCR analyses for DNA sequences will be included here. See refs 24–26 for additional detail.

4.3.1 Giemsa banding

A powerful method for cell identification is karyotype analysis after treatment with trypsin and the Giemsa stain (Giemsa or G-banding). The banding patterns are characteristic for each chromosome pair and permit recognition by an experienced cytogeneticist even of comparatively minor inversions, deletions, or translocations. Many lines retain multiple marker chromosomes, readily recognizable by this method, which serve to identify the cells specifically and positively.

Many modifications of the original technique are available. *Protocol 20* can be applied to metaphase spreads obtained as indicated in *Protocol 6*, step 10.

Protocol 20
Cell identification by Giemsa banding

Equipment and reagents

- Air dried slides
- 0.025 M phosphate buffer (PB): 3.4 g KH_2PO_4/litre adjusted to pH 6.8
- Trypsin solution (1% Difco 1:250 distilled water) made in bulk and stored in aliquots at −70 °C
- Staining solution: 6.5 ml PB, 0.55 ml trypsin solution, 2.5 ml of 100% methanol, and 0.22 ml of stock Giemsa solution (commercial)

Method

1. Use air dried slides within two to seven days.
2. Incubate in 0.025 M phosphate buffer at 60 °C for 10 min. The PB is also used in subsequent steps.
3. Prepare the staining solution just before use (PB, trypsin, methanol, and Giemsa solution).
4. Flood each slide with about 1 ml of staining solution and leave for 15 min.
5. Rinse briefly with distilled water and air dry completely.

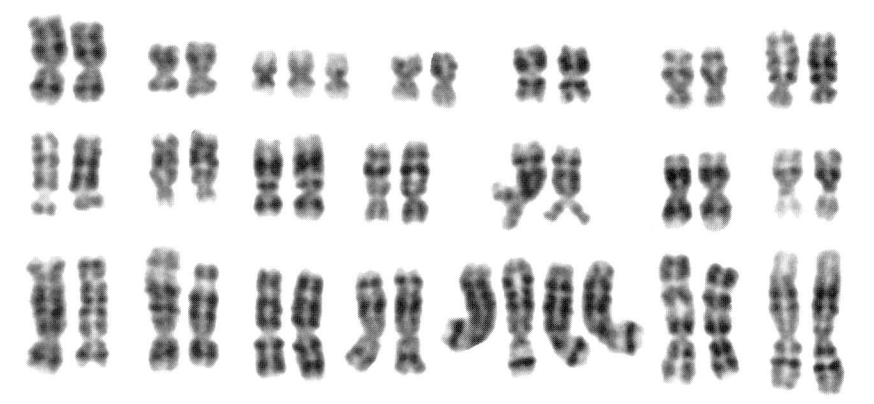

Figure 6. Giemsa banded chromosomes from the cell line ATCC.CCL-208 (4MBr-5) isolated from a rhesus monkey bronchus.

Completely dried slides are used to examine under bright-field, oil-immersion planapochromat objectives without coverslip. Oil can be placed directly on the slide. However, care must be taken not to scratch the cell surface. Oil must be removed as completely as possible immediately after the use of slides. Generally a few changes in xylene should be satisfactory. A typical banded preparation is shown as *Figure 6*.

See refs 27 and 28 for further detail and additional example preparations.

4.3.2 DNA fingerprinting

DNA fingerprinting of human cell lines provides the resolution necessary to confirm the authenticity of a cell line and to rule out the possibility of cell line cross-contamination. DNA fingerprinting became possible with the discovery by Jeffreys *et al.* (29) of hypervariable repetitive minisatellite (base repetitive motif is 10-20 bp) DNA. Nakamura *et al.* (30) expanded the number of hypervariable loci used to characterize DNA, and introduced the term variable number of tandem repeats (VNTR) to describe these regions. Traditional use of these loci involved the laborious, time-consuming procedure of Southern blotting. Two strategies have been used when working with VNTR loci. The first strategy employs a multilocus probe that can simultaneously illuminate the genotype at many loci. Although highly informative, this strategy is problematic as a routine screening technique due to the difficulty in interpreting the complex multiple banding patterns (31). The alternative strategy is to use a cocktail of less informative single locus VNTR probes. When the single locus probe genotypes are combined, a level of discrimination similar to multilocus VNTR fingerprints is possible with the simplicity of interpreting two band patterns. The adaptation of the polymerase chain reaction (PCR) has removed the need to use Southern blotting when examining these VNTR loci (32) which means that loci can be typed in hours rather than days.

Smaller microsatellite (base repetitive motif is 2-6 bp) loci have been identified as well. Edwards *et al.* (33) demonstrated the usefulness of these short tandem repeat (STR) loci in differentiating humans at the DNA level. One significant advantage enjoyed by STR loci over their minisatellite cousins is their small size. The small size of STR loci allows multiplex PCR reactions to be developed in which many loci are simultaneously examined in a single reaction. The authors' laboratory employs a commercially available multiplexed STR system for routinely screening new cell line accessions for authenticity as well as validating any subsequent distribution of an authenticated cell line (34).

When authenticating a new line, it is recommended that DNA be extracted from the cell line using a traditional liquid extraction method as this tends to minimize STR artefacts which may complicate allele assignment. For validation of subsequent passages of the cell line, more expedient DNA techniques may be used, which may produce ambiguous allele assignments. Generally, comparison with the authentic DNA fingerprint easily resolves these ambiguities. We describe a filter paper-based DNA extraction method now used routinely at ATCC.

Although the STR system presented utilizes fluorescent labels and an auto-

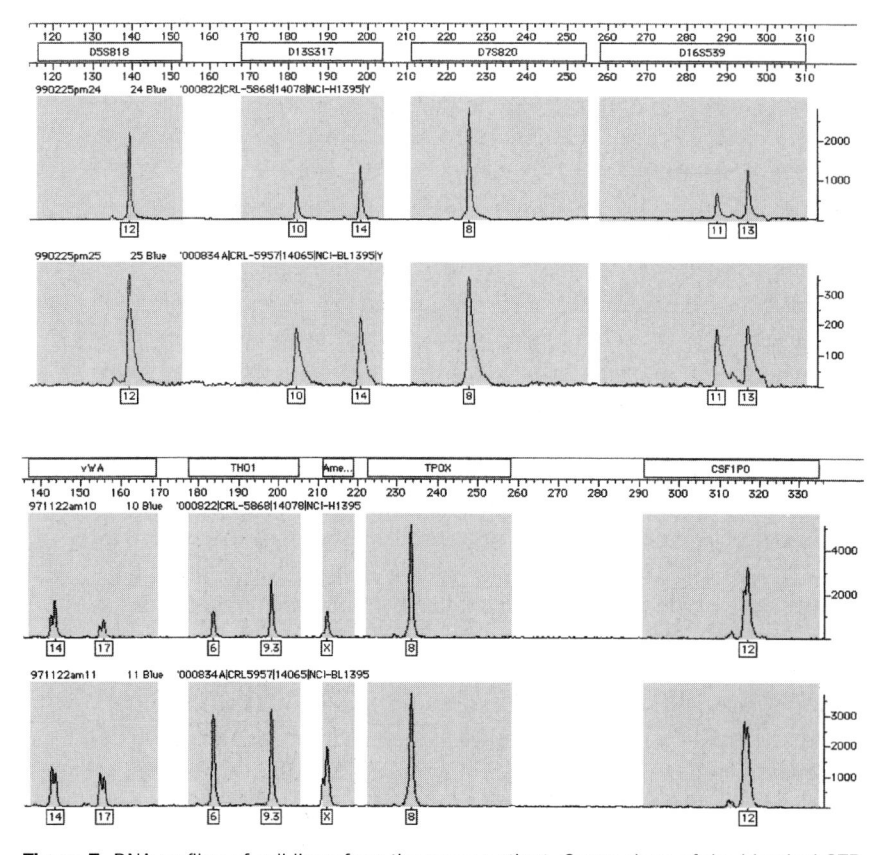

Figure 7. DNA profiles of cell lines from the same patient. Comparison of the identical STR profiles for the EBV-transformed lymphoblast line CRL-5957 and the tumour line CRL-5868 derived from the same female patient. Upper tracings represent the four STR loci D5S818, D13S317, D7S820, and D16S539. The lower tracings represent the four STR loci vWA, TH01, TPOX, and CSF1PO as well as the amelogenin locus used for gender identification.

CELL LINE PRESERVATION AND AUTHENTICATION

mated collection device, it should be noted that various STR formats have been developed that use isotopic, chemiluminescent, or silver staining to detect and type these loci with similar accuracy. One facet of DNA fingerprinting data, which cannot be overemphasized, is the need to convert PCR fragment sizes into alleles when developing genotypes for cell lines. This requires the development of an allelic ladder which contains all of the commonly observed alleles for each STR locus under examination (35). This ladder must be run on every gel, as it will serve to normalize the data and minimize the errors that inevitably crop up when comparing fragment size data between different gels. More importantly, this allows different laboratories, which may use different STR formats, to compare their results in a straightforward, unambiguous manner.

Typical STR profiling results are presented in the electropherograms (*Figures 7* and *8*). *Figure 7* represents the data generated when analysing two cell lines derived from the same patient. *Figure 8* compares the STR profiles generated from two unrelated cell lines.

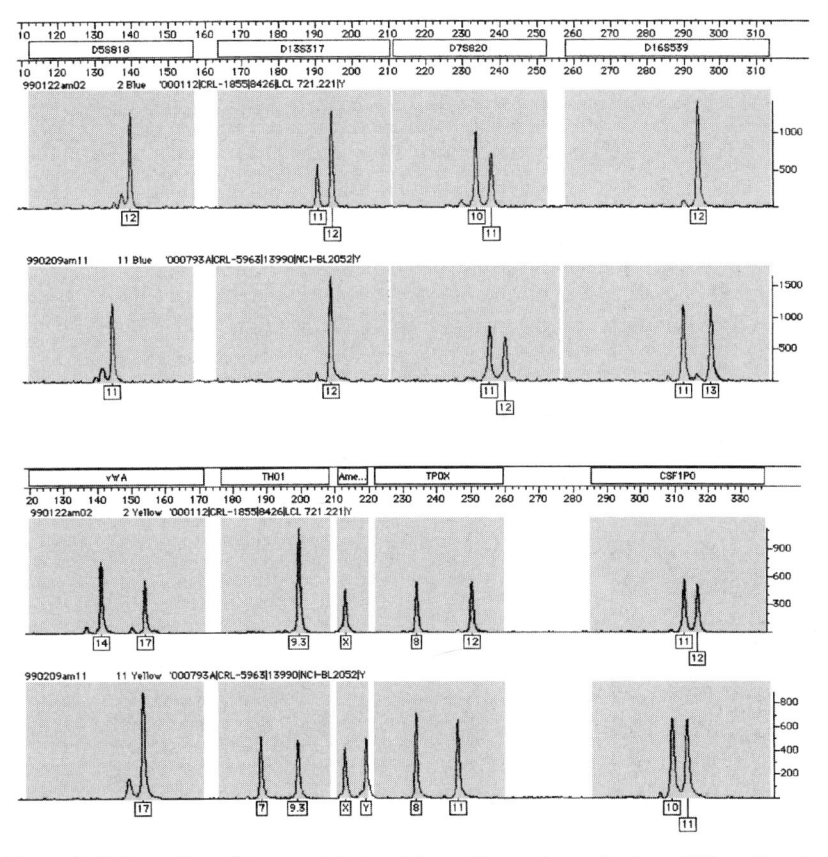

Figure 8. DNA profiles of two unrelated cell lines. Comparison of unique STR profiles for the unrelated male cell line CRL-5963 and female cell line CRL-1855. Upper tracings represent the four STR loci D5S818, D13S317, D7S820, and D16S539. The lower tracings represent the four STR loci vWA, TH01, TPOX, and CSF1PO as well as the amelogenin locus used for gender identification.

97

Protocol 21
Preparation of genomic DNA from cell lines

Equipment and reagents

- Rocking platform
- Eppendorf microcentrifuge
- Puregene® cell lysis solution (Gentra Systems, Cat. No. D-50K2)
- 4 mg/ml stock solution of DNase-free RNase A
- Saturated 6 M NaCl solution
- Isopropanol
- Ethanol
- TE buffer

Method

1. Prepare 1 ml of cell lysis buffer for each cell line to be extracted by mixing 30:1 Puregene® cell lysis solution and 20% SDS stock solution.

2. Starting with a cell pellet containing approx. 10^6 cells in a 1.5 ml microcentrifuge tube (for most cell lines, cell pellet volume should be 3–7 μl), add 600 μl of cell lysis buffer, then thoroughly resuspend pellet by pipetting mixture up and down.

3. Add 3 μl of 4 mg/ml DNase-free RNase A to cell suspension, then mix by inverting tubes several times. Incubate samples at 37 °C on a rocking platform for a minimum of 15 min (incubation can be extended to 2–3 h if convenient).

4. At the completion of the incubation, add 200 μl of a saturated 6 M NaCl solution, mix by vortexing, then pellet cellular debris by centrifuging sample at room temperature in an Eppendorf microcentrifuge at 14 000 r.p.m. for 3 min.

5. Transfer as much of the supernatant as possible to a clean 1.5 ml microcentrifuge tube, being careful not to disturb the flocculent cell debris pellet. Approximate volume of the supernatant transferred should be 500 μl.

6. Add 500 μl (an equal volume) of isopropanol at room temperature to precipitate the DNA from the supernatant, then mix by inverting the tube several times. Collect DNA by centrifuging sample at room temperature in an Eppendorf microcentrifuge at 14 000 r.p.m. for 1 min.

7. Remove supernatant, then carefully wash DNA pellet with 500 μl of 70% ethanol at room temperature.

8. Allow pellet to air dry for 15 min.

9. Redissolve DNA in 100 μl TE buffer, then incubate sample at 37 °C on a rocking platform for a minimum of 2 h (incubation can be extended to overnight if convenient).

10. At the completion of the 37 °C incubation, continue to allow the DNA to redissolve by incubating sample at 4 °C on a rocking platform for a minimum of one day (the longer this incubation is carried out, the more accurate the DNA quantitation will be).

CELL LINE PRESERVATION AND AUTHENTICATION

Protocol 21 continued

11. Quantitate DNA by UV photospectometry. Typical yields vary from 5–100 μg. Yield will depend on the cell line extracted. Yields tend to be consistent between extractions for any given cell line.

Protocol 22

Preparation of genomic DNA from cell lines using FTA™ paper

Equipment and reagents
- FTA classic format card (Life Technologies, Cat. No. 10786010)
- Sealed pouch containing desiccant (UltraBARRIER pouch, Life Technologies, Cat. No. 64030026)
- HARRIS micro-punch (Life Technologies, Cat. No. 10786069)
- PBS
- TE buffer

Method

1. Prepare a 10^6 cells/ml cell suspension. Cells can be washed and resuspended in $1 \times$ PBS at the desired concentration. Alternatively, ampoules at the correct cell concentration, retrieved from liquid nitrogen storage tanks, have been successfully used with no washing; i.e. the cell suspension consisted of growth media and 5% DMSO.

2. Spot one drop from a 200 μl pipette tip of the cell suspension onto an FTA classic format card and allow card to air dry thoroughly for at least 1 h (this can be extended to overnight if convenient). The volume of the drop will vary with the cell line and the viscosity of the fluid in the cell suspension. Dried cards should be stored long-term at room temperature in a sealed pouch containing desiccant.

3. Remove a 2 mm circle from the dried stain using a HARRIS micro-punch and transfer to a 500 μl microcentrifuge tube.

4. Add 200 μl of FTA purification reagent to sample and incubate for 5 min at room temperature.

5. At the end of the incubation, decant or aspirate as much of the purification reagent as possible.

6. Repeat steps 4 and 5 two more times.

7. Add 200 μl of TE to each sample and incubate for 5 min at room temperature.

8. At the end of the incubation, decant or aspirate as much of the TE as possible.

9. Repeat steps 7 and 8 once more.

10. Allow washed 2 mm circle to air dry for at least 1 h (this can be extended to overnight if convenient).

11. Sample is now ready for PCR amplification.

Protocol 23

PCR amplification and STR typing of cell lines

Equipment and reagents

- Thermal cycler
- PE/ABD 373 automated sequencer
- 6% denaturing (8.3 M urea) polyacrylamide gel
- Gold ★ buffer
- PowerPlex 10 × primer pair mix
- AmpliTaq Gold (5 U/µl)
- GS-500 ROX
- Deionized formamide
- See *Tables* 6 and 7

Method

1. Set up PCR reaction in a 500 µl microcentrifuge tube. Input amount of DNA should be 1-2 ng or a 2 mm FTA paper punch. Add the following reagents to the sample:
   - Gold ST☆R buffer   2.5 µl
   - PowerPlex 10 × primer pair mix 2.5 µl
   - AmpliTaq Gold (5 U/µl)   0.45 µl
   - Sterile ddH$_2$O to final volume 25.0 µl

2. Mix samples thoroughly by vortexing, then collect samples by a brief microcentrifuge spin. Overlay samples with 25 µl of mineral oil.

3. Incubate samples in thermal cycler using the following temperature profile:

Cycles	Temperature	Time (min)
1	95 °C	11.0
1	96 °C	2.0
10	94 °C	1.0
	60 °C	1.0
	70 °C	1.5
22[a]	90 °C	1.0
	60 °C	1.0
	70 °C	1.5
1	60 °C	30.0
1	4 °C	∞

4. Prepare samples for gel loading by combining the following reagents:
   - PCR sample   2.0 µl
   - GS-500 ROX   0.5 µl
   - Deionized formamide   2.5 µl
   - Total volume   5.0 µl

5. Heat denature samples at 95 °C for 2 min, then immediately plunge samples into wet ice and incubate for 10 min.

6. Collect sample volume with a brief microcentrifuge spin, then load 1.5 µl of sample onto a 6% denaturing (8.3 M urea) polyacrylamide gel.

CELL LINE PRESERVATION AND AUTHENTICATION

Protocol 23 continued

7. Samples are electrophoresed on a PE/ABD 373 automated sequencer using the following parameters: 2500 V, 40 mA, 40°C, 6 h.[b]
8. Data is collected using PE/ABD *GeneScan 672 v1.0 Data Collection Software*.
9. Sample fragment sizes are calculated using PE/ABD *GeneScan v3.1 Analysis Software*.
10. Allele assignments are made using PE/ABD *Genotyper v2.0 Software*.

[a] When using FTA paper punches as input DNA, cycle number must be reduced down from 22 cycles to 16 cycles.

[b] This run time allows collection of the 350 bp band in the GS-500 internal lane standard.

Table 6. Laboratory stock buffers and reagents

1 × PBS:	135 mM NaCl, 2.5 mM KCl, 10 mM Na_2HPO_4, 2 mM KH_2PO_4 pH 7.4
20% SDS	
6 M NaCl	
Isopropanol	
70% ethanol	
Low TE buffer:	10 mM Tris, 0.1 mM EDTA pH 7.6

Table 7. Vendor reagents and supplies

Cell lysis solution (Gentra Systems, Cat. No. D-50K2)
RNase A solution (Gentra Systems, Cat. No. D-50K6)

FTA classic format card (Life Technologies, Cat. No. 10786010) includes:
    FTA purification reagent
    TE^{-4} (low TE buffer, supplied with kit)

UltraBARRIER pouch (Life Technologies, Cat. No. 64030026)
HARRIS micro-punch (Life Technologies, Cat. No. 10786069)

GenePrint™ PowerPlex™ 1.2 System (Promega, Cat. No. DC6101) includes:
    Gold STAR buffer
    PowerPlex 10 × primer pair mix

AmpliTaq Gold™ (Perkin Elmer, Cat. No. N808-0241)
Mineral oil (Perkin Elmer, Cat. No. O186-2302)
GS-500 ROX (PE/ABD, Cat. No. 401734)
Formamide (Sigma, Cat. No. F-9037)
6% polyacrylamide (8.3 M urea) Burst-Pak™ Gel (Owl Scientific, Cat. No. SEQ-6-10)
5 × TBE liquid concentrate (Amresco, Cat. No. J885–1L)

4.4 Miscellaneous characterizations and cell line availability

A host of additional cell line characterizations may be considered depending upon the tissue type of origin and the intended use for given cell lines. For example, analysis of fine-structure, immunological tests for cytoskeletal and tissue-specific proteins, or assays for monoclonal antibody secreted may be required. For further information see refs 21, 26, 36.

Authenticated cell lines may be obtained from national banks established over the past 10-30 years. Such public repositories include the ATCC, the Human Genetic Mutant Cell Repository at the Coriell Institute for Medical Research in Camden, NJ, USA, the German Collection of Microorganisms and Cell Cultures (DSMZ) in Braunschweig, the European Collection of Animal Cell Cultures, Porton Down, Salisbury, UK, the Japanese Cell Bank for Cancer Research at the National Institute of Hygiene Services in Tokyo, and the Riken Cell Bank in Tsukuba City, Japan. Detailed information on each is now available on the Internet.

Acknowledgements

This work was supported in part under grants (1-R26-CA25635) and (5-RO1-HD32179) from the National Large Bowel Cancer Project of the National Cancer Institute and the National Institute of Child Health and Human Development respectively, and contract (NO 1-RR-9-2105) from the Division of Research Resources of the National Institutes of Health.

References

1. Schaeffer, W. I. (1979). *In Vitro*, **15**, 649.
2. Nelson-Rees, W. A. and Flandermeyer, R. R. (1977). *Science* (Wash.), **195**, 1343.
3. Nelson-Rees, W., Daniels, W. W., and Flandermeyer, R. R. (1981). *Science* (Wash.), **212**, 446.
4. Nelson-Rees, W. A., Flandermeyer, R. R., and Hawthorne, P. K. (1974). *Science* (Wash.), **184**, 1093.
5. Barile, M. F. (1977). In *Cell culture and its application* (ed. R. T. Acton and J. D. Lynn), p. 291. Academic Press, London and New York.
6. McGarrity, G. J. (1982). *Adv. Cell Culture*, **2**, 99.
7. Hay, R. J., Williams, C. D., Macy, M. L., and Lavappa, K. S. (1982). *Am. Rev. Respir. Dis.*, **125**, 222.
8. Consult the ATCC Website at http://www.atcc.org for up-to-date listings.
9. Hay, R. J. (1984). In *Markers of colonic cell differentiation* (ed. S. R. Wolman and A. J. Mastromarino), p. 3. Raven Press, New York.
10. Stulberg, C. S. (1973). In *Contamination in tissue culture* (ed. J. Fogh), p. 1. Academic Press, London and New York.
11. Macy, M. L. (1978). *Tissue Culture Assoc. Man.*, **4**, 833.
12. Hsu, T. C. and Benirschke, K. (1967-1975). *An atlas of mammalian chromosomes*, 9 volumes. Springer-Verlag, New York, Heidelberg and Berlin.
13. Dilworth, S., Hay, R. J., and Daggett, P.-M. (1979). *Tissue Culture Assoc. Man.*, **5**, 1107.
14. Freshney, R. I. (1987). *Culture of animal cells*, 2nd edn. Alan R. Liss, Inc., New York.
15. Cour, I., Maxwell, G., and Hay, R. J. (1979). *Tissue Culture Assoc. Man.*, **5**, 1157.
16. Hay, R. J., Macy, M. L., and Chen, T. R. (1989). *Nature*, **339**, 487.
17. Tang, J., Hu, M., Lee, S., and Roblin, R. (1999). *In Vitro Cell Dev. Biol.*, **35**, 1.
18. Kellogg, D. E. and Kwok, S. (1990). In *PCR protocols: a guide to methods and applications* (ed. M. A. Innis, D. H. Gelfand, J. J. Sninsky, *et al.*), pp. 337-47. Academic Press, Inc.
19. Baginski, I., Perrie, A., Watson, D., *et al.* (1990). In *PCR protocols: a guide to methods and applications* (ed. M. A. Innis, D. H. Gelfand, J. J. Sninsky, *et al.*), pp. 348-55. Academic Press, Inc.

20. Rovozzo, G. C. and Burke, C. N. (1973). *A manual of basic virological techniques*. Prentice-Hall, Inc., Englewood Cliffs, New Jersey.
21. LeDuc, J. W., Smith, G. A., Macy, M. L., and Hay, R. J. (1985). *J. Infect. Dis.*, **152**, 1081.
22. Hay, R. J. (1998). *Testing cell cultures for microbial and viral contaminants, cell biology: a laboratory handbook*, **2** (1), 43–62.
23. Hay, R. J. (ed.) (1992). *ATCC quality control methods for cell lines*, 2nd edn. Rockville, MD.
24. O'Brien, S. J., Shannon, J. E., and Gail, M. H. (1980). *In Vitro*, **16**, 119.
25. Wright, W. C., Daniels, W. P., and Fogh, J. (1981). *J. Natl. Cancer Inst.*, **66**, 239.
26. Hay, R. J. (1988). *Anal. Biochem.*, **171**, 225.
27. Sun, N. C., Chu, E. H. Y., and Chang, C. C. (1973). *Mamm. Chromosome Newsl.*, Jan., 26.
28. Chen, T. R., Hay, R. J., and Macy, M. L. (1982). *Cancer Genet. Cytogenet.*, **6**, 93.
29. Jeffreys, A. J., Wilson, V., and Thein, S. L. (1985). *Nature*, **314**, 67.
30. Nakamura, Y., Leppert, M., O'Connell, P., Wolf, R., Holm, T., Culver, M., *et al.* (1987). *Science*, **235**, 1616.
31. Gilbert, D. A., Reid, Y. A., Gail, M. H., Pee, D., White, C., Hay, R. J., *et al.* (1990). *Am. J. Hum. Genet.*, **47**, 499.
32. Latorra, D., *et al.* (1990). *PCR Methods Appl.*, **3** (6), 351.
33. Edwards, A., Hammond, H. A., Jin, L., Caskey, C. T., and Chakraborty, R. (1992). *Genomics*, **12**, 241.
34. Durkin, A. S. and Reid, Y. A. (1998). *ATCC Q. Newsl.*, **18**, 1.
35. Sajantila, A., Puomilahti, S., Johnsson, V., and Ehnholm, C. (1992). *Biotechniques*, **12** (1), 16.
36. Wigley, C. B. (1975). *Differentiation*, **4**, 25.

Chapter 4
Development of serum-free media

Soverin Karmiol
Bio Whittaker, Inc., PO Box 127, 8830 Biggs Ford Road, Walkersville, MD 21793, USA.

1 Introduction

The removal of serum or other undefined components from cell culture medium involves a multifaceted approach. This is mainly due to the interactive nature of cell culture systems. Cell culture systems require various components: basal nutrients, a set of supplements that convey information, such as growth factors, mitogens, hormones, and cytokines, and for anchorage-dependent cells, a substrate. In designing any cell culture system attention must be paid to all these components and their interactions.

The role of the basal nutrient component needs to be addressed in a systematic manner. For normal eukaryotic cells, the optimization of the basal nutrients is an important aspect in the generation of serum-free media.

The removal of an essential nutrient, such as an amino acid, vitamin, or mineral, would effectively inhibit the activity of the cells in a serum-free culture environment (1-3). It has been demonstrated qualitatively that various cell types require the same nutrients, but that quantitatively there are significant differences. It is reasonable to assume that there is an optimized concentration for a set of nutrients for a particular function for a particular cell type.

2 Role of serum and other undefined tissue extracts in cell culture systems

In media these undefined supplements (serum, pituitary brain, and other extracts) provide not only essential nutrients and attachment factors, but also act as detoxificants. The undefined nature of these supplements and their potential for carrying adventitious agents and determining the phenotype of cells, motivates their avoidance or minimization whenever possible. These supplements can contain inhibitors or promoters of the processes under study (4, 5). In the bioprocessing industry the contribution of these supplements to the final product necessitates extensive downstream purification. In addition, the cell phenotype can be better controlled in the absence of serum (6).

3 Response curves

3.1 Proliferating cultures

The quantitative evaluation of nutrient requirements necessitates titration of the basal components in a systematic manner. The responses evaluated can be any cellular characteristic: proliferation, product secretion, or expression of a differentiated function. Two types of responses are possible: the essential nutrient response, and the non-essential nutrient response (*Figure 1*). The essential response is characterized by a zero response to zero concentration, followed by a positive increase to rising concentrations, a plateau region where no increase in response is observed with higher concentration, and finally an inhibitory region. The slope of the positive response, the length of the plateau, and the slope of the inhibitory regions are dependent on the nutrient and cell under study. The aim of this exercise is to determine the concentration of the nutrient that is represented by the middle of the plateau.

In optimizing the concentration of a nutrient, amino acids represent the largest group where interactions occur, because there are common transport systems for amino acids. These systems interact in a positive or negative manner (7). A consequence of this is that by changing the concentration of one amino acid it is possible to influence the response curve of another amino acid. By placing the concentration of a component in the middle of the plateau region, a shift in any direction will minimize the risk of entering a deficient or inhibitory area.

The non-essential condition is characterized by a significant response at zero concentration of the nutrient. There may also be a response to increasing concentrations of the nutrient. If the desired outcome is maximal proliferation, then the concentration of the non-essential nutrient also needs to be optimized. Examples of this are observed in *Figures 2* and *3*. In *Figure 2*, L-phenylalanine and L-tyrosine are titrated using normal human aortic smooth muscle cells (AOSMC). The L-tyrosine titration demonstrates that in the absence of L-tyrosine, proliferation close to maximum is observed. However, in *Figure 3*, where L-phenylalanine and L-tyrosine are titrated using normal human bronchial/tracheal epithelial cells (NHBE), L-tyrosine appears to be an essential nutrient. In whole animals L-phenylalanine can be converted to L-tyrosine by L-phenylalanine hydroxylase. This metabolism has been observed in various other cell culture systems (8, 9), but not with NHBE. In a similar fashion, titrating the amino acid pair L-methionine and L-cysteine (*Figures 4* and *5*), it is seen that L-cysteine is not an essential amino acid for AOSMC, whereas in NHBE, L-cysteine is essential. L-cysteine is synthesized from L-serine and L-methionine in the liver and brain and in fibroblasts expressing cystathionine synthetase (10).

Nutrient requirements for cells in culture are determined best in clonal

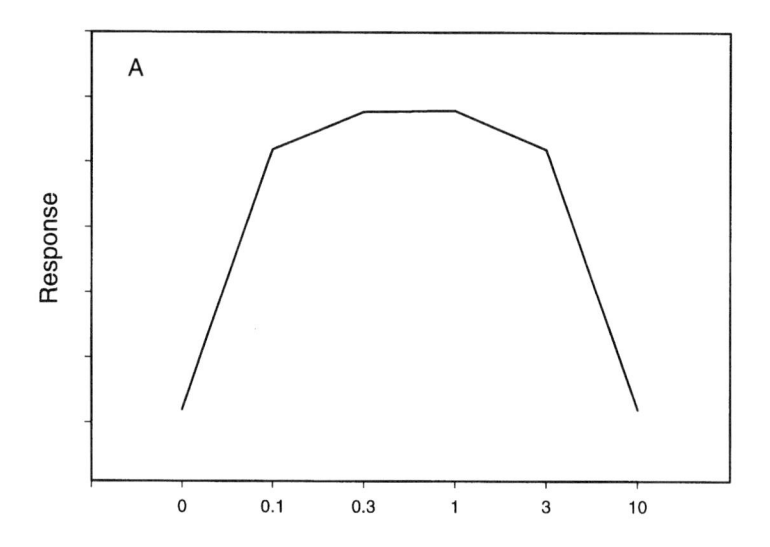

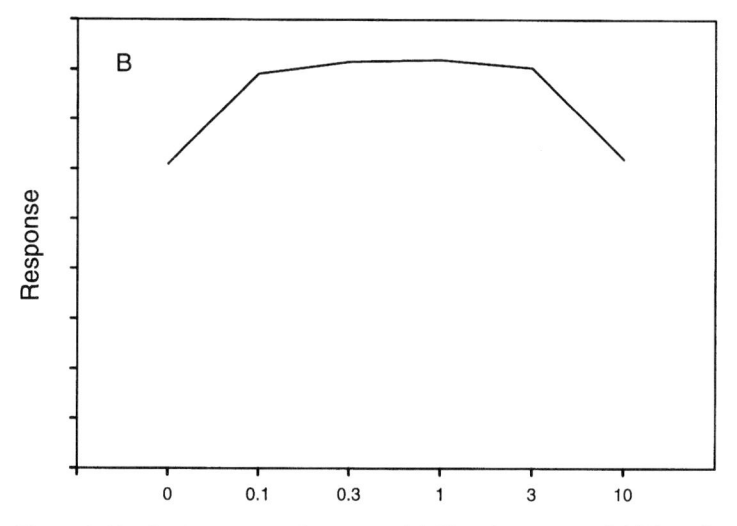

Figure 1. Idealized response of an essential (A) and non-essential (B) nutrient. Based on these responses it is possible to determine the optimized concentration of the nutrient. Note that in the absence of an essential nutrient (one that cannot be synthesized by the cell) no response is observed at zero concentration. On addition of the essential nutrient a dose response is observed: an increase in concentration results in an increase in response. At higher concentrations a plateau is observed where there is no increase in response with increased concentration. The third part of the response is characterized by a decrease in response with increasing concentration of nutrient. In the case of the non-essential nutrient a significant response is observed at zero concentration. However, further addition of the non-essential nutrient may result in an enhanced response. Although the non-essential nutrient can be synthesized by the cell it may not be synthesized at rates sufficient to meet the maximum requirement and/or the addition of the nutrient may relieve the metabolic load required for its synthesis.

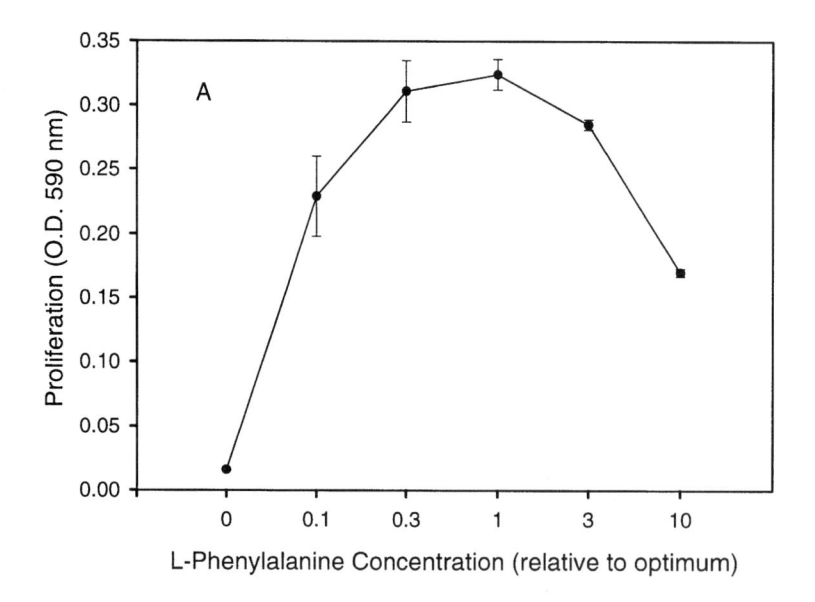

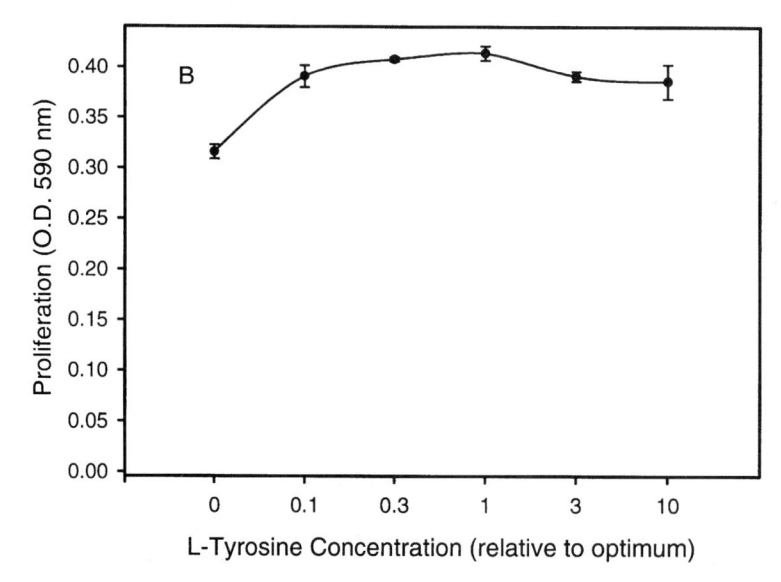

Figure 2. Titration of L-phenylalanine (A) and L-tyrosine (B) using AOSMC. L-phenylalanine is demonstrating essential nutrient behaviour while L-tyrosine is demonstrating non-essential nutrient behaviour. In the absence of L-tyrosine, L-phenylalanine may be contributing to the L-tyrosine pool through the action of phenylalanine hydroxylase, but there may also be protein breakdown contributing to the L-tyrosine pool. The cells were plated at 200 cells/cm^2 and kept in culture for 10–11 days without a medium change and then fixed in formaldehyde and stained with crystal violet. Quantitation was performed by extracting the crystal violet and reading the optical density at 590 nm. The optical density of extracted crystal violet is correlated with cell number during logarithmic growth. No serum or other tissue extract was present in the medium unless otherwise stated. The values on the x axis refer to multiples of the log of the optimal concentration.

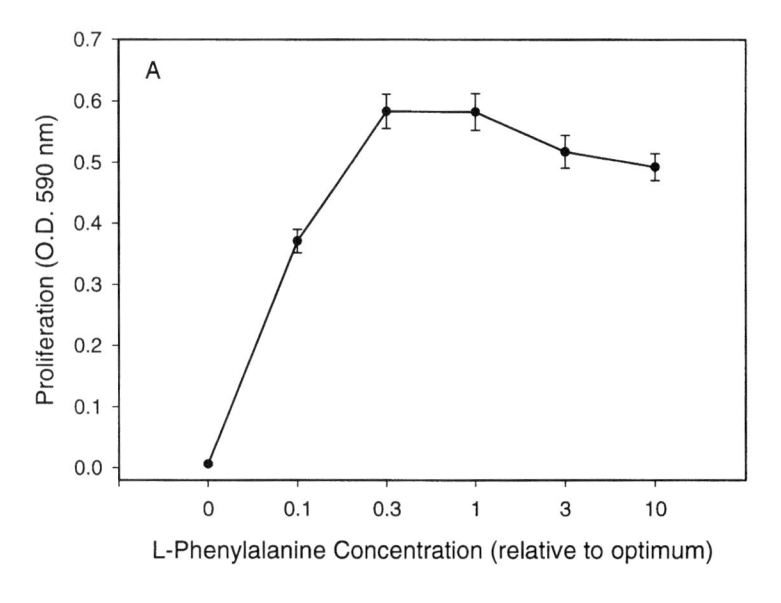

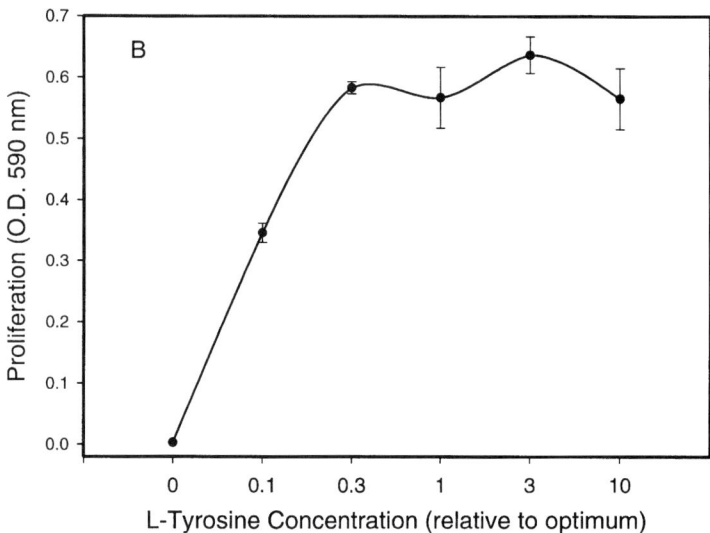

Figure 3. Titration of L-phenylalanine (A) and L-tyrosine (B) in NHBE. Both L-phenylalanine and L-tyrosine behave as essential nutrients. The values on the x axis refer to multiples of the log of the optimal concentration. These cells were plated at 100 cells/cm² and kept in culture for seven days without a medium change. No serum or other tissue extract was present in the medium unless otherwise stated. The values on the x axis refer to multiples of the log of the optimal concentration. Quantitation was performed as described in *Figure 2*.

growth assays. At high density, cells are able to condition the medium by excreting essential factors. In a clonal growth assay the volume of medium is too large for the small number of cells to effectively condition the medium. Consequently, the cells are dependent on the nutrients present in the medium and a deficiency may be observed.

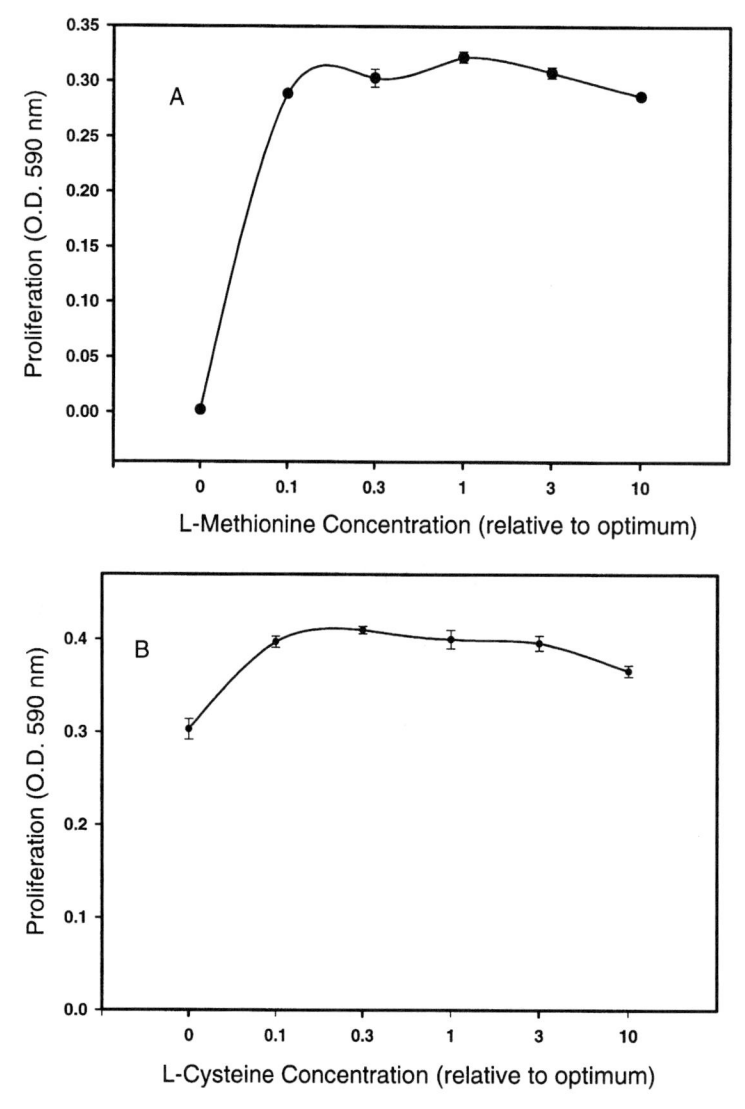

Figure 4. Titration of L-methionine (A) and L-cysteine (B) in AOSMC. L-methionine is demonstrating essential nutrient behaviour while L-cysteine is demonstrating non-essential nutrient behaviour. L-cysteine can be synthesized from L-serine which provides the carbon chain and L-methionine which provides the sulfur atom. No serum or other tissue extract was present in the medium. The values on the x axis refer to multiples of the log of the optimal concentration. Conditions are similar to those stated in *Figure 2*.

One basal component with profound effects on cell proliferation is sodium chloride (NaCl). *Figure 6* demonstrates the response of normal human epidermal keratinocytes (NHEK) to various concentrations of NaCl. The response is characterized by a relatively sharp peak. The sensitivity to NaCl may be due to the energy expenditure associated with the maintenance of the Na^+-K^+-ATPase membrane transport (11, 12).

DEVELOPMENT OF SERUM-FREE MEDIA

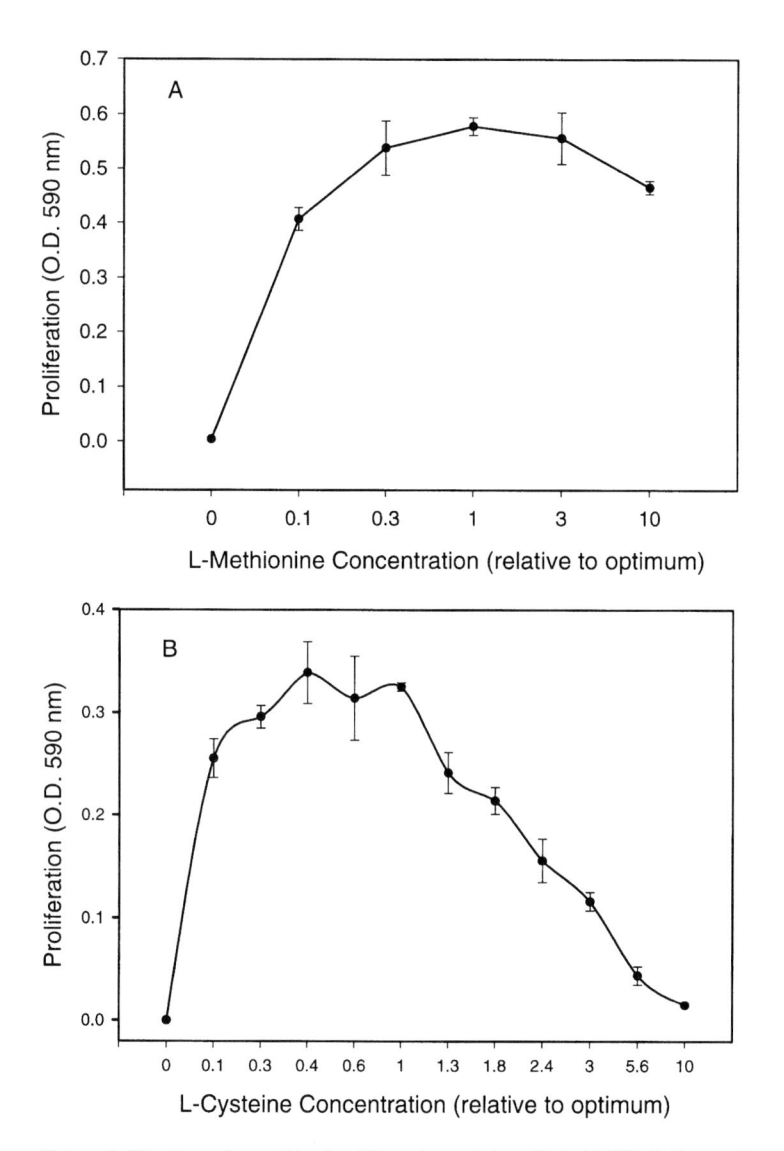

Figure 5. Titration of L-methionine (A) and L-cysteine (B) in NHBE. Both L-methionine and L-cysteine behave as essential nutrients. The values on the x axis refer to multiples of the log of the optimal concentration. These cells were plated at 100 cells/cm² and kept in culture for seven days without a medium change. No serum or other tissue extract was present in the medium unless otherwise stated. The values on the x axis refer to multiples of the log of the optimal concentration. Quantitation was performed as described in *Figure 2*.

3.2 Non-proliferating cultures (hepatocytes)

Normal adult rhesus monkey hepatocytes do not proliferate to any significant extent. A characteristic of hepatocytes is cytochrome P450 activity and the medium optimization uses this activity as an end-point. In *Figure 7* the titration of two amino acids is depicted using hepatocytes exposed for 48 hours to 3-

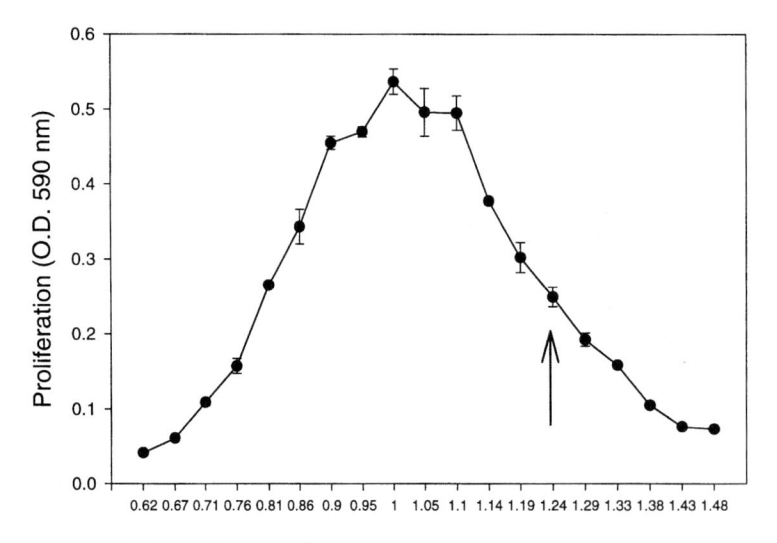

Figure 6. Response of normal human keratinocytes to various concentrations of sodium chloride. Sodium chloride demonstrates essential nutrient behaviour, which is not surprising given the various functions sodium chloride executes. Of the many functions of sodium chloride, influence over the osmolarity of the medium is one of the most profound. As can be seen from the graph, a 15–20% change in concentration has significant effects. The arrow refers to the concentration of NaCl prior to the optimization process. No serum or other tissue extract was present in the medium. The values on the *x* axis refer to multiples of the log of the optimal concentration. Quantitation was performed as described in *Figure 2*.

methylcholanthrene. A homologous series of the *n*-alkyl ethers of phenoxazone (alkoxyresorufins) is available for the study of various cytochrome P450s (13). The phenoxazone ethers are hydroxylated and *O*-dealkylated to a common fluorescent metabolite, resorufin, which fluoresces at 590 nm when excited at 530 nm. The particular phenoxazone ether used to measure cytochrome P450 upon 3-methylcholanthrene induction was ethoxyphenoxazone (14).

The L-leucine titration demonstrates a dose response, but no dose response is observed for L-tyrosine. L-leucine is demonstrating essential amino acid activity while L-tyrosine is not, and this may be due to the conversion of L-phenylalanine to L-tyrosine by phenylalanine hydroxylase.

4 Antimicrobials, phenol red, Hepes, and light

Cell culture systems have a high growth potential for micro-organisms. During primary culture, control of the microbial load is highly desirable. However, once the culture is established, antimicrobials are unnecessary in small scale cultures and potentially could mask occult infection. In attempting to establish defined systems the use of antimicrobials and phenol red is debatable; however, in the bioprocessing industry where thousands of litres are at stake a microbial contamination would be a financial disaster.

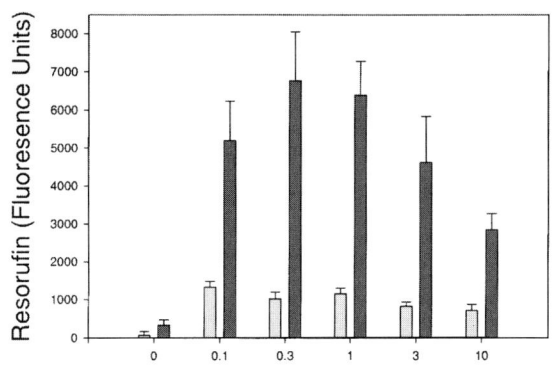

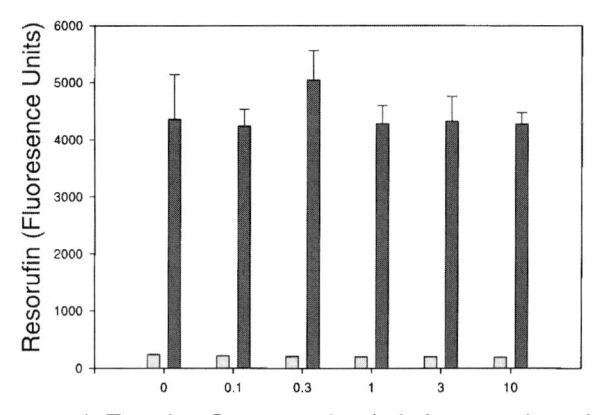

Figure 7. Response of Rhesus monkey hepatocytes to varying concentrations of L-leucine and L-tyrosine. The isolated hepatocytes were inoculated at 10^5 cells/cm^2 on collagen coated plates. After eight days in culture the hepatocytes were stimulated with 2 μM 3-methycholanthrene for an additional two days. 3-Methylcholanthrene is known to increase the expression of cytochrome P450 1A1/2. At that time 7-ethoxyphenoxazone was added for 40 min. After the incubation period the fluorescence of the 7-hydroxyphenoxazone, also known as resorufin (the product of the reaction), was read in the CytoFluor reader set at excitation 530 nm and emission of 590 nm. Each concentration has two bars: the smaller bar represents the fluorescence of the unstimulated hepatocytes and the larger bar represents the fluorescence of the stimulated hepatocytes. L-leucine demonstrates essential nutrient behaviour and L-tyrosine non-essential nutrient behaviour. The liver contains phenylalanine hydroxylase producing L-tyrosine and this is the most likely interpretation of the non-essential behaviour of L-tyrosine in this experiment.

4.1 Phenol red

Phenol red is widely used as a pH indicator in cell culture media. Unfortunately, phenol red or contaminants can interact unfavourably with cultured cells, causing toxicity and oestrogenic activity. Sodium–potassium ion homeostasis is perturbed in serum-free medium containing phenol red and the effect is neutralized by serum or serum albumin (15, 16). Grady et al. (17) described a

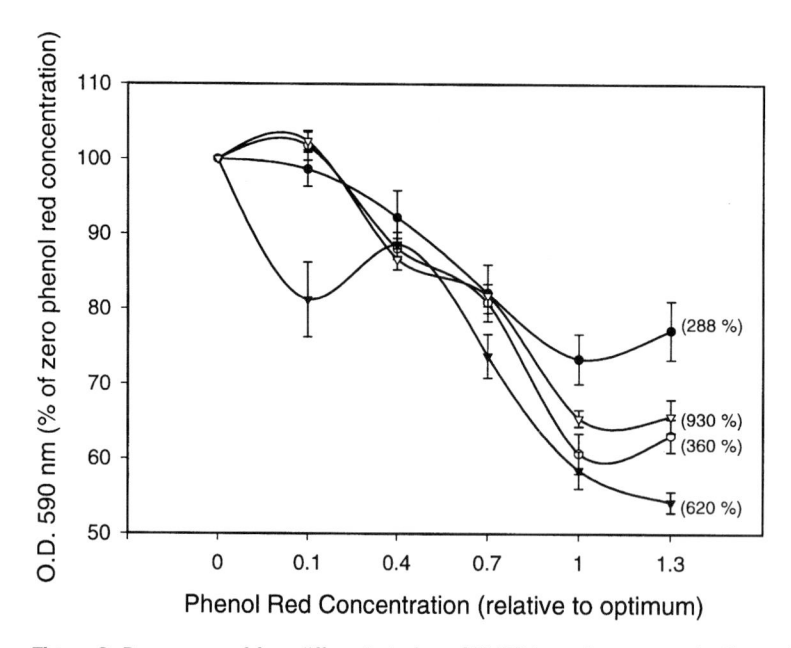

Figure 8. Responses of four different strains of NHBE to various concentrations of one lot of phenol red. These cells were inoculated at 10^3 cells/cm^2 and kept in culture for five to seven days. The cells were cultured in the absence of any undefined components. In each strain phenol red demonstrated a toxic effect at the recommended concentration, 1 on the x axis (3.3×10^{-6} M). The addition of bovine pituitary extract (BPE) at 3.3×10^{-6} M phenol red resulted in an increase in proliferation. This value is depicted in brackets as a per cent of the value in the absence of BPE. Quantitation was performed as described in *Figure 2*.

contaminant that became toxic when the pH rose above 7.4 but was virtually inert at lower pH values. HPLC analysis revealed many contaminating peaks and identified the cytotoxicity with a particular peak that differed from the oestrogenic activity. Anti-oestrogens frequently suppress growth below that of control cells not treated with oestrogen, even after careful steps have been taken to eliminate all oestrogen from the serum (18). The oestrogenic activity is due to a lipophilic impurity that binds to the oestrogen receptor with an affinity 50% that of oestradiol (19). *Figure 8* demonstrates the response of four different strains of normal human bronchial/tracheal epithelial cells (NHBE) to varying concentrations of phenol red. These cells were cultured in the absence of bovine pituitary extract (BPE) and all experienced inhibition of proliferation. In the presence of BPE, at 3.3×10^{-6} M phenol red, the cells demonstrated an apparent neutralization of the phenol red inhibitory effect.

4.2 Gentamicin

Gentamicin belongs to the aminoglycoside family of antibiotics. These antibiotics are nephrotoxins *in vivo* (20) and *in vitro* (21). However, toxicity is not exclusively localized to the kidney: the ear (22), heart (23), and the cornea (24) have all demonstrated susceptibility.

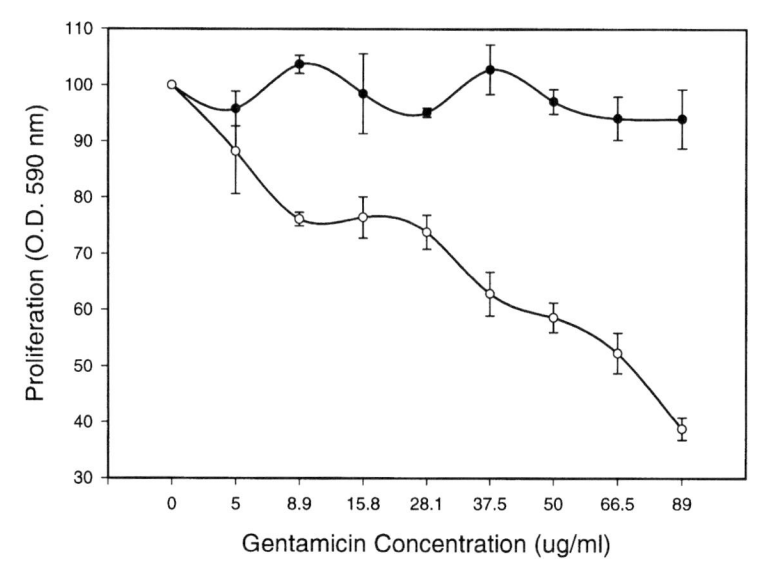

Figure 9. The response of NHBE to various concentrations of gentamicin from two different lots. These cells were inoculated at 100 cells/cm^2 and left in culture for seven days without a medium change. There were no undefined components in the medium. One lot of gentamicin demonstrated toxicity while the other did not. Each lot of gentamicin needs to be evaluated for toxicity. Quantitation was performed as described in *Figure 2*.

The mechanisms of toxicity range from free radical formation (25) to calcium transport inhibition (26). Hepatocyte metabolic conversion of gentamicin to a compound toxic to sensory cells from the inner ear is another mechanism of toxicity (27). Gentamicin is toxic *in vitro* to corneal epithelial cells (28), fibroblasts (29), and hepatocytes (30). The influence of two different lots of gentamicin on the proliferation of NHBE is shown in *Figure 9*, indicating that one lot was inhibitory.

4.3 Hepes

Hepes has been used extensively in cell culture due to its excellent buffering capacity in the physiological range and low binding of cations (31), but it can be toxic. For example, cytoplasmic vacuoles developed in chick embryo epiphyseal chondrocytes (32) and membrane inclusion bodies in human dermal fibroblasts (33) that resolved when Hepes was replaced by another buffer.

The toxicity exhibited by Hepes appears to be mediated by the production of reactive oxygen species. The interaction of Hepes-buffered culture medium with fluorescent light at room temperature resulted in toxicity to V79 Chinese hamster cells. The toxicity was prevented by shielding or the inclusion of catalase in the medium, implicating extracellular hydrogen peroxide as the toxic agent (34). In endothelial cells, Hepes stimulates the production of toxic oxygen metabolites resulting in a decrease in growth (35). Quin2 is a transition metal ion chelator that potentiates iron-driven oxidant formation. Oxidants were detected in

Hepes buffer (36). SIN-1 is an oxide-releasing compound and its toxicity to L929 cells is due to a co-operative action of hydrogen peroxide and reactive nitrogen species in the presence of Hepes (37).

4.4 Light

Fluorescent light can cause the deterioration of tissue culture medium, resulting in toxicity and mutagenicity (38). The deleterious effects are due to prolonged exposure, whereas short exposures can produce a mitogenic effect (39). The phototoxicity is due to wavelengths of light that can be absorbed by riboflavin. The pairing of riboflavin with tryptophan (40), Hepes (41), and guanine (42) results in toxicity mediated by reactive oxygen species (43).

Tryptophan and riboflavin are found in conventional and serum-free optimized media and because they are essential they will be included in all newly designed media. Limited exposure to fluorescent light during manufacture and storage is critical. Specially designed fluorescent tubes not emitting the relevant wavelengths or shields to block those wavelengths are available.

Protocol 1
Procedure for medium component toxicity testing

1. Determine the most defined medium formulation that allows proliferation of the cells. For example, omit the antimicrobials, and work with the lowest possible serum concentration.
2. Determine the length of time the cells can be maintained in culture without a medium change.
3. Determine the lowest possible density the cells can be plated. The actual inoculum will depend on the inherent plating efficiency of the cells in that medium, the level of detection, and the sensitivity of the method used for quantitating the cell number.[a]
4. If the component of interest is a medium component, make the medium from scratch minus that component (for examples on formulating media from scratch see refs 44 and 45). If the component is a supplement leave out the supplement.
5. Introduce the component at concentrations that flank the presumptive or existing concentration. Use logarithmic intervals, the size of which will be determined by the sensitivity of the response (compare *Figures 2-5* with *6*). The intervals need to be determined empirically.
6. Introduce the medium with the various concentrations into the plates and pre-incubate for at least 30 min before introducing the cells.
7. Incubate for the length of time determined in the above step.
8. Determine the relationship of component concentration to cell performance.

[a] Having a large volume to cell number ratio will reduce the effect the cells have on the medium. This condition will increase the sensitivity of the assay.

5 Purity of components

In cell culture the largest component of the system is water. Consequently, a relatively small contaminant in the water can have a significant effect on the cells. On the other hand water may provide essential nutrients that become limiting as the purity of water is increased. One class of nutrients that could account for this phenomenon is the trace metals.

Ideally, in chemically defined media all the components are known and there are no contaminants present. Though the availability of recombinant proteins and peptides has made the pursuit of chemically defined media more attainable, most fall short of this ideal. Cells may need albumin or other serum fractions for the desired performance or stability. Also, manufacturers may not necessarily screen for all potential contaminants in their organic compounds, such as selenium contamination of a thyroxine preparation (46). The many surfaces with which the medium components come in contact during manufacturing may contribute contaminants.

The bioprocessing industry is redefining the notion of purity due to the potential infectious agents that may be present in medium components. It is desirable that all components are of non-animal origin. For example methionine and cysteine are generally isolated from animal hair, so new sources are required.

6 Fatty acids

Serum provides lipids in various forms to cultured cells. These forms include cholesterol, phospholipids, triglycerides, fatty acids, fat soluble vitamins, and various esterified forms of these lipids.

The essential fatty acids that cannot be synthesized by animals are 18:2n-6 and 18:3n-3 ('18' refers to the number of carbon atoms and the number following the colon refers to the total number of double bonds in the molecule; refer to *Table 1*). Animals do not contain the enzymes necessary to place a double bond in these positions. Consequently, the acquisition of these molecules occurs only through an exogenous supply; however, animals (excluding true carnivores) contain desaturation and elongation/shortening enzymes that result in polyunsaturated fatty acids, most notably arachidonic acid, 20:4n-6, and eicosapentenoic acid, 20:5n-3, and fatty acids of higher degrees of unsaturation and elongation (47–49).

The removal of serum can result in an essential fatty acid deficiency since serum is the predominant lipid source for cultured cells. In the absence of a lipid source cells will synthesize, desaturate, and elongate fatty acids of the n-9 and n-7 families. Under these deficiency conditions there is a characteristic fatty acid profile represented by the polyunsaturated fatty acids of the n-9 and n-7 families (50). Cultured cells will utilize exogenous fatty acids preferentially to endogenously synthesized ones (51, 52). The various fatty acid families are metabolized by the same enzymes but at different rates. The essential fatty

Table 1. Fatty acid composition of phosphatidylcholine of normal human fibroblasts supplemented with various fatty acids[a]

	Supplementation[b]		
Fatty acid[c]	18:1n-9	18:2n-6	18:3n-3
16:0	25.80[d]	32.70	33.40
16:1n-9	7.50	5.90	7.00
16:1n-7	1.80	2.20	3.20
18:0	3.70	7.20	7.30
18:1n-9	48.80	18.10	22.30
18:1n-7	4.60	4.00	4.30
18:2n-6	0.60	17.20	1.50
18:3n-3	0.00	0.00	6.00
20:1n-9	0.70	0.20	0.10
20:3n-9	1.80	1.40	0.60
20:3n-6	0.20	2.20	0.20
20:4n-6	0.60	4.60	0.70
20:3n-3	0.00	0.00	0.90
20:4n-3	0.00	0.00	1.90
20:5n-3	0.00	0.00	5.30
22:3n-9	0.60	0.20	0.30
22:4n-6	0.00	0.50	0.00
22:5n-3	0.00	0.00	0.50
22:6n-3	0.40	0.40	0.50
Degrees of saturation[e]			
Saturated	32.2	43.2	44.8
Monounsaturated	63.7	30.4	36.9
Polyunsaturated	4.1	26.5	18.3

[a] Normal human fibroblast numbers were expanded in medium containing 0.5% fetal bovine serum (FBS). This level of FBS was insufficient to supply the required fatty acids for these fibroblasts under these conditions. Consequently, the cells displayed an essential fatty acid deficiency profile (not shown), similar to that displayed by the cells supplemented with the 18:1n-9 fatty acid. Subsequently, the cells were supplemented with pure fatty acids complexed to fatty acid depleted bovine serum albumin for several days.

[b] After the fibroblasts were cultured in an essential fatty acid deficient environment the medium was supplemented for several days with 18:1n-9, 18:2n-6, and 18:3n-3 fatty acids complexed to fatty acid depleted bovine serum albumin. The phospholipids were extracted, phosphatidylcholine purified, and the fatty acids of that phospholipid evaluated as fatty acid methyl esters by GLC.

[c] Fatty acids are designated by the number of carbon atoms (first two numbers), the number of double bonds in the molecule (the number after the colon), and on which carbon atom the first double bond is from the last 'nth' carbon (n-position of carbon atom) in the fatty acid molecule, previously named the omega carbon.

[d] Values are expressed as the nmole % of total fatty acids.

[e] This represents the sum of the nmole % of the fatty acids for each designation.

acids are preferentially metabolized over the non-essential fatty acid families. This situation can be exploited to generate cells with a fatty acid profile favouring a particular fatty acid family (see *Table 1*).

Polyunsaturated fatty acids have an extensive metabolism; generating eicosanoids, epoxides with vascular function, anandamide (a ligand of the cannabinoid receptors), and non-enzymatic oxidation products of 20:4n-6, the isoprostanes, which are isomers of prostaglandins (53, 54). Also, polyunsaturated fatty acids have functions that result from the fatty acid and not its metabolic product. For example, intracellular arachidonic acid opened a potassium-selective channel in neonatal rat atrial cells (55) and under some conditions extracellular arachidonic and docosahexanoic (22:6n-3) acids blocked the major voltage-dependent potassium channel (Kv1.5) in cardiac cells (56). Fatty acids can also influence the physical characteristics of the membrane by influencing its fluidity (57). Polyunsaturated fatty acids can increase the membrane fluidity, as in J774A.1 cells, and decrease the uptake of acetylated low density lipoprotein (49).

The fatty acid composition of normal human fibroblasts *in vitro* is demonstrated in *Table 1*. The supplemented fatty acid is represented to a large extent in the phosphatidylcholine (PC) of fibroblasts. The cells are capable of metabolizing the precursor fatty acid by elongation and desaturation to other fatty acids. Cells with different fatty acid compositions may be functionally different. It may be tempting to consider the removal of serum the final step in the process of generating a serum-free medium, but care must be taken to consider the fatty acid composition of the cells.

Protocol 2
Complexing fatty acids to albumin

1. Either bovine or human serum albumin can be used. The albumin needs to be close to fatty acid-free, and not just fatty acid-poor.

2. Run a toxicity curve to determine non-toxic concentrations (see *Protocol 1*).

3. Use the highest albumin concentration possible. For the chosen albumin concentration, calculate the fatty acid concentration that results in a mole ratio of fatty acid to albumin that is not greater than 2. This calculation determines the highest concentration of fatty acid possible in the system.[a] Make albumin stocks that are 50–100 × in a physiological buffer.

4. Dissolve the fatty acid in 100% ethanol of the highest purity. Calculate a fatty acid concentration so that once it is complexed to the albumin the final concentration of the ethanol the cells experience does not exceed 0.1%; however, each cell culture system needs to be tested.

5. Add the fatty acid dissolved in ethanol to the albumin solution in the smallest possible volume. Initially the solution will be cloudy but will clear upon gentle rotation at room temperature for approx. 15–20 min.[b]

6. Maintain the smallest possible headspace and shield from fluorescent light. Ideally, flush the surface of the solution with nitrogen through a 25 mm 0.2 μm sterile filter disk.

> **Protocol 2** continued
>
> 7. Aliquot the fatty acid–albumin complex and store at −70 °C. Maintain low headspace.
>
> [a] Albumin possess two high affinity binding sites. Beyond the binding capacity of albumin the fatty acids are more toxic.
>
> [b] Albumin is soluble in ethanol, so no precipitation of the albumin will occur due to the presence of ethanol presented with the fatty acid.

Acknowledgements

I thank Dr R. Ham for his advice, Jerry Hammond for the work on NHBE, NHEK, and monkey hepatocyte titrations, Sorian Damian for the cytochrome P450 assay development, and Dr Hoda Elgendy for the AOSMC titrations.

References

1. Pongrac, J. L. and Rylett, R. J. (1998). *J. Neurosci. Methods*, **84**, 69.
2. Ham, R. G., Hammond, S. L., and Miller, L. L. (1977). *In Vitro*, **13**, 1.
3. McKeehan, W. L. and Ham, R. G. (1978). *Nature*, **275**, 756.
4. Schwartz, Z., Nasatzky, E., Brooks, B. P., Soskolne, W. A., and Boyan, B. D. (1994). *Endocrinology*, **134**, 1640.
5. Hakvoort, A., Haselbach, M., and Galla, H. J. (1998). *Brain Res.*, **795**, 247.
6. Hakvoort, A., Haselbach, M., Wegener, J., Hoheisel, D., and Galla, H. J. (1998). *J. Neurochem.*, **71**, 1141.
7. Christensen, H. N. (1990). *Physiol. Rev.*, **70**, 43.
8. Petruschka, L., Rebrin, I., Grimm, U., and Herrman, F. H. (1990). *Clin. Chim. Acta*, **193**, 65.
9. Cartwright, E. C. and Danks, D. M. (1972). *Biochim. Biophys. Acta*, **264**, 205.
10. Uhlendorf, B. W. and Mudd, S. H. (1968). *Science*, **160**, 1007.
11. Adeola, O., Young, L. G., McBride, B. W., and Ball, R. O. (1989). *Br. J. Nutr.*, **61**, 453.
12. McBride, B. W. and Early, R. J. (1989). *Br. J. Nutr.*, **62**, 673.
13. Burke, M. D. and Mayer, R. T. (1983). *Chem. Biol. Interact.*, **45**, 243.
14. Burke, M. D., Thompson, S., Weaver, R. J., Wolf, C. R., and Mayer, R. T. (1994). *Biochem. Pharmacol.*, **48**, 923.
15. Hopp, L. and Bunker, C. H. (1993). *J. Cell. Physiol.*, **157**, 594.
16. Lubin, M. (1993). *In Vitro Cell Dev. Biol. Anim.*, **29A**, 597.
17. Grady, L. H., Nonneman, D. J., Rottinghaus, G. E., and Welshons, W. V. (1991). *Endocrinology*, **129**, 3321.
18. Katzenellenbogen, B. S., Norman, M. J., Eckert, R. L., Peltz, S. W., and Mangel, W. F. (1984). *Cancer Res.*, **44**, 112.
19. Bindal, R. D. and Katzenellenbogen, J. A. (1988). *J. Med. Chem.*, **31**, 1978.
20. Cronin, R. E. (1979). *Clin. Nephrol.*, **11**, 251.
21. Monteil, C., Leclere, C., Fillastre, J. P., and Morin, J. P. (1993). *Ren. Fail.*, **15**, 475.
22. Hinojosa, R. and Lerner, S. A. (1987). *J. Infect. Dis.*, **156**, 449.
23. Hendry, P. J., Taichman, G. C., Taichman, S. J., and Keon, W. J. (1988). *Can. J. Cardiol.*, **4**, 219.

24. Hasany, S. M. and Basu, P. K. (1989). *Lens Eye Toxic Res.*, **6**, 93.
25. Conlon, B. J., Aran, J. M., Erre, J. P., and Smith, D. W. (1999). *Hear Res.*, **128**, 40.
26. Schwertz, D. W., Kreisberg, J. I., and Venkatachalam, M. A. (1986). *J. Pharmacol. Exp. Ther.*, **236**, 254.
27. Crann, S. A., Huang, M. Y., McLaren, J. D., and Schacht, J. (1992). *Biochem. Pharmacol.*, **43**, 1835.
28. Lass, J. H., Mack, R. J., Imperia, P. S., Mallick, K., and Lazaru, H. M. (1989). *Curr. Eye Res.*, **8**, 299.
29. Tulkens, P. and Von Hoof, F. (1980). *Toxicology*, **17**, 195.
30. Weinberg, J. M., Simmons, F. Jr., and Humes, H. D. (1980). *Res. Commun. Chem. Pathol. Pharmacol.*, **27**, 521.
31. Good, N. E., Winget, G. T., Winter, W., Connolly, T. N., Izawa, S., and Singh, R. M. (1966). *Biochemistry*, **5**, 467.
32. Poole, C. A., Reilly, H. C., and Flint, M. H. (1982). *In Vitro*, **18**, 755.
33. Verdery, R. B., Nist, C., Fujimoto, W. Y., Wight, T. N., and Glomset, J. A. (1981). *In Vitro*, **17**, 956.
34. Zieger, M. A., Glofcheski, D. J., Lepock, J. R., and Kruuv, J. (1991). *Cryobiology*, **28**, 8.
35. Bowman, C. M., Berger, E. M., Butler, E. N., Toth, K. M., and Repine, J. E. (1985). *In Vitro Cell Dev. Biol.*, **21**, 140.
36. Sandstrom, B. E., Granstrom, M., and Marklund, S. L. (1994). *Free Radic. Biol. Med.*, **16**, 177.
37. Lomonosova, E. E., Kirsch, M., Rauen, U., and de Groot, H. (1998). *Free Radic. Biol. Med.*, **24**, 522.
38. Bradley, M. O. and Sharkey, N. A. (1977). *Nature*, **266**, 724.
39. Parshad, R. and Sanford, K. K. (1977). *J. Cell. Physiol.*, **92**, 481.
40. Edwards, A. M., Silva, E., Jofre, B., Becher, M. I., and De Ioannes, A. E. (1994). *J. Photochem. Photobiol.*, **24**, 179.
41. Zigler, J. S., Lepe-Zuniga, J. L., Vistica, B., and Gery, I. (1985). *In Vitro Cell Dev. Biol.*, **21**, 282.
42. Bessho, T., Tano, K., Nishmura, S., and Kasai, H. (1993). *Carcinogenesis*, **14**, 1069.
43. Silva, E., Ugarte, R., Andrade, A., and Edwards, A. M. (1994). *J. Photochem. Photobiol.*, **23**, 43.
44. Peehl, D. M. and Ham, R. G. (1980). *In Vitro*, **16**, 526.
45. Shipley, G. D. and Ham, R. G. (1981). *In Vitro*, **17**, 656.
46. Hamilton, W. G. and Ham, R. G. (1977). *In Vitro*, **13**, 537.
47. Grammatikos, S. I., Subbaiah, P. V., Victor, T. A., and Miller, W. N. (1994). *Ann. N. Y. Acad. Sci.*, **745**, 92.
48. Salem, N. Jr., Wegher, B., Mena, P., and Uauy, R. (1996). *Proc. Natl. Acad. Sci. USA*, **93**, 49.
49. Shichiri, G., Kinoshita, M., and Saeki, Y. (1993). *Arch. Biochem. Biophys.*, **303**, 231.
50. Sardesai, W. (1992). *Nutr. Clin. Pract.*, **7**, 179.
51. Stoll, L. T. and Spector, A. A. (1984). *In Vitro*, **20**, 732.
52. Rosenthal, M. D. (1987). *Prog. Lipid Res.*, **26**, 87.
53. Galli, C., Marangoni, F., Galella, G. (1993). *Prostaglandins Leukotrienes and Essential Fatty Acids*, **48**, 51.
54. Galli, C. and Marangoni, F. (1997). *Nutrition*, **13**, 978.
55. Kim, D. and Clapham, D. E. (1989). *Science*, **244**, 1174.
56. Honore, E., Barhanin, J., Attali, B., Lesage, F., and Lazdunski, M. (1994). *Proc. Natl. Acad. Sci. USA*, **91**, 1937.
57. Cribier, S., Morrot, G., and Zachowski, A. (1993). *Prostaglandins Leukotrienes and Essential Fatty Acids*, **48**, 27.

Chapter 5
Three-dimensional culture

L. A. Kunz-Schughart and W. Mueller-Klieser[†]
Institute of Pathology, University of Regensburg, D-93042 Regensburg, Germany.
[†] Institute of Physiology and Pathophysiology, Johannes Gutenberg University Mainz, Duesbergweg, D-55099 Mainz, Germany.

1 Introduction

Three-dimensional (3D) cell cultures have been widely used in biomedical research since the early decades of this century. Holtfreter (1) and later Moscona (2, 3) pioneered the field by their research on morphogenesis using spherical re-aggregated cultures of embryonic or malignant cells. Numerous subsequent *in vitro* studies on organogenesis or expression of malignancy were based on these early investigations. Substantial novel input into research on cell aggregates came from fundamental studies of Sutherland and associates (4–6). They pioneered multicellular tumour spheroids (MCTS) as an *in vitro* model for studies on tumour cell response to therapy. MCTS have also been used to study basic biological mechanisms, including the regulation of proliferation, differentiation, cell death, invasion, angiogenesis, and the immune response (for reviews see refs 7–11). This chapter will mainly focus on techniques for growing spheroids, with limited information on 3D cultures other than spheroids.

One major advantage of 3D cell cultures is their well-defined geometry—whether planar or spherical—which makes it possible to directly relate structure to function, and which enables theoretical analyses, for example of diffusion fields. Combining such approaches with molecular analysis has demonstrated that, in comparison to conventional cultures, cells in 3D culture more closely resemble the *in vivo* situation with regard to cell shape and cellular environment, and that shape and environment can determine gene expression and the biological behaviour of the cells. One impressive example is the ectopic implantation of embryonic cells, which can result in malignant transformation, whereas the same cells undergo normal embryogenesis in the uterus. Conversely, teratocarcinoma cells may undergo normal development when implanted into an embryo (12). One further example is the relative resistance of cancer cells to drugs in 3D culture compared to the same cells grown as conventional monolayer or in single cell suspension (for a recent review see ref. 13).

2 Multicellular tumour spheroids (MCTS)

2.1 MCTS monocultures

2.1.1 The history of MCTS and their relevance to solid tumours

Multicellular tumour spheroids (MCTS) mimic solid tumours better than monolayer cultures. MCTS have significantly increased our knowledge of radiation response of mammalian cells (14), intercellular communication, cell invasion, angiogenesis, and neovascularization (reviewed by refs 15–17).

Although attempts to initiate spheroids directly from biopsies for routine individual predictive drug testing had little success, MCTS from established cell lines provide an *in vitro* model to study mechanisms controlling drug penetration, binding, and action (for reviews see refs 8–10, 14, 18–21).

In terms of three-dimensional growth, MCTS behave like the initial, avascular stages of solid tumours *in vivo*, unvascularized micrometastatic foci, or intercapillary tumour microregions, as shown in *Figure 1*. Growth kinetics can be described by the Gompertz equation (22–24) which has been shown to model three-dimensional *in vivo* and *in vitro* growth reasonably well (25). MCTS develop discrete cell populations similar to those found in microregions of solid tumours with three major groups: actively cycling cells closest to the blood and nutrient supply; quiescent, yet intact and viable cells as intermediates; and necrotic cells/areas furthest from the blood supply (*Figure 2*). Beyond a critical size (≥ 500 μm), most spheroids from established cell lines develop massive necrosis in the centre, surrounded by a viable rim of cells with a thickness of 100–300 μm.

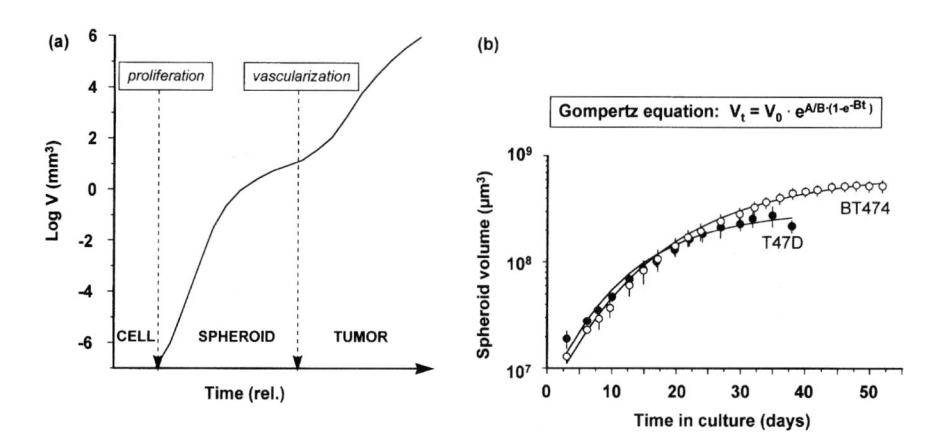

Figure 1. (a) Schematic illustration of the different stages of solid tumour growth that can be mathematically described by the Gompertz function. (b) Representative spheroid volume growth kinetics of two different human breast cancer cell lines grown in liquid-overlay culture in conventional, supplemented media. Lines show non-linear least squares best fits to the Gompertz function with $V_0 = 0.498 \times 10^7$, $A = 0.298$, and $B = 0.060$ for BT474 cells, and $V_0 = 0.578 \times 10^7$, $A = 0.331$, and $B = 0.083$ for T47D cells.

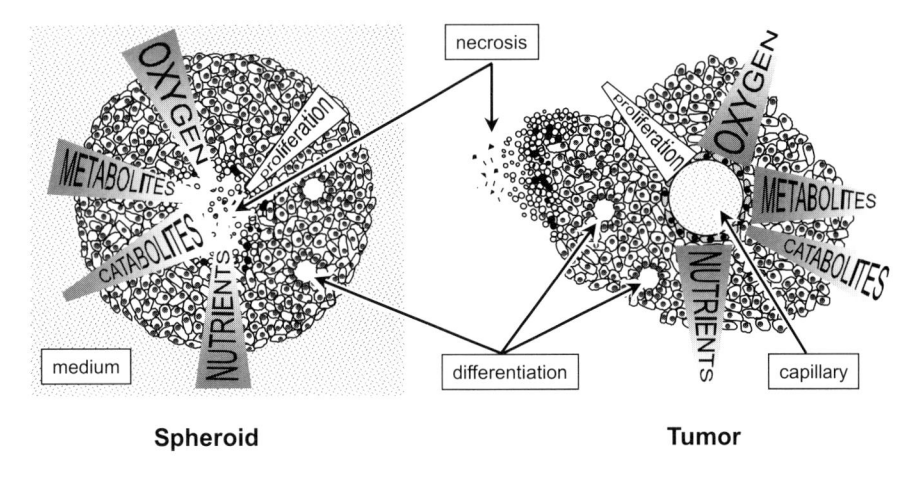

Figure 2. Schematic illustration of the analogy between tumour microregions and multicellular tumour spheroids.

2.1.2 Current research with MCTS monocultures

Many of the early findings with the V79 (Chinese hamster lung cells) and the EMT-6 (mouse sarcoma) spheroid models were supported by subsequent studies with human tumour cell lines. Necrotic cell death in the spheroid centre mainly relates to the limited inward and outward diffusion (*Figure 2*). Thus, in some tumour spheroid types, such as WiDr (human colon adenocarcinoma) and Rat1-T1 (*ras*-transfected rat embryo fibroblasts), necrosis and hypoxia are coincident (26, 27), and necrotic cell death might be explained by a single limiting factor (28). In contrast, there is experimental evidence from other MCTS types that necrosis is a complex, multifactorial event (29–31).

Investigations on microenvironmental and epigenetic mechanisms involved in the regulation of cell proliferation, differentiation, and gene expression have been intensified enormously within the last ten years (7, 10, 32). Many cell types seem capable of maintaining intracellular homeostasis in 3D culture until shortly before necrotic cell death, despite environmental stress (21). There are almost no structural or metabolic markers for this 'pre-necrotic' stage except for a reduced oxygen uptake and/or respiratory activity, and a decrease in mitochondrial function. Cell cycle arrest in the nutrient deprived inner regions of large *myc/ras* co-transformed rat fibroblast spheroids (MR1) is accompanied by an upregulation of the CDK inhibitor p21$^{Kip1/Cip1}$, while cell cycle arrest in monolayer cultures and in *myc*-transfected, non-tumorigenic aggregates is associated with enhanced expression of p18^{Ink4} (33). These data showed for the first time that cell cycle control differs mechanistically in monolayer and spheroid culture and may change at different stages of malignant transformation.

Due to the massive central necrotic cell death in MCTS, programmed cell death in 3D aggregates has long been ignored. However, recent data show that apoptosis occurs in spheroids of intestinal epithelial cells (34). Growth inhibition via stimulation of apoptosis has also been described by Fujiwara *et al.* (35, 36) for p53-defective human lung cancer spheroids transfected *in situ* with a viral wild-type p53 vector. Mueller-Klieser reported apoptotic cell death of multinucleated giant cells in a highly differentiated rhabdomyosarcoma spheroid and in 3D cultures of V79 cells using the TUNEL assay (10, 37).

3D cultures undergo morphological and functional reorganization (for review see ref. 7). Amongst these activites are:

(a) Modified deposition and/or assembly of biomatrix such as the extracellular matrix molecules fibronectin, collagen types I, III, and IV, or laminin (38).

(b) Expression and function of ECM receptors (integrins), receptor subsets and subunits (39–41), and cadherins (42).

(c) The expression and function of biological response modifiers such as growth and angiogenic factors (i.e. EGF, TGFα, VEGF) and/or their receptors (38, 39, 43–48).

(d) The development of gap junctional communication (49, 50).

2.1.3 MCTS culture

A common principle of all spheroid culture methods is the prevention of cellular attachment to a substrate such as surfaces of Petri dishes. Many non-transformed and most malignant cells tend to form spherical aggregates under these conditions.

There are two strategies for cultivating MCTS: in stirred medium and in static medium. In the following section, both these technical approaches are described, and the advantages and disadvantages for specific applications are discussed.

The most widely used method for growing MCTS from established cell lines is the spinner flask culture (*Figure 3*). With this technique a large number of

Spheroid Culturing Techniques

Roller tube — Spinner flask — Gyratory shaker — Soft agar and liquid overlay

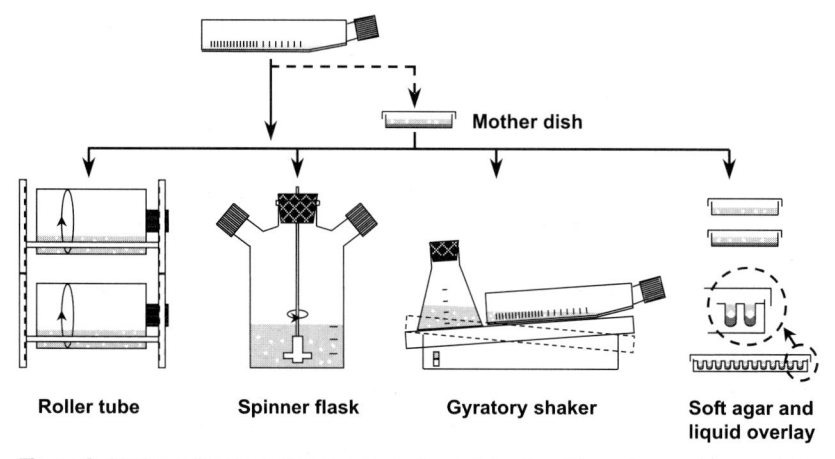

Figure 3. Methods for the cultivation of multicellular spheroids under continuous stirring (spinner flask, roller tube, gyratory shaker, etc.) or static on non-adherent surfaces (liquid-overlay in 96-well plates).

MCTS may be generated simultaneously in large volume cultures, and these can be scaled-up to mass culture. While some tumour cell types are capable of producing cell aggregates directly from single cell suspensions in spinner flasks, cell–cell interaction, and thus aggregate formation, is often more efficient under static conditions. As a consequence, bacteriological Petri dishes or agar/agarose-coated culture dishes, to which cells do not adhere, are often used to initiate spheroid formation.

Cell number and initiation period are critical variables which depend on the cell type and culture conditions (medium, serum pH, etc.). In general, the cell number inoculated varies between 1×10^5 and 2×10^6 exponentially growing cells per 100 mm dish. Once aggregates have emerged in the unstirred medium within two to five days, they can be transferred into spinner flasks. Medium is not refreshed during the initiation period, whereas it is renewed routinely in spinner cultures. Since spheroid initiation may lead to the formation of aggregates with a large variability in size, and since spheroid-to-spheroid clusters may occur, selection of a specific spheroid population prior to the transfer into spinner flask is recommended. If several selective sedimentation steps, for example using Falcon tubes, are not sufficient to eliminate large cell clusters and single cells, hand-selection or rapid harvesting of spheroids by sieving through nylon screens can be performed. The coefficient of variation of the average spheroid diameter within the selected spheroid population should be ≤ 10%.

Other frequently used techniques to generate large numbers of spheroids are roller tubes and the 'gyratory shaker' method shown in *Figure 3*. As for spinner flask cultures, an initiation period under non-stirring conditions may be required and the medium should be renewed routinely after transfer into the roller tubes or onto the gyratory shaker. Also, in order to guarantee optimum constant supply conditions throughout spheroid growth, the number of cells per volume of medium should be kept relatively constant by gradually reducing the number of spheroids per flask. For many tumour cell types, the optimum cell number does not extend beyond 3×10^5 cells per ml of medium, and two-thirds of the medium has to be refreshed every 24–48 hours.

The most convenient method of growing spinner flask cultures is in a humidified CO_2 incubator at 37°C, if a stirring device suited for incubator atmospheres is available. Alternatively, spinner flasks may be sealed and kept in a non-humidified, temperature controlled atmosphere or in a thermostatically controlled water-bath positioned over a magnetic stirrer. If the volumes of the gas phase and of the culture medium are approximately the same, the medium of the sealed cultures may be refreshed every 24 hours without generating unfavourable environmental conditions due to cellular production of CO_2 and metabolic waste. The size of commercially available spinner flasks for research purposes ranges from 100 ml to 1 litre with a recommended maximum filling level of approximately 50% of the total volume. The choice of size should be determined by the number of spheroids needed or the (minimum) volume of supplemented medium to be applied. Different types of spinner flasks are available from various companies. While the handling of these spinner flasks, in particular the effort required for cleaning, may differ, the authors have experienced no relevant difference between them in the growth kinetics of MCTS.

An additional parameter that should be optimized is the stirring frequency and thus the shear force in the spinner culture. Optimization should minimize the number of cells that are shed from the spheroids into the medium. Some cell types exhibit an intrinsically high cell shedding in spheroid culture that may not be decreased by the growth conditions. As a result, shed single cells may form small clusters and/or adhere to the vessel wall. In general, these complications can be alleviated, if not eliminated, by changing the culture medium more frequently and by siliconization of the inner surfaces of the spinner flasks.

Some cell types do not form ideally spherical spheroids. Indeed, some tumour cells do not even exhibit cell aggregation, such as some variants of the MCF-7 breast cancer cell line. If no or only irregular cell clusters are obtained, a number of parameters can be modified to achieve satisfactory cell–cell aggregation and aggregate growth, including:

(a) Initiation of aggregate formation:
- serum type (batch)
- serum content
- culture medium (glucose and glutamine concentrations)
- non-adhesive dish (microbiological, agar-coated)
- CO_2 in incubator (pH in culture medium)
- cell concentration

(b) Initial phase of spinner flask culturing:
- same factors as in (a); each factor may be different from those in (a)
- stirring frequency
- spinner flask geometry

Protocol 1
Initiation and cultivation of tumour spheroids in spinner flasks

Equipment and reagents
- 100 mm agarose-coated culture dishes (see *Protocol 2*) or bacteriological Petri dishes (Falcon)
- 50 ml tubes (e.g. Falcon)
- Different types of pipettes (Pasteur, Eppendorf, 10–25 ml glass or one-way plastic pipettes)
- Spinner flasks (Techne or Bellco)
- Stirring drive unit
- CO_2 incubator
- Inverted microscope with eyepiece reticule that can be moved with a micrometer screw
- Trypsin/EDTA solution
- MEM, DMEM, RPMI, or other standard medium
- FCS (fetal calf serum)
- Penicillin/streptomycin

THREE-DIMENSIONAL CULTURE

Protocol 1 continued

A. Initiation of 'mother dishes'

1. Enzymatically dissociate exponentially growing monolayer cultures.
2. Stop enzymatic reaction by adding appropriate volume of supplemented medium (e.g. DMEM containing 5% or 10% (v/v) FCS).
3. Transfer single cell suspension into Falcon tube and spin down (1000–1500 r.p.m.).
4. Resuspend cell suspension, count cells with automated cell counter or hemocytometer, and prepare cell suspension with a concentration of 1×10^5 cells/ml.
5. If the optimum cell concentration to be inoculated per dish is not known, seed 1×10^5, 2×10^5, 5×10^5, and/or 1×10^6 cells/dish each into three or four bacteriological Petri dishes or agar-coated culture dishes with a final volume of 15 ml medium.
6. Grow for four days under standard culture conditions (e.g. 5% (v/v) CO_2 in air in a humidified atmosphere at 37 °C). Avoid any motion of dishes—this is important.
7. Check spheroid formation and shape of the aggregates under an inverted microscope and estimate the average size of the round-shaped subpopulation. Spheroids may then be collected and transferred into an one litre spinner flask containing 300–500 ml of supplemented culture medium.

B. Transfer of aggregates into spinner flask

1. Collect aggregates from four dishes and transfer into 50 ml medium in Falcon tubes with a Pasteur pipette or an Eppendorf pipette with cut-off tip or a 10 ml glass pipette. The type of pipette may determine the purity of the collection procedure. For example, a smaller tip will not transfer large cell clusters.
2. Let aggregates of appropriate size sediment for 1–3 min and remove 80% of the medium.
3. Add 40 ml of fresh medium.
4. Repeat steps 2 and 3 once or twice to reduce the number of extremely small aggregates and single cells in the suspension.
5. Transfer all or an aliquot of the selected spheroids into a bacteriological or agar-coated dish for accurate sizing.
6. Size a representative population of 25–50 spheroids by measuring two orthogonal diameters for each individual spheroid, either using an inverted microscope equipped with a calibrated reticule or by applying a calibrated image processing apparatus (51). Calculate the average spheroid volume.
7. Transfer 50–100% of the selected spheroid population into an one litre spinner flask containing 300–500 ml of supplemented medium. If all spheroids are to be used for further culturing, all steps (1–6) have to be accomplished under sterile conditions.

Protocol 1 continued

C. Cultivation of MCTS in spinner flasks

1. Transfer the one litre spinner flasks into humidified 37°C temperature controlled CO_2 incubator and place onto stirring drive unit with a speed of 75-100 r.p.m. (smaller spinner flasks will need a lower speed).
2. Routinely feed spheroids at day 2 and every first or second day thereafter by allowing spheroids to sediment in the spinner flask for a few minutes and by replacing two-thirds of the medium.
3. In parallel, size a representative population of 25-50 spheroids according to part B, step 6.
4. Estimate the number of spheroids in the spinner flask. Dissociate 10-50 spheroids using a 5 × concentrated enzymatic solution (5 × compared to that used for monolayer cultures of the same cell line), in order to determine the number of cells/spheroid and the number of cells per medium volume unit.
5. Reduce the number of spheroids if the cell concentration exceeds 2×10^5 cells/ml or if pH in the medium drops substantially below 7.0.
6. Prior to a final experiment, hand-selection with a Pasteur pipette or an Eppendorf pipette with cut-off tip or rapid harvesting of spheroids of approximately the same size by pipetting them through a nylon screen may be required.

Protocol 2
Coating of 96-well plates or culture dishes with agarose

Equipment and reagents

- 100 mm culture dishes and/or 96-well plates (e.g. Falcon)
- 10-25 ml glass or one-way plastic pipettes and/or multistep pipette with dispenser tips
- High gelling agarose
- MEM, DMEM, RPMI, or other standard medium
- Penicillin/streptomycin
- Distilled H_2O

Method

1. Prepare up to 200 ml (for 96-well plates) and up to 600 ml (for culture dishes) of 1.5% (w/v) agarose in serum-free medium (e.g. DMEM).
2. Autoclave agarose solution for at least 20 min.
3. Let agarose solution cool down to approx. 60°C.

Alternative method

1. Prepare 40 ml (for 96-well plates) and 100 ml (for culture dishes) of 3% (w/v) agarose in distilled H_2O.
2. Autoclave agarose solution for at least 20 min.

Protocol 2 continued

3. Warm up (to 37°C) 160 or 500 ml of serum-free medium (e.g. DMEM) and dilute freshly autoclaved 3% (w/v) agarose in H_2O with appropriate amount of medium to 0.5%.
4. (a) Use 20 ml or 25 ml glass pipette to spread 10 ml agarose solution into each culture dish and check for complete surface coating; 60 dishes may be prepared at a time.
   (b) Use multistep pipette with appropriate dispenser tip to spread 100 µl agarose in each well of a 96-well plate; coating of 20 plates is feasible.
5. Agarose-coated dishes and 96-well plates may be stored for approx. one week (depending upon the quality/strength of the agarose gel) in a humidified atmosphere containing 5% (v/v) CO_2 in air at 37°C. For long-term storage (up to two weeks), dishes/plates may be sealed and stored at 4°C. These dishes/plates should be incubated under standard culture conditions for 30–60 min prior to spheroid initiation.

Spinner culture is preferable to growth of MCTS on stationary, non-adhesive surfaces, if large numbers of spheroids with a maximum size are needed. Spheroids continuously grown on agar/agarose-coated surfaces may not reach their maximum size, and the thickness of the viable cell rim is often relatively small. On the other hand, various modifications for culturing spheroids in so-called agar- or liquid-overlay (Figure 3) have been established. The application of 96-well plates is of particular interest, since individual spheroids can be monitored repeatedly and may be manipulated throughout growth. Following an initiation period under static conditions, liquid-overlay cultures may be kept on a gyratory shaker to guarantee optimum nutrient supply. As a prerequisite, 96-well plates are coated with 0.5–1.5% agar/agarose as described in Protocol 2. A novel semi-adhesive substrate used for spheroid cultivation takes advantage of the spontaneous detachment of small cell aggregates formed on semi-adhesive substrates such as proteoglycans or positively charged polystyrene (52).

Protocol 3

Cultivation of liquid-overlay tumour spheroid cultures in 96-well plates

Equipment and reagents

- Agarose-coated 96-well plates (see Protocol 2)
- Various pipettes including an 8- or 12-channel Eppendorf pipette
- Inverted microscope with an eyepiece reticule that can be manipulated with a micrometer screw
- Gyratory shaker (optional)
- 0.25% trypsin
- 1% EDTA solution
- MEM, DMEM, RPMI, or other standard medium
- FCS
- Penicillin/streptomycin

Protocol 3 continued

Method

1. Enzymatically dissociate exponentially growing monolayer cultures (e.g. with 0.25% trypsin/1% EDTA).
2. Stop enzymatic reaction by adding appropriate volume of supplemented medium (e.g. DMEM containing 5% or 10% (v/v) FCS).
3. Transfer single cell suspension into Falcon tube and spin down at 300 g.
4. Resuspend cell suspension, count cells with automated cell counter or hemocytometer, and prepare cell suspension at a concentration of 2×10^4 cells/ml.
5. If the optimum cell concentration to be inoculated per well is not known, seed 5×10^2, 1×10^3, 2×10^3, and/or 4×10^3 cells/well each into 24 agar-coated wells with a final volume of 200 μl medium.
6. Incubate liquid-overlay cultures under standard culture conditions (e.g. 5% (v/v) CO_2 in air in a humidified atmosphere at 37°C). Avoid any motion of plates for a minimum of three to four days.
7. Check cell aggregation under an inverted microscope at days 3-4 and size a representative population of 12-24 spheroids by measuring two orthogonal diameters for each individual spheroid either with a reticule manipulated by a micrometer screw in the eyepiece of an inverted microscope or using a calibrated image processing apparatus.
8. Eventually place 96-well plates on gyratory shaker positioned in the CO_2 incubator for further culturing.
9. Feed spheroids routinely every 24-48 h by refreshing 100 μl of the medium using 8- or 12-channel Eppendorf pipettes and size a representative spheroid population until maximum size is reached.

2.1.4 Additional methodological aspects

In addition to the volume of the spheroids, which should be monitored routinely, analysis of the proliferative activity is useful. The proliferation status can be determined autoradiographically by [^{3}H]thymidine labelling (TLI = thymidine labelling index) or immunohistologically utilizing the BrdUrd antibody technique or antibodies detecting specific proliferation-associated antigens such as PCNA or Ki67. Alternatively, spheroids may be dissociated enzymatically and fixed or unfixed single cell suspensions can be analysed for their cell cycle distribution using flow cytometry and DNA-specific fluorochromes such as propidium iodide, 7-AAD (7-aminoactinomycin D), mithramycin, or Hoechst DNA dyes.

Sections of frozen or fixed, paraffin-embedded spheroids can be stained by any conventional labelling technique, or examined by scanning or transmission electron microscopy.

A method for isolating specific cell subpopulations from spheroids (53) is

THREE-DIMENSIONAL CULTURE

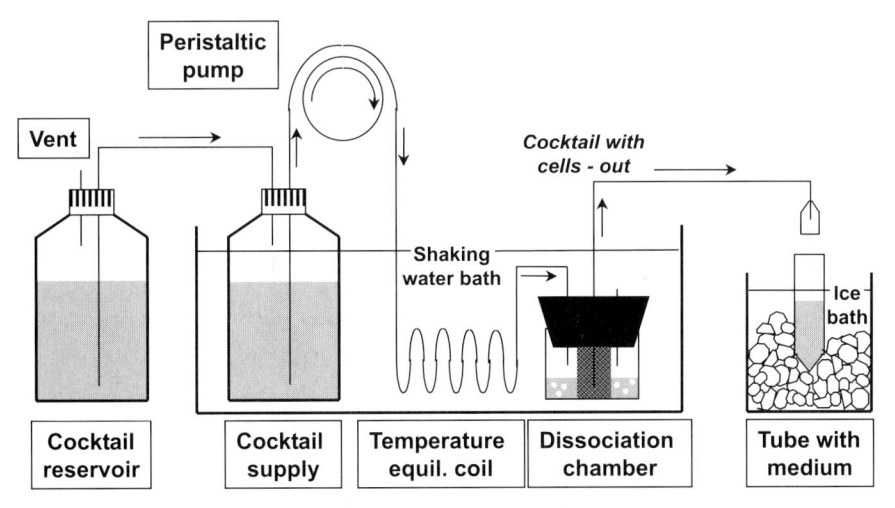

Figure 4. Experimental set-up for automated selective dissociation of multicellular spheroids consisting of an enzyme cocktail reservoir and supply unit, a peristaltic pump for enzyme transport through a temperature equilibration coil into a specially designed dissociation chamber, both placed in a 37°C temperature controlled shaking water-bath. The dissociation chamber contains an outer and an inner chamber which are separated by a 75 μm mesh, and is sealed by a stopper. Spheroids are placed in the outer chamber and the enzyme cocktail is pumped through the chamber so that dissociated cells are transferred into a tube containing ice-cold medium.

described below in detail. This automated dissociation technique is based on the sequential exposure of the cell aggregates to an enzyme solution (54). The experimental set-up consists of autoclavable, standard laboratory equipment mounted on a gyratory platform in a 37°C water-bath as shown in *Figure 4*. The enzyme solution is continuously pumped through a temperature equilibration coil into a 'racetrack' dissociation chamber which is filled with 15 ml of enzyme solution prior to the dissociation procedure. The chamber consists of two compartments which are separated by a nylon screen with a pore size of 75 μm. Pumping enzyme solution into the outer compartment and out of the inner part generates a flow of liquid through the screen. Thus, the spheroids are kept in suspension in the outer compartment, while dissociated single cells reach the inner compartment, following the fluid flow. The cell suspension is constantly removed from the inner compartment and collected on ice. Sequential aliquots are sampled from different levels within the spheroids.

2.2 MCTS co-cultures

2.2.1 General aspects of tumour heterogeneity

The presence of histologically different cell types within tumours, including host defence cells, fibroblasts and endothelium contributes to their heterogeneity. This heterogeneity can to some extent be mimicked *in vitro* by heterologous spheroids consisting of tumour cells and host cells (for a review see ref. 21).

Such spheroids may be initiated directly from mixtures of suspended host cells, such as fibroblasts or monocytes and tumour cells. One disadvantage of this approach is the risk of non-reproducible formation of intraspheroidal clusters of one cell type.

Alternatively, preformed MCTS of defined sizes and histomorphological characteristics can be incubated with suspensions of immune cells such as monocytes or LAK (lymphokine-activated killer) cells, to study immune cell infiltration and its cytostatic and/or cytotoxic effects on the tumour cells (55–57).

2.2.2 Co-cultures of MCTS and immune cells

Mononuclear cells from peripheral blood can be prepared by standard density centrifugation over Ficoll/Hypaque, and human blood monocytes can be separated from this cell mixture by their intrinsic activity to adhere to plastic or glass surfaces. A more convenient technique uses leukaphoresis and countercurrent elutriation (58). Differentiated macrophages can be generated by long term culture in Teflon bags, and these can be used for co-cultivation experiments (58, 59).

Protocol 4

Co-cultivation of MCTS and monocytes in liquid-overlay culture

Equipment and reagents

- Agarose-coated 96-well plates (see *Protocol 2*)
- Various pipettes including an 8- or 12-channel Eppendorf pipette
- Inverted microscope with an eyepiece reticule that can be manipulated by a micrometer screw
- 0.25% trypsin/EDTA solution
- MEM, DMEM, RPMI, or other standard medium
- FCS
- Penicillin/streptomycin

A. Initiation and cultivation of MCTS and preparation of monocyte suspension

1. Initiate tumour spheroids in agarose-coated 96-well plates as described in *Protocol 1*.
2. Measure the average spheroid diameter and volume over time, and decide when to add the monocyte suspension (e.g. small size spheroids without necrosis versus large spheroids with necrotic core).
3. Make fresh preparation of human blood monocytes (e.g. reference 58).
4. Prepare monocyte/macrophage suspension in medium containing 10% FCS (later adding 2–5% AB serum which may be required for macrophage differentiation; the medium and serum should be low in or free of LPS) at a concentration of $1-4 \times 10^5$ cells/ml.

Protocol 4 continued

B. Co-cultivation of MCTS and monocytes/macrophages

1. Remove 100 µl of the medium.
2. Add $1-4 \times 10^4$ monocytes/macrophages (100 µl of the immune cell suspension) to each well of a 96-well plate containing individual tumour spheroids.
3. Freeze or fix an aliquot of spheroids after 24–48 h of co-cultivation to investigate immune cell migration capacity using immunohistochemistry and/or dissociate co-cultures to obtain single cell suspensions for further flow cytometric or molecular analyses.
4. Freeze or fix another aliquot of spheroids after six to eight days of co-cultivation to investigate macrophage differentiation using either immunohistochemistry or flow cytometry or molecular analysis following enzymatic dissociation of co-cultures.

2.2.3 Co-cultures of MCTS and fibroblast aggregates

The extracellular matrix (ECM), mainly produced by stromal fibroblasts, has a crucial role in epithelial cell differentiation, growth, and gene expression. In order to investigate interactions between stromal fibroblasts and tumour cells, MCTS can be cultured on confluent fibroblast monolayers (60) or with fibroblast aggregates (33), as described with human bladder and breast cancer cells (21, 61). While some tumour cell types overgrow fibroblast aggregates and may show some specific cell–cell interactions in the contact zone, others, in particular highly metastatic tumour cells, seem to invade the fibroblast aggregates.

Protocol 5

Co-cultivation of MCTS and fibroblast aggregates in liquid-overlay culture

Equipment and reagents

- Agarose-coated 96-well plates (see *Protocol 2*)
- 100 mm dishes and 50 ml tubes (e.g. Falcon)
- Various pipettes (see *Protocol 3*)
- Sterile scalpel and a pair of scissors
- 0.25% trypsin/1 mM EDTA solution
- DMEM with 4.5 g/litre glucose
- DMEM with 1.0 g/litre glucose
- FCS
- Penicillin/streptomycin and amphotericin
- PBS (Mg^{2+}- and Ca^{2+}-free phosphate-buffered saline)
- Acetone (for fixation of monolayers)
- Liquid nitrogen (for freezing)
- 4% (v/v) formalin in PBS for fixation

A. Preparation of fibroblasts from primary (breast) tumour biopsies

1. Remove adipose and non-tumour tissue from biopsy.

Protocol 5 continued

2. Wash tissue with PBS.
3. Incubate tissue for 30–60 min in a high concentration of penicillin/streptomycin (5000 IU/ml penicillin, 5000 µg/ml streptomycin) in PBS.
4. Place tissue in medium and slice into 1–4 mm³ pieces under sterile conditions.
5. Transfer 20–30 tumour fragments into one 25 cm² culture flask.
6. Incubate for approx. 1 h until fragments have adhered to the surface of the culture flask.
7. Cover fragments with DMEM containing 20% (v/v) FCS, 25 mM glucose, penicillin/streptomycin, and amphotericin.
8. Monitor cell outgrowth after two weeks and routinely thereafter.
9. Transfer cells into 75 cm² culture flask after sufficient outgrowth using trypsin and reduce serum to 10% (v/v) and glucose concentrations to 5 mM.
10. Routinely prepare slides with low passage monolayer cultures for fixation and immunohistochemical detection of contaminating cells such as tumour or endothelial cells.

B. Initiation of MCTS and fibroblast aggregates

1. Initiate MCTS and fibroblast aggregates separately according to the routine initiation protocol; if possible use same medium. In general, cell numbers to be inoculated differ for tumour cells (5×10^2 to 2×10^3/well) and fibroblasts (3×10^3 to 6×10^3) by a factor of two to four in order to obtain aggregates within the same size range after an initiation interval of three to five days.
2. Routinely size MCTS and fibroblast aggregates after three to four days and daily thereafter until transfer.

C. Co-cultivation of MCTS and fibroblast aggregates

1. When tumour spheroids are slightly larger than fibroblast aggregates (e.g. 350 µm versus 300 µm), transfer individual 3D fibroblast cultures into agarose-coated wells, each containing one spheroid, using a Pasteur pipette or an Eppendorf pipette with a cut-off tip (8- or 12-channel Eppendorf pipette may be applied if cut-off tips are available).
2. Monitor growth behaviour every day and feed co-cultures routinely, until fibroblast aggregate is overgrown.
3. Freeze or fix co-culture at different growth stages for immunohistochemical analysis and histomorphological observation of cell–cell interactions. To avoid cell death, do not culture fibroblast aggregates for more than 16 days.

2.2.4 Co-cultures of MCTS and endothelial cells

MCTS are also used in studies of angiogenesis and invasion, such as the embryonic chick heart fragment confrontation culture system in semi-solid medium (for reviews see refs 15, 62–66). Culture of MCTS on confluent endothelial cells with underlying ECM can be used to study tumour cell invasion, but not angiogenesis. Data obtained with melanoma cell spheroids have led to the hypothesis that free radical-mediated endothelial cell damage is one mechanism contributing to melanoma metastasis (67).

Protocol 6
Co-cultivation of MCTS and endothelial cells

Equipment and reagents

- 100 mm bacteriological Petri dishes and culture dishes (e.g. Falcon)
- Sterile scalpel, pair of scissors, and cannulas
- Inverted phase-contrast microscope
- 70% (v/v) ethanol
- PBS
- Ham's F12/Iscove's medium (1:1)
- FCS

A. Initiation and cultivation of MCTS

1. Initiate MCTS according to *Protocols* 1 and 3.
2. Routinely feed spheroids.
3. Size spheroids after a defined initiation interval.

B. Preparation of umbilical cord vein

1. Wash surface of freshly isolated umbilical cord vein with 70% (v/v) ethanol, place into Petri dish, and cover with PBS.
2. Cut umbilical cord longitudinally.
3. Section umbilical cord vertically to obtain 1.5–2 cm pieces and check for integrity.
4. Transfer pieces into new Petri dish and fix tissue with endothelial cell layer upward, e.g. by using the tips of sterile cannulas.
5. Cover umbilical cord tissue with Ham's F12/Iscove's (1:1) medium containing 10% FCS.
6. Incubate under standard culture conditions (5% (v/v) CO_2 in air, humidified, 37°C).

C. Co-cultivation of MCTS and umbilical cord vein endothelial cells

1. Place spheroids of an appropriate size (e.g. small MCTS without necrosis versus larger spheroids with necrosis) onto freshly prepared umbilical cord vein using a Pasteur pipette or an Eppendorf pipette with cut-off tip.
2. Systematically monitor migration of tumour cells using an inverted phase-contrast microscope.
3. Fix co-culture at different growth stages for routine immunohistochemical analysis and histomorphological observation.

Protocol 7
Co-cultivation of MCTS and endothelial cells on basal membrane

Equipment and reagents

- 100 mm bacteriological Petri dishes, culture dishes, 6-well plates, and 25 cm² culture flasks (e.g. Falcon)
- Sterile scalpel, pair of scissors, and cannulas
- Inverted phase-contrast microscope
- 70% (v/v) ethanol
- PBS
- **Hepes** solution: 50 mM **Hepes** pH 7.5, 685 mM NaCl, 20 mM KCl, 1% (w/v) glucose
- 0.02% collagenase in **Hepes**
- 0.2% (w/v) gelatin in PBS
- DMEM (Ham's F12/Iscove's medium, 1:1)
- FCS
- 5% (w/v) dextran
- 0.2 M NH_4OH

A. Preparation of basement membrane (BM) (68)

1. Wash freshly isolated cattle eyes with 70% (v/v) ethanol.
2. Isolate cornea under sterile conditions using a pair of scissors and a scalpel.
3. Place cornea in sterile Petri dish and scrape off endothelial cells from the inner surface with a scalpel.
4. Transfer endothelial cells into a 5 cm culture dish and cover with 2–5 ml of DMEM containing 10% (v/v) FCS.
5. Culture cells to confluence under standard culture conditions (5–8% (v/v) CO_2 in air, humidified atmosphere, 37 °C).
6. Routinely refresh medium every 48 h.
7. Transfer confluent endothelial cells into a 25 cm² culture flask and subculture for several passages.
8. Seed $1-5 \times 10^5$ cells into each well of a 6-well plate and culture until confluence.
9. Feed confluent monolayers with DMEM containing 10% (v/v) FCS and 5% (w/v) dextran at days 0 and 4.
10. After an additional incubation interval of four days remove medium, rinse with PBS, and lyse cells with 0.2 M NH_4OH (1–2 ml for 10 min).
11. Remove cell debris with PBS, seal 6-well plates, and store at 4 °C until use.

B. Isolation and cultivation of human umbilical vein endothelial cells (HUVEC) (69)

1. Wash surface of umbilical cord with 70% (v/v) ethanol.
2. Rinse the inside of the umbilical cord with 10–20 ml **Hepes** solution.
3. Inject 4 ml of a 0.02% (w/v) collagenase solution in **Hepes** into the umbilical cord vein and carefully seal the edges.
4. Incubate in 37 °C water-bath for 10 min.

Protocol 7 continued

5. Gently massage the umbilical cord.
6. Open both ends and collect endothelial cell suspension into 50 ml Falcon tube by injection of 10 ml Hepes solution.
7. Spin down cells (5 min, 300 g) and resuspend in 5 ml Ham's F12/Iscove's (1:1) medium containing 10% (v/v) FCS.
8. Inoculate cells into a 25 cm^2 culture dish coated with 0.2% (w/v) gelatin in PBS.
9. Incubate cells for 30–60 min under standard culture conditions, monitor cell adherence, wash carefully, and refresh medium.
10. Feed cells every other day and subculture at confluence using 0.02% (w/v) collagenase in Hepes.

C. Co-cultivation of MCTS and HUVEC on BM

1. Inoculate 1–5 × 10^5 HUVEC onto each well of a BM-coated 6-well plate and culture until confluence using serum-supplemented Ham's F12/Iscove's (1:1) medium.
2. Initiate and culture MCTS according to the routine initiation protocol (see *Protocols 1 and 3*).
3. Transfer individual spheroids with defined sizes onto confluent HUVEC monolayers (two spheroids per well).
4. Systematically monitor migration of tumour cells using an inverted phase-contrast microscope.

2.2.5 Additional technical aspects

Separation of cells from defined locations within sections from 3D cultures can be achieved using microdissection techniques, such as laser microbeam combined with an optical tweezer (e.g. 70, 71). Dissociation of co-cultures by enzymatic or mechanical means allows for flow cytometric analyses and for separation of cellular subpopulations by various sorting techniques (for review see ref. 21). As a prerequisite for cell detection and separation, cell subpopulations can be permanently labelled by the introduction of specific marker genes, for example using β-galactosidase (72) or green fluorescent protein (GFP) or labels for cell membranes (73). GFP can be detected in routine flow cytometry (see for example refs 10, 74–76).

3 Experimental tissue modelling

3.1 Current research on tissue modelling

Substantial progress has been made towards the development of an artificial liver. Strategies include culture of hepatocyte spheroids (77–80) or of heterospheroids consisting of hepatocytes and fibroblasts or parenchymal cells (81–83). The application of artificial support systems, such as porous gelatin sponges (84), agarose (85), or collagen (86), as well as the induction of aggregation by exogenous adhesion molecules (87) is advantageous for long-term culture of liver cells. Reconstitution of the specific geometry, microenvironment, and metabolism of the hepatic sinusoid in a complex 3D culture system has been reported (88) and pig hepatocytes have been grown in co-culture in an artificial 3D capillary system (89, 90).

Development of an artificial pancreas has focused on spheroids from insulin-secreting recombinant mouse pituitary AtT 20 cells and mouse insulinoma beta TC3 cells (91). Long-term viability and function of human fetal islet-like cell clusters implanted under the kidney capsule of athymic mice have been achieved (92). Another promising strategy is the use of isolated piscine islets, so-called Brockmann bodies, as 'natural' pancreatic spheroids which can be implanted as xenografts after microencapsulation with alginate (93). Oxygen supply seems to play a crucial role for the survival and maintenance of function of such islets (93, 94).

The pituitary gland has been modelled using multicellular spheroids to study hormone release, including luteinising hormone (LH), following stimulation with the releasing hormone LHRH (95, 96). Three-dimensional cultures of melatonin secreting cells from the pineal gland have also been described (97). In contrast to pineal cell monolayers, 3D cultures remain functional for more than three weeks secreting melatonin when challenged with isoproterenol. Despite their tendency to spontaneously form follicles in aggregation culture, there are few reports on isolated adult thyroid cells (98). Thyroid cell spheroids allow study of cell motility, cell adhesion, and E-cadherin expression in thyroid follicle biogenesis (99).

3D brain cultures has been reviewed comprehensively (11, 95) and have been used to study neural myelination and demyelination (100–103), neuronal degeneration (104), and the neurotoxicity of lead (105). Neuronal behaviour following HIV infection (106, 107), Alzheimer's disease (108), and Morbus Parkinson (109) has been studied using aggregated brain cell cultures. Retinal development has also been modelled (110).

3D chondrocyte research has focused on molecular aspects of matrix formation (77, 111, 112) and differentiation (113, 114).

Aggregated cell cultures of heart cells have been used to study aspects of cardiac development and physiology, including sinoatrial node cell preparations (115), atrial cell preparations (116, 117), embryonic chick heart cell aggregates (118–120), and ventricular cell aggregates (121).

Other tissues that have been modelled in 3D cultures include mesangial cells (122, 123), urothelial cells (124), and nasopharyngeal cells (125).

3.2 Tissue modelling of skin and mucosa

Heterotypic 3D cultures including fibroblasts and keratinocytes have been used for studies on the formation of skin (126) and hair follicles (127, 128), and the influence of mesenchyme on the differentiation of keratinocytes (127, 129). A method is described in *Protocol 8* and shown diagrammatically in *Figure 5*.

THREE-DIMENSIONAL CULTURE

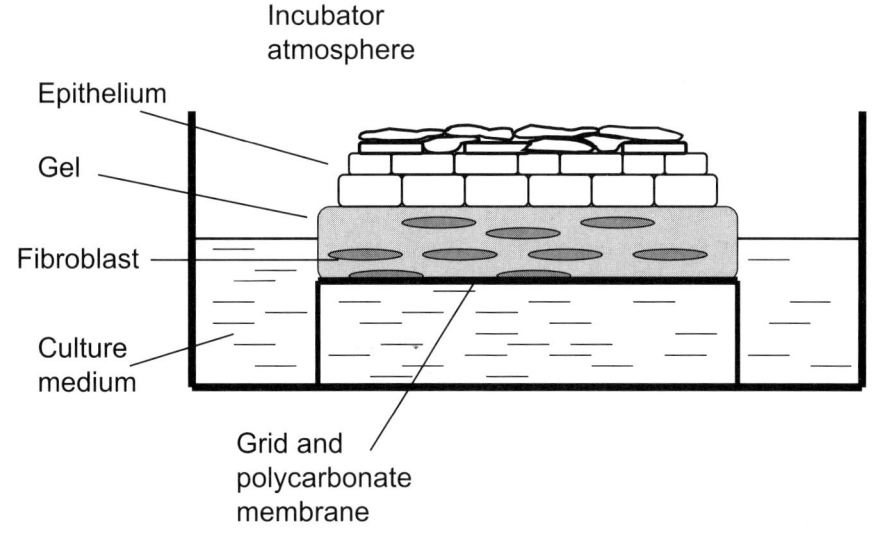

Figure 5. Tissue culture chamber for cultivation of oral mucosa cells. The epithelial cells are exposed to the incubator atmosphere, and may not grow if covered by culture medium.

Protocol 8
Isolation and cultivation of different cell populations from oral mucosa

Equipment and reagents

- Standard cell culture equipment (including laminar flow hood, incubator, centrifuge, etc.)
- Tissue strainer (70 μm pore size; Falcon)
- Nucleopore track-etch membrane (polycarbonate membrane; Corning Costar)
- Microsurgical scissors and tweezers
- 100 ml dispase grade II (240 U; Boehringer Mannheim)
- DMEM with 4.5 g/litre glucose
- FCS
- Penicillin/streptomycin
- Trypsin/EDTA solution
- PBS (e.g. Sigma-Aldrich)
- Collagen type I
- 1 M NaOH, sterile
- FAD_{spec} medium: 1:3 (v/v) of Ham's F12 and DMEM supplemented with 18.2 μg/ml adenine, 8.33 μg/ml cholera toxin, 10 ng/ml EGF, 0.4 μg/ml hydrocortisone, 5 μg/ml insulin

A. Isolation of cells

1. Wash tissue in 70% (v/v) alcohol.
2. Rinse tissue in sterile phosphate-buffered saline (PBS). Repeat steps 1 and 2.
3. Separate connective tissue (pink) from epithelium (white) with tweezers, scalpel, and micro-scissors.

141

Protocol 8 continued

4. Mince epithelium with micro-scissors to homogeneity.
5. Mix epithelium with 3 ml dispase II and transfer to a sterile Petri dish.
6. Incubate for 30 min at 37°C.
7. Inactivate enzyme by adding 30 ml DMEM with 10% fetal calf serum (FCS).
8. Spin down for 10 min at 1200 r.p.m. (300 g) at room temperature.
9. Resuspend pellet in 1.2 ml FAD_{spec} medium with 5% (v/v) FCS.
10. Transfer suspension to small (25 cm^2) tissue culture flasks and incubate under standard conditions.
11. Mince connective tissue with micro-scissors and suspend in 1 ml DMEM (4.5 g/litre glucose) with 10% FCS.
12. Transfer suspension to small size tissue culture flasks and incubate under standard conditions.

B. Cell culture: keratinocytes

1. Incubate epithelial cell suspension at 37°C in a humidified atmosphere containing air and 5% (v/v) CO_2.
2. Replenish FAD_{spec} medium with 5% (v/v) FCS twice a week until cellular outgrowth from adhered cellular islets can be observed.
3. Continue with three complete renewals of the medium.
4. At near confluence, trypsinize cells and pass the suspension through a cell strainer to remove cell clusters.

Fibroblasts

1. Incubate suspension of cells and clusters from connective tissue at 37°C in a humidified atmosphere containing air and 5% (v/v) CO_2.
2. Replenish medium (DMEM, 4.5 g/litre glucose, 10% FCS) on a weekly schedule, until cellular outgrowth from adhered cellular islets can be observed.
3. Continue with replenishment of the medium two to three times per week thereafter.
4. At near confluence, trypsinize cells and remove clusters, as described for the keratinocytes.

C. Organotypic co-culture

(a) Preparation of fibroblasts

1. Trypsinize fibroblasts and resuspend in FAD_{spec} or DMEM.
2. Determine cell number.
3. Sediment cells by centrifugation (10 min, 300 g, room temperature).
4. Resuspend pellet in an equal volume of FCS.

Protocol 8 continued

(b) Preparation of collagen gel

All work to be done on ice near 0 °C, unless stated otherwise.

1. Mix 1:10 (v/v) 80% (v/v) collagen type I (4 mg/ml) and 10% (v/v) Hank's buffered salt solution.
2. Neutralize this mixture with 5 M NaOH (to achieve pink colour).
3. Add FCS containing cells to the gel mixture and adjust cell concentration to 1.5–3 $\times 10^5$ fibroblasts/ml gel and FCS content to 10% (v/v).
4. Fill gel suspension into wells of a multititre plate (24-well, 1 ml).
5. Let the gel solidify for 1–2 h at 37 °C, and cover it with FAD$_{spec}$ medium.
6. During incubation, fibroblasts should change their shape from a spherical to an elongated form within two days.

(c) Preparation of the keratinocytes

1. Trypsinize keratinocytes and suspend in FAD$_{spec}$ medium.
2. Determine cell number.
3. Overlay 1.5–3 $\times 10^5$ keratinocytes in a drop of medium on top of the fibroblast gel.

(d) Cultivation of the co-cultures

1. Incubate co-cultures for 24–48 h (humidified, 37 °C, 5% CO_2 and air).
2. Replenish medium (careful suction) two to three times per week.

3.3 Embryoid bodies

Embryoid bodies (EB) are derived from embryonic stem cell (ES) cell lines (130) which have retained their capacity to generate cells of the haematopoietic (131), endothelial (132–135), muscle (136), and neuronal lineages (137). Development of skeletal muscle myocytes in embryoid bodies is similar to that *in vivo* (136) and beating EB with cardio-specific receptors, ionic channels, and action potentials have been produced (138, 139). Oxygen-regulated gene expression during embryonic development has also been studied (140).

References

1. Holtfreter, J. A. (1947). *J. Exp. Zool.*, **106**, 197.
2. Moscona, A. (1952). *Exp. Cell Res.*, **3**, 535.
3. Moscona, A. (1961). *Exp. Cell Res.*, **22**, 455.
4. Inch, W. R., McCredie, J. A., and Sutherland, R. M. (1970). *Growth*, **34**, 271.
5. Sutherland, R. M., Inch, W. R., McCredie, J. A., and Kruuv, J. (1970). *Int. J. Radiat. Biol.*, **18**, 491.
6. Sutherland, R. M., McCredie, J. A., and Inch, W. R. (1971). *J. Natl. Cancer Inst.*, **46**, 113.
7. Knuechel, R. and Sutherland, R. M. (1992). In *Spheroid culture in cancer research* (ed. R. Bjerkvig), p. 159. CRC Press Boca Raton.

8. Sutherland, R. M. (1988). *Science*, **240**, 177.
9. Mueller-Klieser, W. (1987). *J. Cancer Res. Clin. Oncol.*, **113**, 101.
10. Mueller-Klieser, W. (1997). *Am. J. Physiol.*, **273**, C1109.
11. Bjerkvig, R., Lund-Johansen, M., and Edvardsen, K. (1997). *Curr. Opin. Oncol.*, **3**, 223.
12. Martin, G. R. (1980). *Science*, **209**, 768.
13. Desoize, B., Gimonet, D., and Jardiller, J. C. (1998). *Anticancer Res.*, **18**, 4147.
14. Freyer, J. P. (1992). In *Spheroid culture in cancer research* (ed. R. Bjerkvig), p. 217. CRC Press Boca Raton.
15. Bracke, M., Romijn, H., Vakaet, L., Vyncke, B., De Mets, M., and Mareel, M. (1992). In *Spheroid culture in cancer research* (ed. R. Bjerkvig), p. 73. CRC Press Boca Raton.
16. Engebraaten, O. (1992). In *Spheroid culture in cancer research* (ed. R. Bjerkvig), p. 107. CRC Press Boca Raton.
17. Neeman, M., Abramovitch, R., Schiffenbauer, Y. S., and Tempel, C. (1997). *Int. J. Exp. Pathol.*, **78**, 57.
18. Knuechel, R. and Sutherland, R. M. (1990). *Cancer J.*, **3**, 234.
19. Carlsson, J. (1992). In *Spheroid culture in cancer research* (ed. R. Bjerkvig), p. 277. CRC Press Boca Raton.
20. Carlsson, J. and Nederman, T. (1992). In *Spheroid culture in cancer research* (ed. R. Bjerkvig), p. 199. CRC Press Boca Raton.
21. Kunz-Schughart, L. A., Kreutz, M., and Knuechel, R. (1998). *Int. J. Exp. Pathol.*, **79**, 1.
22. Winsor, C. P. (1932). *Proc. Natl. Acad. Sci. USA*, **18**, 1.
23. Laird, A. K. (1964). *Br. J. Cancer*, **18**, 490.
24. Laird, A. K. (1965). *Br. J. Cancer*, **19**, 278.
25. Marusic, M., Bajzer, Z., and Vuk-Pavlovic, S. (1994). *Bull. Math. Biol.*, **56**, 617.
26. Kunz-Schughart, L. A., Groebe, K., and Mueller-Klieser, W. (1996). *Int. J. Cancer*, **66**, 578.
27. Monz, B., Karbach, U., Groebe, K., and Mueller-Klieser, W. (1996). *Oncol. Rep.*, **1**, 1177.
28. Groebe, K. and Mueller Klieser, W. (1996). *Int. J. Radiat. Oncol. Biol. Phys.*, **34**, 395.
29. Acker, H., Carlsson, J., Mueller-Klieser, W., and Sutherland, R. M. (1987). *Br. J. Cancer*, **56**, 325.
30. Acker, H., Carlsson, J., Holtermann, G., Nederman, T., and Nylen, T. (1987). *Cancer Res.*, **47**, 3504.
31. Carlsson, J. and Acker, H. (1988). *Int. J. Cancer*, **42**, 715.
32. Acker, H. and Carlsson, J. (1992). In *Spheroid culture in cancer research* (ed. R. Bjerkvig), p. 135. CRC Press Boca Raton.
33. LaRue, K. E., Bradbury, E. M., and Freyer, J. P. (1998). *Cancer Res.*, **58**, 1305.
34. Rak, J., Mitsuhashi, Y., Erdos, V., Huang, S. N., Filmus, J., and Kerbel, R. S. (1995). *J. Cell Biol.*, **131**, 1587.
35. Fujiwara, T., Grimm, E. A., Mukhopadhyay, T., Cai, D. W., Owen-Schaub, L. B., and Roth, J. A. (1993). *Cancer Res.*, **53**, 4129.
36. Fujiwara, T., Grimm, E. A., Mukhopadhyay, T., Zhang, W. W., Owen-Schaub, L. B., and Roth, J. A. (1994). *Cancer Res.*, **54**, 2287.
37. Romero, F. J., Zukowski, D., and Mueller Klieser, W. (1997). *Am. J. Physiol.*, **272**, C1507.
38. Knuechel, R., Keng, P., Hofstaedter, F., Langmuir, V., Sutherland, R. M., and Penney, D. P. (1990). *Am. J. Pathol.*, **137**, 3.
39. Waleh, N. S., Gallo, J., Grant, T. D., Murphy, B. J., Kramer, R. H., and Sutherland, R. M. (1994). *Cancer Res.*, **54**, 838.
40. Paulus, W., Huettner, C., and Tonn, J. C. (1994). *Int. J. Cancer*, **58**, 841.
41. Hauptmann, S., Denkert, C., Löhrke, H., Tietze, L., Ott, S., Klosterhalfen, B., *et al.* (1995). *Int. J. Cancer*, **61**, 819.
42. Byers, S. W., Sommers, C. L., Hoxter, B., Mercurio, A. M., and Tozeren, A. (1995). *J. Cell Sci.*, **108**, 2053.

43. Acker, H., Pietruschka, F., and Deutscher, J. (1990). *Br. J. Cancer*, **62**, 376.
44. Laderoute, K. R., Murphy, B. J., Short, S. M., Grant, T. D., Knapp, A. M., and Sutherland, R. M. (1992). *Br. J. Cancer*, **65**, 157.
45. Mansbridge, J. N., Knüchel, R., Knapp, A. M., and Sutherland, R. M. (1992). *J. Cell. Physiol.*, **151**, 433.
46. Mansbridge, J. N., Ausserer, W. A., Knapp, M. A., and Sutherland, R. M. (1994). *J. Cell. Physiol.*, **161**, 374.
47. Murphy, B. J., Laderoute, K. R., Vreman, H. J., Grant, T. D., Gill, N. S., Stevenson, D. K., et al. (1993). *Cancer Res.*, **53**, 2700.
48. Neeman, M., Abramovitch, R., Tempel, C., Meir, G., Itin, A., Shweiki, D., et al. (1993). *Proc. Soc. Magn. Reson. Med.*, **1**, 194.
49. Huelser, D. F. (1992). In *Spheroid culture in cancer research* (ed. R. Bjerkvig), p. 172. CRC Press Boca Raton.
50. Knuechel, R., Siebert-Wellnhofer, A., Traub, O., and Dermietzel, R. (1996). *Am. J. Pathol.*, **4**, 1321.
51. Kunz-Schughart, L. A., Habbersett, R. C., and Freyer, J. P. (1997). *Am. J. Physiol.* **273**, C1487.
52. Koide, N., Sakaguchi, K., Koide, Y., Asano, K., Kawaguchi, M., Matsushima, H., et al. (1990). *Exp. Cell Res.*, **186**, 227.
53. Freyer, J. P. and Schor, P. L. (1989). *In Vitro Cell Dev. Biol.*, **25**, 9.
54. Kunz-Schughart, L. A. and Freyer, J. P. (1997). *In Vitro Cell Dev. Biol.*, **33**, 73.
55. Jääskeläinen, J., Kalliomäki, P., Paetau, A., and Timonen, T. (1989). *J. Immunol.*, **142**, 1036.
56. Jääskeläinen, J., Lehtonen, E., Heikkilä, P., Kalliomäki, P., and Timonen, T. (1990). *J. Natl. Cancer Inst.*, **82**, 497.
57. Konur, A., Kreutz, M., Knuechel, R., Krause, S. W., and Andreesen, R. (1998). *Int. J. Cancer*, **78**, 648.
58. Krause, S. W., Rehli, M., and Andreesen, R. (1998). *Immunol. Rev.*, **161**, 119.
59. Andreesen, R., Gadd, S., Brugger, W., Löhr, G. W., and Atkins, R. C. (1988). *Immunobiology*, **177**, 186.
60. Brouty-Boyé, D., Mainguene, C., Magnien, V., Israel, L., and Beaupain, R. (1994). *Int. J. Cancer*, **56**, 731.
61. Schuster, U., Büttner, R., Hofstaedter, F., and Knuechel, R. (1994). *J. Urol.*, **151**, 1707.
62. Folkman, J. (1990). *J. Natl. Cancer Inst.*, **82**, 4.
63. Folkman, J. (1992). *Semin. Cancer Biol.*, **3**, 65.
64. Folkman, J. (1997). *Experientia Supplement*, **79**, 1.
65. Mareel, M. M., Van Roy, F. M., and De Baetselier, P. (1990). *Cancer Metastasis Rev.*, **9**, 45.
66. Vermeulen, S., Vanmarck, V., VanHoorde, L., VanRoy, F., Bracke, M., and Mareel, M. (1996). *Pathol. Res. Pract.*, **192**, 694.
67. Offner, F. A., Wirtz, H., Schiefer, R. J., Bigalke, I., Klosterhalfen, B., Bittinger, F., et al. (1992). *Am. J. Pathol.*, **141**, 601.
68. Gospodarowicz, D., Vlodavsky, J., and Savion, N. (1980). *J. Supramol. Struct.*, **13**, 339.
69. Jaffé, E. A., Nachman, R. L., Becker, C. G., and Minick, G. R. (1973). *J. Clin. Invest.*, **52**, 2745.
70. Becker, I., Becker, K. F., Rohrl, M. H., Minkus, G., Schuetze, K., and Hofler, H. (1996). *Lab. Invest.*, **75**, 801.
71. Schuetze, K. and Clement-Sengewald, A. (1994). *Nature*, **368**, 667.
72. Bradley, C. and Pitts, J. (1994). *Br. J. Cancer*, **70**, 795.
73. Teare, G. F., Horan, P. K., Slezak, S. E., Smith, C., and Hay, J. B. (1991). *Cell. Immunol.*, **134**, 157.
74. Cubitt, A. B., Heim, R., Adams, S. R., Boyd, A. E., Gross, L. A., and Tsien, R. Y. (1995). *Trends Biochem. Sci.*, **20**, 448.

75. Lybarger, L., Dempsey, D., Franek, K. J., and Chervenak, R. (1996). *Cytometry*, **25**, 211.
76. Heim, R. and Tsien, R. Y. (1996). *Curr. Biol.*, **6**, 178.
77. Li, A. P., Colburn, S. M., and Beck, D. J. (1992). *In Vitro Cell Dev. Biol.*, **28A**, 673.
78. Tong, J. Z., De Lagausie, P., Furlan, V., Cresteil, T., Bernard, O., and Alvarez, F. (1992). *Exp. Cell Res.*, **200**, 326.
79. Peshwa, M. V., Wu, F. J., Follstad, B. D., Cerra, F. B., and Hu, W. S. (1994). *Biotechnol. Prog.*, **10**, 460.
80. Yuasa, C., Tomita, Y., Shono, M., Ishimura, K., and Ichihara, A. (1993). *J. Cell. Physiol.*, **156**, 522.
81. Endoh, K., Ueno, K., Miyashita, A., and Satoh, T. (1993). *Res. Commun. Chem. Pathol. Pharmacol.*, **82**, 317.
82. Takezawa, T., Mori, Y., Yonaha, T., and Yoshizato, K. (1993). *Exp. Cell Res.*, **208**, 430.
83. Yagi, K., Sumiyoshi, N., Yamada, C., Michibayashi, N., Nakashima, Y., Kawase, M., et al. (1995). *J. Ferment. Bioeng.*, **80**, 575.
84. Lin, K. H., Maeda, S., and Saito, T. (1995). *Biotechnol. Appl. Biochem.*, **21**, 19.
85. Shiraha, H., Koide, N., Hada, H., Ujike, K., Nakamura, M., Shinji, T., et al. (1996). *Biotechnol. Bioeng.*, **50**, 416.
86. Nishikawa, Y., Tokusashi, Y., Kadohama, T., Nishimori, H., and Ogawa, K. (1996). *Exp. Cell Res.*, **223**, 357.
87. Dai, W. G. and Saltzman, W. M. (1996). *Biotechnol. Bioeng.*, **50**, 349.
88. Bader, A., Knop, E., Kern, A., Boeker, K., Fruehauf, N., Crome, O., et al. (1996). *Exp. Cell Res.*, **226**, 223.
89. Gerlach, J. C., Schnoy, N., Encke, J., Smith, M. D., Muller, C., and Neuhaus, P. (1995). *Hepatology*, **22**, 546.
90. Gerlach, J. C. (1994). *Adv. Exp. Med. Biol.*, **368**, 165.
91. Tziampazis, E. and Sambanis, A. (1995). *Biotechnol. Prog.*, **11**, 115.
92. Beattie, G. M., Butler, C., and Hayek, A. (1994). *Cell Transplant.*, **3**, 421.
93. Schrezenmeir, J., Gerö, L., Laue, C., Kirchgessner, J., Müller, A., Hüls, A., et al. (1992). *Transplant. Proc.*, **24**, 2925.
94. Schrezenmeir, J., Kirchgessner, J., Gerö, L., Kunz, L. A., Beyer, J., and Mueller-Klieser, W. (1994). *Transplantation*, **57**, 1308.
95. Shacoori, V., Saiag, B., Girre, A., and Rault, B. (1995). *Res. Commun. Mol. Pathol. Pharmacol.*, **87**, 115.
96. VanBael, A., Proesmans, M., Tilemans, D., and Denef, C. (1995). *J. Mol. Endo.*, **14**, 91.
97. Khan, N. A., Shacoori, V., Havouis, R., Querne, D., Moulinoux, J. P., and Rault, B. (1995). *J. Neuroendocrinol.*, **7**, 353.
98. Yap, A. S. and Manley, S. W. (1994). *Exp. Cell Res.*, **214**, 408.
99. Yap, A. S., Stevenson, B. R., Keast, J. R., and Manley, S. W. (1995). *Endocrinology*, **136**, 4672.
100. Kelero de Rosbo, N., Honegger, P., Lassmann, H., and Matthieu, J. M. (1990). *J. Neurochem.*, **55**, 583.
101. Loughlin, A. J., Honegger, P., Woodroofe, M. N., Comte, V., Matthieu, J. M., and Cuzner, M. L. (1994). *J. Neurosci. Res.*, **37**, 647.
102. Matthieu, J. M., Comte, V., Tosic, M., and Honegger, P. (1992). *J. Neuroimmunol.*, **40**, 231.
103. Tosic, M., Torch, S., Comte, V., Dolivo, M., Honegger, P., and Matthieu, J. M. (1992). *J. Neurochem.*, **59**, 1770.
104. Chatterjee, S. S. and Noldner, M. (1994). *J. Neural Transm. Suppl.*, **44**, 47.
105. Zurich, M. G., Monnet Tschudi, F., and Honegger, P. (1994). *Neurotoxicology*, **15**, 715.
106. Kolson, D. L., Buchhalter, J., Collman, R., Hellmig, B., Farrell, C. F., Debouck, C., et al. (1993). *AIDS Res. Hum. Retroviruses*, **9**, 677.

107. Yeung, M. C., Pulliam, L., and Lau, A. S. (1995). *AIDS*, **9**, 137.
108. Gebicke-Haerter, P. J., Appel, K., Honegger, P., and Berger, M. (1994). *J. Neurosci. Res.*, **38**, 32.
109. Wiese, C., Cogoli-Greuter, M., Argentini, M., Mader, T., Weinreich, R., and Winterhalter, K. H. (1992). *Biochem. Pharmacol.*, **44**, 99.
110. Willbold, E., Reinicke, M., Lance Jones, C., Lagenaur, C., Lemmon, V., and Layer, P. G. (1995). *Eur. J. Neurosci.*, **7**, 2277.
111. Mallein-Gerin, F., Ruggiero, F., Quinn, T. M., Bard, F., Grodzinsky, A. J., Olsen, B. R., et al. (1995). *Exp. Cell Res.*, **219**, 257.
112. Poliard, A., Nifuji, A., Lamblin, D., Plee, E., Forest, C., and Kellermann, O. (1995). *J. Cell Biol.*, **130**, 1461.
113. Denker, A. E., Nicoll, S. B., and Tuan, R. S. (1995). *Differentiation*, **59**, 25.
114. Dozin, B., Quarto, R., Campanile, G., and Cancedda, R. (1992). *Eur. J. Cell Biol.*, **58**, 390.
115. Boyett, M. R., Kodama, I., Honjo, H., Arai, A., Suzuki, R., and Toyama, J. (1995). *Cardiovasc. Res.*, **29**, 867.
116. Li, G. R., Feng, J., Shrier, A., and Nattel, S. (1995). *J. Physiol. (Lond.)*, **484**, 629.
117. Wang, Z., Fermini, B., Feng, J., and Nattel, S. (1995). *Am. J. Physiol. (Heart Circ. Physiol.)*, **268** (**37**), H1992.
118. Clay, J. R., Kristof, A. S., Shenasa, J., Brochu, R. M., and Shrier, A. (1994). *Prog. Biophys. Mol. Biol.*, **62**, 185.
119. Rabkin, S. W. (1993). *Gen. Pharmacol.*, **24**, 699.
120. Zhang, J., Rasmusson, R. L., Hall, S. K., and Lieberman, M. (1993). *J. Physiol.*, **472**, 801.
121. Tatsukawa, Y., Arita, M., Kiyosue, T., Mikuriya, Y., and Nasu, M. (1993). *J. Mol. Cell Cardiol.*, **25**, 707.
122. Ayo, S. H. and Kreisberg, J. I. (1991). *J. Am. Soc. Nephrol.*, **2**, 1153.
123. He, C. J., Striker, L. J., Tsokos, M., Yang, C. W., Peten, E. P., and Striker, G. E. (1995). *Am. J. Physiol. (Cell Physiol.)*, **269** (**38**), C554.
124. Boxberger, H. J., Sessler, M. J., Maetzel, B., Mosleh, I. M., Becker, H. D., and Meyer, T. F. (1994). *Epithelial Cell Biol.*, **3**, 85.
125. Boxberger, H. J., Sessler, M. J., Maetzel, B., and Meyer, T. F. (1993). *Eur. J. Cell Biol.*, **62**, 140.
126. Ihara, S., Watanabe, M., Nagao, E., and Shioya, N. (1991). *Cell Tissue Res.*, **266**, 65.
127. Limat, A., Breitkreutz, D., Hunziker, T., Klein, C. E., Noser, F., Fusenig, N. E., et al. (1994). *Cell Tissue Res.*, **275**, 169.
128. Limat, A., Breitkreutz, D., Stark, H. J., Hunziker, T., Thikoetter, G., Noser, F., et al. (1991). *Ann. N. Y. Acad. Sci.*, **642**, 125.
129. Limat, A., Hunziker, T., Breitkreutz, D., Fusenig, N. E., and Braathen, L. R. (1994). *Skin Pharmacol.*, **7**, 47.
130. Keller, G. M. (1995). *Curr. Opin. Cell Biol.*, **7**, 862.
131. Hole, N., Graham, G. J., Menzel, U., and Ansell, J. D. (1996). *Blood*, **88**, 1266.
132. Wang, R., Clark, R., and Bautch, V. L. (1992). *Development*, **114**, 303.
133. Doetschman, T., Shull, M., Kier, A., and Coffin, J. D. (1993). *Hypertension*, **22**, 618.
134. Krah, K., Mironov, V., Risau, W., and Flamme, I. (1994). *Dev. Biol.*, **164**, 123.
135. Heyward, S. A., Dubois-Stringfellow, N., Rapoport, R., and Bautch, V. L. (1995). *FASEB J.*, **9**, 956.
136. Rohwedel, J., Maltsev, V., Bober, E., Arnold, H. H., Hescheler, J., and Wobus, A. M. (1994). *Dev. Biol.*, **164**, 87.
137. Strubing, C., Ahnert-Hilger, G., Shan, J., Wiedenmann, B. H. J., and Wobus, A. M. (1995). *Mech. Dev.*, **53**, 275.

138. Wobus, A. M., Kleppisch, T., Maltsev, V., and Hescheler, J. (1994). *In Vitro Cell Dev. Biol.*, **30A**, 425.
139. Wobus, A. M., Rohwedel, J., Maltsev, V., and Hescheler, J. (1995). *Ann. N. Y. Acad. Sci.*, **752**, 460.
140. Gassmann, M., Fandrey, J., Bichet, S., Wartenberg, M., Marti, H. H., Bauer, C., *et al.* (1996). *Proc. Natl. Acad. Sci. USA*, **93**, 2867.

Chapter 6
Tissue engineering

Robert A. Brown and Rebecca A. Porter
Tissue Repair Unit, University College London, Charles Bell House, 67 Riding House Street, London W1P 7LD, U.K.

1 Introduction

Tissue engineering (TE) is the application of engineering principles to cell culture for the purpose of constructing functional anatomical units, normally for reconstructive surgery. There are three main components: cell biology, materials science (polymer/protein biochemistry), and biological engineering.

TE aims to supply body parts for repair of damaged tissues and organs, without causing an immune response or infections, or using cadaveric tissue, or mutilating other parts of the patient. The dream is to blend the advantages of biological integration from living grafts with the ease, stability, and safety of prosthetic implants and as such it applies to most surgical specialities (1, 2).

Two contrasting philosophies can be identified behind current approaches to tissue engineering (3). The first assumes that living cells possess an innate and self-sufficient potential for biological regeneration. The implication is that addition of suitable cell types to a suitable support matrix which allows proliferation and movement will result in an organized and functional tissue, resembling the tissue of origin (2). This is likely to be the simplest, most economic approach, where it can be applied. The second approach assumes that cells require a greater degree of control to produce new and functioning tissue structures. This approach reflects the understanding that, far from regenerating, mature tissues repair rather poorly *in vivo* and many cells do not organize in culture. This view leads to more complex solutions, generally supplying organizational cues to regulate resident cell function and provide spatial and synthetic control cues. Based on what is understood about tissue repair and regeneration biology it seems likely that epithelial (4) and endothelial cell layers will prove better at organizing themselves through innate cell behaviours. In particular, cells such as keratinocytes and vascular endothelial cells make strong cell–cell interactions and form coherent sheets. On the other hand, connective tissues or mesenchymal layers, in which extensive deposition and remodelling of collagen matrix is needed (central to many TE applications), tend not to form appropriate structures spontaneously (5, 6).

The contribution of tissue repair processes is important in TE developments, since these are almost all surgical implants, including skin, blood vessels, heart valves, nerves, joint surfaces, and ligaments. Consequently, there is a spectrum of stages at which TE constructs can be designed to be implanted. At one end of this spectrum is the engineered tissue which is designed to guide, control, and improve the repair function, effectively acting as an implanted provisional tissue, for example to support peripheral nerve repair. At the other end of the spectrum is a fully functioning tissue ready to work in the patient.

It is presently inconceivable that a functional peripheral nerve could be engineered to repair a defect or gap in a nerve tract. What is needed is a cellular repair tissue to guide and encourage regeneration of existing axons across the gap. This support function tends to be simpler to produce than a fully functioning implant, needing simpler and shorter *in vitro* tissue maturation periods, as this process occurs *in vivo*. However, tissue function will recover far more slowly and such constructs need to be functionally sophisticated to control local repair processes long after implantation.

Fully functioning implants include constructs designed to operate as large blood vessel implants or heart valves, where significant periods of patient recovery are impossible or impractical. Tissue applications falling between these two extremes can be tackled in a mixture of ways. Current forms of TE skin, for example, tend to resemble repair or granulation tissue more than a skin graft, although graftable dermis would have many advantages.

2 Design stages for tissue engineering

A generalized scheme for the design of a TE construct is given in *Table 1* with some specific examples. The surgical criteria are frequently based on previous procedures using grafts and prostheses, including ease of fixation, minimal patient discomfort, and the rapid restoration of function.

The source of donor cells is critical to the design. However, this is often overshadowed by the question of whether autologous cells are essential. Where a mature implant is needed and little remodelling is expected, use of the patient's own cells is favoured. However, allogeneic cells become more attractive where the TE construct is designed as a temporary repair tissue. The implanted cells have a temporary role and are expected to be gradually lost in remodelling. Allogeneic and xenogeneic cells can be immunogenic and need special precautions to screen for infection. However, they are more amenable for use in the mass production of consistent, rapidly available, low cost constructs than is the case where the patient's own cells must be used. In this respect the use of cultured cells appears to provide a fortunate advantage, in that prolonged culture or cryopreservation seems to reduce the antigenicity of allogeneic cells (reviewed in ref. 19). Early experiences with new commercially available TE skin (Apligraf™), using neonatal human foreskin cells, appear to support this view (20).

Design criteria related to the support material include their porosity and structure. The survival time of the material is important, as are its degradation

Table 1. Analysis of five major design stages useful to follow in TE constructions

Design stage	Demonstration examples[a]	
	Repair support	Fully functioning implant
	Nerve repair implant	Large vessel implant
1. List and rank the imperatives	Rapid, aligned axon regeneration Limited life of support material Ease of surgical fixation Non-fibrogenic	Immediately functional Mechanically strong (fluid pressure and suturing) Non-thrombogenic lumen Strong elastic wall
2. Select and optimize materials	PLA/PGA synthetic polymer Collagen sponge Fibronectin fibrous tube (crosslinked)	Collagen gel or collagen-GAG sponge PLA or PGA synthetic polymer
3. Select and isolate the cells	Schwann cells (SC) Perineural fibroblasts or crude cell extracts from nerve	Smooth muscle cells (SMCs) Endothelial cells (ECs) Fibroblasts
4. Assembly and culture stages	Seed SCs at staging points along material or seed SCs at distal end of construct Seed outer surface with fibroblasts Minimal/zero culture prior to implantation	Seed outer layers (SMCs) Seed lumenal surface (ECs) Long-term culture (under flow)
5. Incorporation of control and monitoring processes	Minimally inflammatory substrate matrix Contact guidance substrate surfaces Material surface for rapid migration Graded substrate degradation Slow release growth factors (e.g. NGF, NT3)	Fluid flow through lumen Growth factors (e.g. VEGF, FGF) Monitor for continuous endothelial layer Monitor for mechanical properties of wall Monitor for platelet adhesion to lumen

[a] These are illustrated by examples for repair support (peripheral nerve implants) and fully functioning implant (large blood vessel) TE applications, compiled from various published attempts at each tissue type. PLA = poly(lactic acid) and PGA = poly(glycolic acid) resorbable synthetic polymers (7). Collagen-GAG = insoluble reconstituted sponges of bovine collagen and glycosaminoglycans (chondroitin sulfate) (8). Fibronectin + orientated fibrous materials from aggregated plasma fibronectin (9), crosslinked with copper salts to regulate tissue half-life (10). Incorporation of Schwann cells into PGA nerve guidance implants was reported by Hadlock et al. (11), SMCs and endothelial lumen cells to blood vessels (12). Contact guidance substrates using fibronectin (13) and added NGF or NT3 (14, 15) or collagen (16, 17). Endothelial cell alignment by fluid flow described by Nerem et al. (18).

products, cell adhesion characteristics, and ability to propagate surface guidance and mechanical cues. These features will, by accident or design, control the types of local cells which are recruited (macrophages, polymorphs, fibroblasts, smooth muscle cells, epithelial, or endothelial cells), their spatial organization, and the nature of the matrix they assemble.

Three tissue examples are analysed (skin, urothelium, and peripheral nerve) in terms of their TE requirements and design features.

2.1 Tissue engineered skin

Tissue engineering of skin became feasible in 1975 with the demonstration that sheets of human keratinocytes could be grown in the laboratory in a suitable form for grafting. This was a simple, cohesive sheet of cells cultured from the donor on a feeder layer of fibroblasts (21). This technique has been extensively modified and applied clinically, but as a skin treatment without a dermal layer, its uses are limited. The epithelial component is able to regenerate in culture, since the cells grow as a continuous sheet over a suitable denuded surface, producing a continuous layer which progresses to form cornified layers. However, it is the underlying dermal layer which presents more difficulties, with its regular collagenous architecture, blood capillaries, nerves, and accessory organs such as sweat glands.

There are various forms of implantable skin substitutes which can be considered as TE constructs. The first and simplest is a basic collagen-glycosaminoglycan sponge known as Integra™. Although Integra alone is a bioartificial material, rather than a TE construct, it has been used to carry seeded cells, as have a number of other collagen sponges and hyaluronan films (22, 23). Integra consists of insoluble bovine collagen type I and the glycosaminoglycan chondroitin sulfate in a ratio of 98:2 (24). Dermagraft™ consists of PGA polymer mesh of suitable pore size, seeded with human dermal fibroblasts from neonatal foreskins. As with Integra, this can be covered in a keratinocyte sheet at the time of implantation (25). Apligraf™ consists of human dermal fibroblasts seeded into a type I collagen gel and allowed to contract under tension. A layer of human keratinocytes is then seeded over the upper surface at the air–liquid interface (26). Both cell types in this construct are again derived from neonatal foreskins and so are allogeneic. Such TE constructs are available for surgical use, though with a limited shelf-life, presently in the order of five days. Clinical assessment of the performance and fate of Apligraf suggests that implants normally integrate well into surrounding tissues, forming a good skin cover. Importantly, there is no evidence of antibody formation to the bovine collagen, and little sign of rejection of the allogeneic cells in the construct (20), which are likely to disappear as the construct is remodelled.

2.2 Tissue engineered urothelium

Since human urothelial cells (27) and bladder smooth muscle cells (28) can be cultured, it is likely that construction of tissue engineered urothelial implants will be possible. The criteria are that the final structures need to form elastic tubes or bladders able to remain patent (i.e. without strictures), and the implant should not allow crystal formation from urine or harbour local infections. The structural requirements of the tissue are relatively simple in that an outer muscle layer should be lined on the lumenal surface by an intact, differentiating sheet of urothelium (29).

Support materials tested have included resorbable polymers [poly(glycolic acid) and poly(lactic-co-glycolic acid) co-polymer: PGA and PLGA] (28) and crosslinked collagen sponges (30, 31). Isolated urothelial cells cultured on collagen sponges formed differentiated sheets of urothelium over the surface of the material, with minimal tendency to promote crystal deposition. Urothelial and bladder muscle cells seeded onto PGA scaffolds formed urothelium-like, vascularized bilayered tissues when implanted into rabbits (28, 32). Recently, this technique has been applied to the tissue engineering of a functional bladder in dogs using a fibrous PGA polymer base, shaped into a bladder, and coated in PLGA 50:50 co-polymer. Muscle cells were seeded onto the outer surface of the bladder after which the lumenal surface was coated with pre-cultured urothelial cells, prior to implantation (33). Implanted bladders achieved near-normal performance and maintained this for up to 11 months.

2.3 Tissue engineered peripheral nerve implants (*Figure 1*)

Peripheral nerve injury is a common consequence of trauma and tumour resection surgery, with hundreds of thousands of reconstructive operations performed each year. Two criteria dominate approaches to assisted repair of peripheral nerve injuries. The first is that regeneration of axons should be guided as tightly as possible from their sprouting at the proximal stump to where they rejoin the degenerating distal stump on the far side of the defect. The second is that axonal regeneration across the nerve defect must be as fast as possible (reviewed in ref. 34). Prolonged delay of reinnervation leads to irreversible muscle atrophy.

Nerve guidance was achieved using silicone conduits attached between the nerve stumps (35) and also reported using tubes of bioresorbable materials such as PLGA co-polymer (36), collagen type I (16), and polymerized hyaluronan (37). The tube walls provide gross guidance, but no spatial cues are available to axons or Schwann cells away from the tube wall. Support for neurite outgrowth into the tube lumen has been provided by collagen and fibrin, with or without added

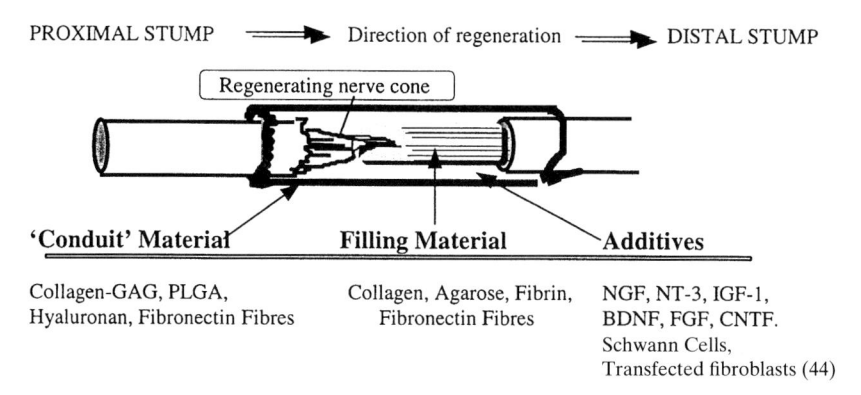

Figure 1. Scheme illustrating the basic design and some variations for peripheral nerve implants. Three elements are commonly used. The conduit outer layer (often the primary source of guidance). The filling material can also incorporate guidance elements but more often is for support of neural cells. Additives include a huge possible variety of growth and neurotrophic factors in many forms, combinations, ratios, and release modes.

growth factors (24, 38), but again such gels provide minimal directional information.

Initial outgrowth for short distances from the proximal stump can be rapid, but for gaps of > 10–15 mm the process can slow or stop, probably due to a lack of supporting growth factors and Schwann cells (34). Continued growth can be achieved by adding purified growth factor or by seeding implants with Schwann cells (14, 15, 39–41). In another approach, optimization of the migration speed (as well as direction) has been described for orientated fibronectin implants, by varying the proportion of fibrinogen in the polymer material, thus altering its surface adhesion properties (13–15, 42, 43).

3 Cell substrates and support materials

The value of a tissue engineering cell support depends on the information and suitability for the required adherent cell type or types. The information provided by the material is most commonly simple spatial cues, providing support surfaces with, for example, sufficient spacing for good cell growth. Biochemical information, such as surface reactive groups able to promote or reduce adherence or to activate cell membrane receptors (e.g. integrins) can be provided. More complex orientational (guiding cell responses, motion, and direction of deposition of extracellular matrix) and mechanical (tension, compression, or shear) information can be expressed through the supporting matrix of the construct.

The types of support materials available can be divided into discrete groups:

- Traditional: abiotic materials; metals, plastic, ceramics.
- Bioprostheses: natural materials modified to become biologically inert.
- Synthetic: resorbable polymers.
- Semi-natural: modified natural materials.
- Natural polymers: proteins, polysaccharides.

Composite devices can be constructed using more than one of these materials.

For the purposes of tissue engineering constructs it is possible to largely omit the first group of traditional materials, since they are not designed to resorb or become biologically integrated within a reasonable period. Hence constructs based on these are more likely to be regarded as bioactive modifications of conventional prosthetic implants. In many cases, the same is also true of bioprosthetic materials in current use. These are materials formed by extensive chemical crosslinking of natural tissues, such as porcine heart valves and tendon (44–46). The native, collagen-based connective tissue is stabilized by glutaraldehyde treatment to produce non-immunogenic substrates which will survive largely unchanged at the implantation site for many years. Although cell layers and even new connective tissue can eventually grow over the surface of such bioprostheses *in vivo* (45, 47), they have been designed and fabricated primarily to function as long as possible independently and without modification by surrounding host tissues. The resistance to cellular infiltration and remodelling of bioprostheses is counter to the basic aims of tissue engineering and largely discounts their use.

A variety of synthetic bioresorbable materials are degraded by hydrolysis and then phagocytosed. The advantage of such materials is that production is relatively cheap and easy, in a controllable and reproducible manner at large scale. Disadvantages lie in their cell compatibility, which is often not as good as for natural polymers, and their degradation products, which can have unwanted cellular effects.

Polymer composition is critical. Varieties include PGA, PLA, polycarbonate, poly ε-caprolactone (reviewed in refs 48 and 49). The most widely applied polymers in tissue engineering are PGA and PLA and co-polymer PLGA (7, 50, 51). The composition, dimensions, and formation of these polymers can be adjusted to control their survival *in vivo* (stability), their gross mechanical properties (important in surgical handling and in replacing structural function), as well as their ability to support cell growth. In the case of PLGA this control is through adjustment of the ratios of PLA and PGA (48). PGA is a crystalline, hydrophilic polyester, typically losing its mechanical strength through hydrolysis over two to four weeks (52). In contrast, PLA is more amorphous and hydrophobic, degrading to release lactic acid and losing mechanical strength after eight weeks, though with much slower total resorption *in vivo* (53). PGA, PLA, and PLGA have been used to support cells in a range of tissue engineering models including cartilage, urothelium, smooth muscle, and skin (32, 54, 55).

The format of the support material is also important for different TE applications and relatively simple to modify using these polymers. The solid polymer screws and pins used in orthopaedics are less applicable to TE but early forms of polymer sutures were easily woven to give meshes and braids with porosities suitable for cell growth (7, 51). Non-woven foams are made by incorporation of salt crystals into the polymer casting and subsequent dissolution by aqueous washing (56). Pore sizes are controlled by the size of the seeded crystals, typically between 150–300 μm. These materials can be moulded and extruded into a range of shapes including tubes for nerve guides (48).

Semi-natural and natural substrates are derived from natural macro-molecular polymers or whole tissues. The distinction between them lies in the extent to which materials are modified (to achieve aggregation or stabilization) and how much this leads to frank denaturation. The most critical and pertinent test for TE applications is the extent to which a material participates in the natural remodelling process with local cells.

An example of a chemically crosslinked polysaccharide is mammalian hyaluronan, stabilized by benzyl esterification of increasing numbers of side chains (57). Hyaluronan is a charged polysaccharide (glycosaminoglycan) found naturally at cell surfaces and in the extracellular matrix. It is an important lubricant between gliding surfaces in soft tissues and has been implicated in a range of cellular functions, including angiogenesis and tissue repair (58–60). This aggregate, known as HYAFF (marketed as 'laser skin™'), is progressively more stable as crosslinking increases, but less of the material is biologically native and functional. Materials can be made with varying levels of substitution and these survive for progressively longer periods *in vivo*, eliciting only modest inflammation and local connective tissue response (61). HYAFF has been developed for use in skin grafting, particularly as a carrier for keratinocytes (23).

Collagen sponges are prepared from various forms of insoluble or aggregated collagen (62, 63), acid extracted, and crosslinked with agents such as carbodiimide (64, 65), tannic acid (66), or diphenylphosphorylazide (67, 68). Some of these treatments produce substrates which can be utilized by cells *in vivo* and *in vitro*. A method for aggregating basement membrane into a cell substrate, termed TypeIV/TypeIV$_{ox}$, has been described using oxidatively crosslinked collagen purified from human placenta. It was designed principally to support growth of epithelial and endothelial cell layers and tested for the repair of skin and dura (68-71).

The most natural polymer materials are those in which protein stabilization is produced by drastic dehydration. This maximizes intermolecular interaction within polymers through complete removal of hydration shells, to promote intimate molecular packing (72). The most widely studied of these materials is the collagen-chondroitin sulfate aggregate material of Yannas and Burke (73-77), which is available as 'Integra'. A range of forms have been tested with different added matrix components (78, 79).

Aggregates of plasma fibronectin, produced as aligned fibrous materials, are derived from native protein solutions under directional shear and stabilized by dehydration (9, 80, 81). Their stability and attachment properties can be altered by treatment with trace levels of copper salts (10). For example, trace levels of copper differentially regulated growth and migration of Schwann cells and fibroblasts (82, 83). This raises the possibility of producing modified substrates which have cell selectivity, potentially useful in segregating cell types between anatomical zones or layers.

The most natural forms of cell support are polymers whose aggregation can be achieved in culture as it occurs *in vivo*. Examples include Matrigel, fibrin glue, collagen gels, and some polysaccharides. Of the polysaccharides, chitosan (84) and hyaluronan have been used in the form of a hydrated gel (85). Agarose gels are used to support a chondrogenic phenotype in chondrocytes (86). Matrigel is derived from tumour cells as a thick extracellular matrix, containing type IV collagen, laminin, proteoglycans, and growth factors (87). Despite being a natural substrate, its complex composition and tumour cell origin are limitations.

There are commercially available forms of fibrin glue or fibrin sealant designed for surgical use. These are prepared from fibrinogen and thrombin from human plasma, stabilized by a bovine protease inhibitor, aprotinin (reviewed in ref. 88). One composite form, Neuroplast, comprising elastin and fibrin, has been proposed as a substitute for dura mater and tympanic membranes (89, 90). Fibrin materials have also been used as implantable growth factor depots and agents promoting angiogenesis (91, 92), forming a simple slow release depot for trapped agents. Growth factor depot formulation in fibrin stabilizes FGF activity (93).

Collagen gels or 'lattices' were developed to study contraction as the gel shrinks or tension is generated by fibroblasts (94, 95). Untethered lattices shrink, producing smaller and denser matrices, sometimes regarded as tissue-like materials (96), whilst tethered lattices generate substantial endogenous forces (97, 98). Collagen gels have been seeded with other cell types, such as keratinocytes (99) and endothelial cells (100). Numerous model tissues have been attempted, based on contracted collagen gels (reviewed in ref. 101) as bioartificial grafts for skin replacement (102). The most advanced application is a bilayered tissue engineered allograft, 'Apligraf™' (20) based on a fibroblast-contracted lattice and keratinocyte layer. Orientated forms of collagen lattice have been developed to provide spatial information to cells either mechanically (103) or more deliberately, using magnetic fields (104), to provide contact guidance cues (105).

4 Cell sources

The first requirement for cell culture in any TE application is to generate sufficient cells to seed. The stage at which cells are to be introduced into the implant, the amount of remodelling of the implant substrate required, the degree of function, and the timing are critical.

One of the most important questions which must be addressed is whether the source of cells should be autologous, allogeneic, or xenogeneic. The use of the patient's cells (autologous transplant) is often the most straightforward option. A biopsy of the appropriate tissue is taken from the patient, enzymatically digested or explant cultured, and the cells grown to the required numbers. Whilst this method has the advantage of avoiding the immune response and the possible transfer of infection inherent in the use of allogeneic cells, it does have its drawbacks. Depending on the type of tissue required and the condition of the patient (limiting factors here include disease state and age of patient), it may not be possible to obtain adequate biopsy material.

The cell type may pose problems, for example articular cartilage, which has a relatively low density of cells with a low mitotic capacity (106). Cosmetic issues influence the choice of skin for tissue engineering. Studies using keratinocytes derived from the sole of the foot (107) indicate that the implanted cells resume their original phenotype and continue to produce a thickened epidermis. Genetically altering cells may alleviate this problem, but more needs to be known about the origins and maintenance of these regional cell differences within tissues.

Allogeneic sources of cells have advantages over autologous cells. It can be relatively easy to obtain healthy donor tissue, the cells may be cultured on a large scale at central 'factory' sites, with none of the time constraints of autologous cells. The product will be cheaper, of a consistent quality, and available as and when required by the patient. The major problem associated with allogeneic cells is graft rejection. The level of immune response generated by cells differs between cell types. For example, endothelial cells can induce a large reaction (108, 109), whilst fibroblasts, keratinocytes, and smooth muscle cells (108, 110) are less immunogenic. The reaction may also be dependent upon the donor age of the cells used, as fetal or neonatal cells can elicit little or no immune response (19, 111). There have been many studies reporting methods for reducing the immunogenicity of transplanted cells, though with limited success (112). The current strategies involve segregation of the donor cells from the host using cell encapsulation techniques which allow only the passage in of oxygen and metabolites and the exit of cell-generated hormones (113, 114). There must also be scrupulous staged analyses of cells for possible infections, such as hepatitis, HIV, and CJD, as well as screening for potential hazards, such as abnormal karyotype, or tumorigenic capacity.

Gene therapy has a role to play in tissue engineering. Cell populations can be altered by adding genes either to increase their output of a certain protein or add to their repertoire of expressed genes. Examples include the alteration of keratinocytes to produce transglutaminase-1, which is deficient in patients with lamellar icthyosis, a disfiguring dermal disorder. The engineered cells were transplanted in athymic mice and produced a normal epidermis (115). The technique has also been used to induce fibroblasts and keratinocytes to release factors they would not normally produce. Fibroblasts have been made to produce proteins such as transferrin, factor IX, and factor VIII (116–118). In the vascular system modified endothelial cells have been used to produce anti-thrombogenic factors such as tissue plasminogen activator (119). Chondrocytes have been engineered to produce an antagonist for IL-1, a major cause of degradation of cartilage ECM.

The method of culture of cells for TE implants will depend on the function of the cells. For many cell types, the classical monolayer (or equivalent large scale culture in roller bottles or spinners—see Chapter 2) may allow proliferation of cells, but also increases the possibility of phenotypic changes. Many cells revert to a fibroblast-like morphology after long-term monolayer culture (e.g. chondrocytes) or if allowed to become too confluent may lose their proliferative capacity (e.g. endothelial cells). This change may result in a loss of function due to the inability to produce, for example, a certain hormone or the capacity to contract. It must be assessed whether this alteration is temporary, i.e. cells revert to original phenotype when replaced in a 3D format or under the influence of soluble factors *in vivo*, or if the changes are irreversible. In either case it may be preferable to culture cells in an environment similar to the *in vivo* situation.

Perfusion is a critical issue in 3D culture. A number of factors interact to determine when nutrient and gaseous perfusion of a cell mass become limiting, including:

- cell density
- metabolic rate
- mean effective diffusion distance (cell to mixed nutrient source)
- density and anisotropy of diffusion path material

Culture medium is exhausted more quickly by near-confluent cultures whilst some cell types, for example chondrocytes, survive at nutrient levels which would be critical to other cell types. In many experimental models, cells (e.g. fibroblasts or chondrocytes) grow in low density 3D aqueous gels. Collagen gels typically start at 95–98% water and these need to contract to less than 20% of their starting volume before they approach the cell and matrix densities of native tissues (96). Diffusion into such low density gels is rapid (120). In addition, 3D collagen lattices have a random, homogeneous fibril structure almost never encountered in native collagenous tissues. The native tissues have organized and aligned fibrous structures which will produce pronounced anisotropic diffusion properties. In particular, diffusion will be far more rapid parallel to rather than perpendicular to aligned collagen fibre bundles (121). Since matrix density and alignment are characteristics of mature rather than repair tissues, perfusion will be a greater issue in TE of mature tissues. The principal means of addressing this problem to date has been to incorporate a form of external pumping to generate a fluid flow around or through the tissue construct (122).

Throughout this section it is assumed that adequate perfusion of the construct will also optimize the rate of growth and the deposition of ECM. Control cues can also be optimized by adding growth factors and cytokines. Examples are the use of mitogenic growth factors to promote fibroblast proliferation (PDGF and FGF) (123) or VEGF or FGF to optimize endothelial cell proliferation (124, 125). Similarly, rates of matrix production may be regulated by 'control cues' such as the TGFβ and IGF families of growth factor (126). However, the incorporation of specific growth factors is extremely cell system- and stage-specific and this is made more complex by the need for multi-component cocktails.

5 Orientation

Cells require specific shape, directional, and spatial cues from their environment (3, 127, 128), including:

- contact or substrate guidance
- chemical gradients
- mechanical cues

In its simplest form, contact guidance uses the topographical features of the substrate (128–131). Topographical features are most frequently in the form of aligned fibres or ridges of appropriate dimensions (normally < 100 μm for fibres). Cell types which are aligned in this way include fibroblasts, tenocytes, neurites, macrophages, and Schwann cells (43, 48, 130, 132, 133). Bioresorbable guidance templates suitable for tissue engineering are based on synthetic polymer substrates (127, 128), collagen fibrils (97, 105, 134), and fibronectin (9, 10, 80, 81). Alignment of collagen fibrils within native collagen gels has utilized mechanical or magnetic forces (94, 105), whilst tethering points (103), holes, or defects in the gel (135) and the local effects of contractile cells (136) can produce

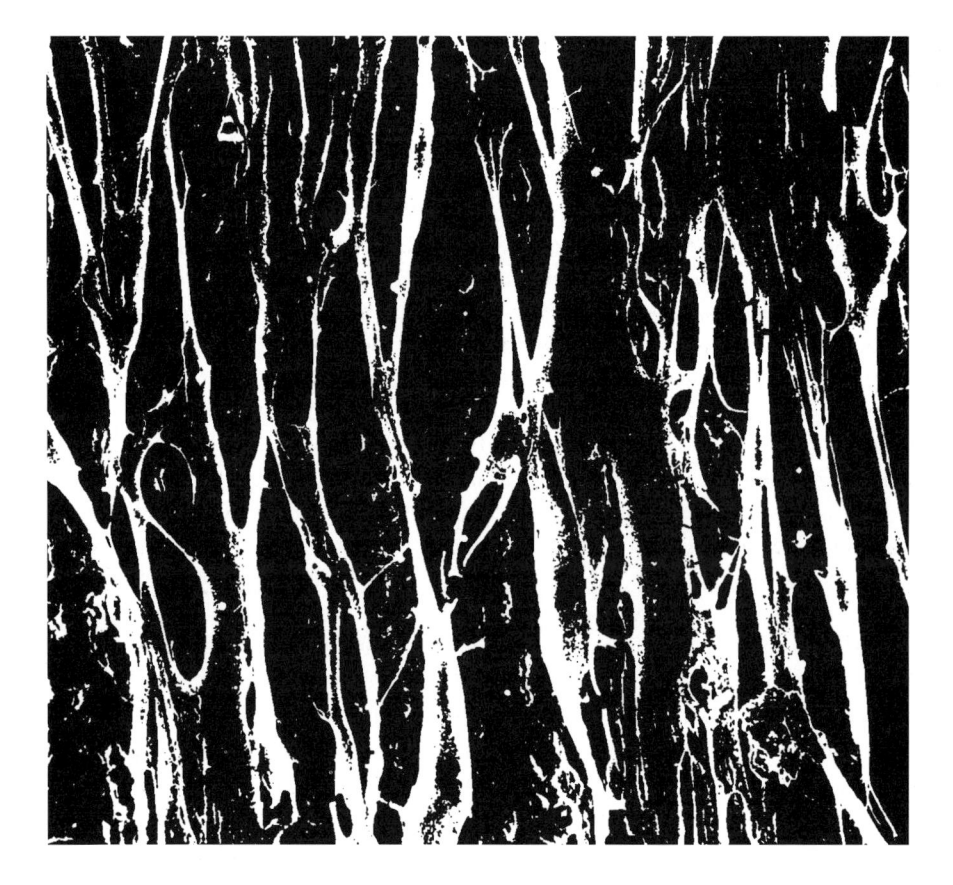

Figure 2. Scanning electron micrograph showing human dermal fibroblasts aligned on a fibronectin cable (alignment top to bottom). Mag. × 900.

local orientation. Orientation of fibronectin fibres involves application of directional shear as part of the fibre aggregation process (9, 80, 81). *Figure 2* shows a form of orientated fibronectin cable with a surface layer of aligned dermal fibroblasts, which migrate rapidly along the structure. It is possible to use differential attachment to substrates as a means of producing alignment in cells (137, 138).

The speed of cell locomotion is cell type- and substrate-dependent, as demonstrated by the ability to alter cell speed on more or less adhesive substrates (43, 139, 140). Fibroblast speeds of 20–40 μm/h are common. It is possible to optimize cell speeds by suitable design of the substrate composition (81, 82), but a far greater impact is made on recruitment by optimizing cell velocity. Random direction of movement will have a minimal velocity until some guiding cue is provided. In the case of cell guidance substrates, such as fibronectin fibre substrates or patterned resorbable polymers, cell speed and persistence of motion can be linked with directional cues in order to optimize the cellular velocity (43, 82, 128, 133).

Delivery of spatial or directional cues to cells in the form of chemical gradients through 3D constructs requires that the chemical agent not only stimulates the cell function, but also that it is presented to cells as a directional gradient. A number of growth factors are chemotactic and certain extracellular macromolecules, such as fibronectin (141, 142) and collagen (143), produce chemotactic fragments on breakdown. Such diffusible agents can be important in tissue organization. Angiogenesis, for example, is based on the concept that new vessels move along gradients of diffusible factors. These can be towards sites of low oxygen tension or high lactate concentration, or towards local sources of growth factor production. For example, vascular endothelial cell growth factor (VEGF) (124), or FGF (125), or matrix components such as oligosaccharide fragments of hyaluronan (60) influence angiogenesis.

Whilst such gradients may be useful in special circumstances, their use in control of architecture is likely to be limited *in vitro*. The most plausible circumstances would be for:

(a) Control over short ranges (e.g. between distinct cell layers in a tissue construct).

(b) During early stages of matrix development and maturation.

Use of hydrated 3D matrices for support of cultured cells, based on collagen (96, 144, 145), fibrin (146), or fibronectin (9, 147) is helpful in minimizing convection mixing of gradients. However, as discussed earlier there are likely to be major practical difficulties in producing and maintaining effective gradients in a controlled manner as the matrix becomes dynamic, dense and anisotropic.

5.1 Mechanical cues

A wide range of cell types are sensitive to mechanical loading. The mechanisms by which cells respond to mechanical signals are complex and include modulation of cytosolic-free calcium (148), stretch-sensitive ion channels (149, 150), and

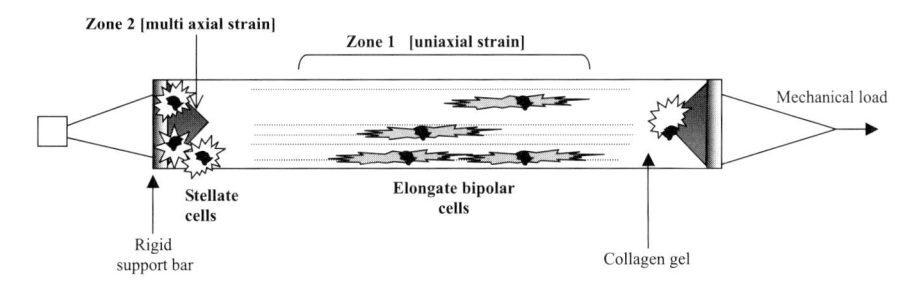

Figure 3. Schematic illustration of fibroblast alignment produced in collagen gels by uniaxial loading (156). Strain was aligned parallel with the long axis of the gel (and applied load in zone 1) and cells here took on an elongate, bipolar morphology, parallel to the strain. Strain in zone 2 was poorly aligned (represented by the dark triangle) due to the shielding effect of the rigid support bars, and cells here were stellate in shape showing no alignment.

deformation of cytoskeletal or integrin components (151). Responses can include changes in cell alignment (136, 152–154), synthesis of active regulatory molecules such as growth factors or hormones (123, 149, 155), altered matrix synthesis (152, 156–158), and enzyme release (159, 160).

There are at least three different forms of mechanical cue: tension, compression, and shear. Fibroblasts are normally considered in terms of tensional forces, vascular endothelial cells predominantly under fluid shear (153, 161), and chondrocytes are adapted to compression forces (156, 162). It is possible to further differentiate each of these forms of mechanical cue, for example into cyclical and static force or on the basis of loading velocity (163, 164). However, our understanding of the effects of these forms of loading or their combinations is limited.

A further, and potentially critical complexity, is that mechanical stimuli used in *in vitro* experimental models frequently have little or no directional component. Systems which give either non-orientated, or highly complex, multiple direction cues can give little information on the control of architecture and tissue polarity. They are valuable primarily for investigation of the role of mechanical forces on tissue production, turnover, and composition (152, 162).

A number of shear force models are being developed for tissue engineering of tube tissues such as blood vessels. Fluid shear forces have been found to operate through cytoskeletal changes (165), cell–cell adhesion sites (166, 167), and protein kinase and G protein signalling in vascular endothelial cells (168). These pathways are reported to affect heat shock protein phosphorylation (169), Cu/Zn superoxide dismutase activity (170), and apoptosis (171). Alteration of extracellular matrix production (172) and cell shape and alignment (18, 153) are more likely to be important for microarchitecture control cues. Current examples of tissue engineered tube structures for use in fluid drainage are bioartificial blood vessels (12) and bladder (33).

The ability of fibroblasts to become orientated along a uniaxial mechanical load affects their pattern of matrix remodelling (98, 158, 173). Similar orientations have been reported in loaded skeletal muscle cells (174). Multi-axial complex loading patterns, produced by the flexcell analytical model, produces complex cell alignments, which are difficult to interpret, though collagen production is quantitatively increased by loading (152). In uniaxial, cyclically loaded collagen lattices, fibroblasts take on an alignment parallel with the maximum strain induced in the material at any given location. This has been used in one experimental model to identify two distinct zones of mechanical loading in the same gel (*Figure 3*).

Most studies of compressional loading have used chondrocytes in a 3D agarose gel matrix (86, 162) or whole cartilage (175, 176). The agarose matrix provides little or no attachment until chondrocytes have elaborated their own pericellular matrix. In addition, agarose is essentially homogeneous in structure, providing uniform mechanical support in all dimensions. Consequently, newly synthesized matrix is easily detected and applied loads act uniformly and predictably on almost all the cells. In contrast, cartilage in organ culture contains cells in a non-homogeneous matrix (177) with mechanically distinct zones (178–180).

6 Protocols

6.1 Cell seeding of implantable materials

There are two main types of cell seeding strategy, depending on the final function of the implant:

(a) Materials that have cells throughout their structure.

(b) Materials that have cells on one or more surfaces.

There will also be situations which call for combinations of both (a) and (b).

The method of cell seeding throughout an implant will depend on the properties of the substrate used. These fall into two basic categories:

(a) 3D substrates which are preformed, e.g. polymers (PGA and PLA), collagen sponges, fibronectin mats.

(b) Substrates which require some form of preparation, such as mixing or gelling, e.g. fibrin sealant, collagen gel.

Protocol 1

Human dermal fibroblast (hDF) seeding within collagen sponges

Equipment and reagents

- Sterile scalpels and forceps
- DPPA non-crosslinked collagen sponges (Coletica S.A.)[a]
- DMEM supplemented with 10% fetal calf serum, 2 mM L-glutamine
- hDF cells: 10^4 cells suspended in 0.5 ml DMEM/1 cm^2 surface area

Method

1. Cut sponge to desired size, shape under sterile conditions, and place each piece in a separate well in the multiwell plate.

2. Slowly add the medium to hydrate the sponge.

3. When the sponge is completely saturated (approx. 2 h), gently squeeze out the medium with a pair of flat forceps without disturbing the architecture of the sponge.

4. Add cell suspension dropwise onto the sponge, which will soak up the cell suspension and disperse the cells within the sponge.

Protocol 1 continued

5. Incubate at 37°C in 5% CO_2 in air for 30 min, to facilitate attachment of hDF.
6. After 30 min add medium (DMEM) containing 50 µg/ml ascorbate.
7. Change the medium every three days. The cultures can then be kept for 8–12 weeks, either free-floating or attached to devices that enable loading regimes to be tested.

[a] Many materials, due to the manufacturing process, will have slightly different 'top' and 'bottom' surfaces. In the example above, the collagen sponge used has a glossy and a matt side. The latter has larger pores than the glossy side and has been found to be unsuitable for culturing epithelial cell sheets, as the cells migrate into the material instead of remaining on the surface.

Protocol 2

Human dermal fibroblast (hDF) seeding within collagen lattices

Reagents

- Rat tail collagen type I, native, acid soluble: 2.3 mg/ml in acetic acid (First Link UK Ltd.)
- 10 × MEM
- 1 M sodium hydroxide
- hDF cells: 10^6 cells/ml of final collagen gel volume, suspended in 0.5 ml DMEM containing 10% fetal calf serum, 2 mM L-glutamine, and 100 U/ml/100 µg/ml penicillin/streptomycin

All of the lattice reagents should be used at 4°C and mixed on ice (prior to the gelling incubation stage) to obtain good collagen polymerization.

Method

1. Mix 4 ml of collagen solution with 0.5 ml of 10 × MEM in a Universal.
2. Add 1 M NaOH solution dropwise with gentle agitation, until the pH of solution reaches about 7.5. Do not shake the solution as the gel will not polymerize.
3. Pour collagen solution into tube containing cell suspension and mix gently to disperse cells.
4. Transfer solution to Petri dish or multiwell plate, and allow to polymerize at 37°C/5% CO_2. Gelling should occur within 15 min.
5. After gelling the collagen lattices can be cultured either as free-floating, where the lattice is released from the edges of the dish with a needle, or tethered (i.e. left attached to the dish).

TISSUE ENGINEERING

Protocol 3
Urothelial cell seeding on collagen sponges (30)

Equipment and reagents

- DPPA crosslinked collagen sponges (Coletica S.A.)
- 4-well multiwell dishes
- Sterile scalpels and forceps
- RPMI 1640 supplemented with 10% fetal calf serum and 2 mM L-glutamine
- Urothelial cells (2×10^5) in 1 ml medium suspension

Method

1. Cut sponge to 1×1 cm squares under sterile conditions and place each piece in a separate well in the multiwell dish.
2. Slowly add approx. 100 µl RPMI 1640 to the well to hydrate the sponge.
3. Add cell suspension to top of hydrated sponge and incubate overnight at 37°C in 5% CO_2 in air.
4. After 24 h remove medium from well and transfer sponge with sterile forceps to new multiwell dish. Add fresh medium and incubate for up to 21 days, with a medium change every three days. Within six days a thickening, stratified layer of cells will develop and after 21 days a layer six to eight cells deep should form.

6.2 Slow release systems for local control of TE constructs or repair sites

Storage and release of diffusable agents from tissue engineering materials can be useful. Phenytoin (5,5-diphenylhydantoin, PHT) is an anti-convulsive agent which influences tissue matrix deposition (181). The drug is deposited into a fibronectin mat and, since it has very poor water solubility, it is trapped in crystal form (182).

Protocol 4
Phenytoin release from fibronectin mats

Equipment and reagents

- Cut materials (e.g. fibronectin mats) into small squares of at least 5 mg mass[a]
- 20 mg/ml PHT (Sigma Chemicals) dissolved in 40% ethanol in distilled water
- Phenol red-free DMEM, to monitor the release of PHT from the material

Method

1. Place dried, weighed pieces of material on the bottom of suitable cell culture wells.

165

> **Protocol 4** continued
>
> 2. Apply 20 μl of the PHT solution to the surface of the dried material and allow it to soak in and rehydrate the material, sticking it to the culture plastic (1 h at 37 °C).
> 3. Rinse in a large volume of fresh medium to remove excess, surface PHT and ethanol. Removal of the ethanol solvent leads to rapid precipitation of the PHT as fine needle-like crystals which become physically trapped in the material structure (*Figure 4*).
> 4. Seed the PHT material.
>
> [a] The fibronectin mats here are examples of almost any porous materials with suitable fibre/pore diameter characteristics. Collagen sponges would also be suitable.

Figure 4. Micrograph of a PHT-loaded fibronectin mat showing the fine crystals of PHT trapped within the structure of the material as they come out of solution (bright birefringent structures seen here under crossed polarizing illumination). Fibronectin mats alone produce little birefringence and are not associated with crystal structures.

The dynamics of PHT release can be measured by covering the material with phenol red-free DMEM (200–500 μl). After various times, the medium is recovered and analysed for PHT by a routine quantitative HPLC assay (183). PHT release profiles from fibronectin mats are shown in *Figure 5*.

TISSUE ENGINEERING

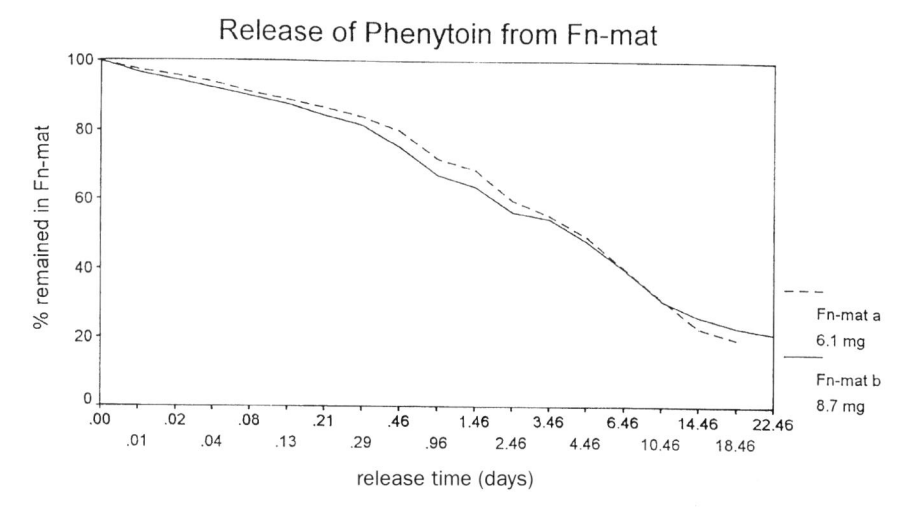

Figure 5. Graph showing the release of PHT from fibronectin mats over a period of 26 days as determined by HPLC analysis in the conditioned medium, at daily intervals.

In an alternative protocol, PHT can be delivered from a fibrin depot using a similar procedure, except that solid PHT is mixed with the fibrinogen component of a commercial fibrin glue preparation (SNBTS, Plasma Fractionation Centre) followed immediately by the normal volume of thrombin solution component, at which point a fibrin clot is formed, trapping the PHT.

Protocol 5
Slow release growth factor depot formation

Reagents

- Growth factors such as basic fibroblast growth factor (unmodified or ^{125}I-labelled, to measure rate of release; Amersham Radiochemicals), nerve growth factor, transforming growth factor β, neurotrophin-3, dissolved to appropriate concentrations in DMEM or PBS
- DMEM supplemented with 2% fetal calf serum, penicillin/streptomycin (50 U/ml and 50 µg/ml respectively)
- PBS: 10 mM sodium phosphate buffer pH 7.4, containing 0.15 M NaCl

Method

1. Cut suitably sized, freeze-dried fibronectin mats, add the growth factor in solution for 24 h at 37 °C; 500 ng/ml NT-3 (184), 3–5 ng/ml NGF (15).
2. Rinse with fresh DMEM to remove unattached growth factor.
3. For release monitoring, sample the supernatant at 24 h intervals.

Soluble growth factors can be incorporated into materials by impregnation into dry materials. This technique relies on the rapid ingress of water, carrying the soluble factors, into the dry porous material, followed by slow diffusional release of the solutes into the surrounding aqueous medium or implant site. Governing factors include the dimensions of the material (i.e. diffusion path length), its mean pore diameter, and the molecular weight of the growth factor.

Release can be monitored by dorsal root ganglion outgrowth bioassay for the neurotrophic factors (14, 15, 184) or release of radiolabelled growth factor (9). In most cases, there is little further release of factors after three to five days incubation at 37 °C.

Acknowledgements

The authors are grateful to the Forte Trust, the Augustus Newman Trust, and the Welton Trust for financial support for this work, to Sarah Underwood and Sarah Harding for the electron micrograph, Mr Xavier Brouha and Dr Vivek Mudera for generously providing their experimental protocols.

References

1. Langer, R. and Vacanti, J. P. (1995). *Sci. Am.*, **273**, 130.
2. Langer, R. and Vacanti, J. P. (1993). *Science*, **260**, 920.
3. Brown, R. A., Smith, K. D., and McGrouther, D. A. (1997). *Wound Repair Regen.*, **5**, 212.
4. Woodley, D. T. (1996). In *The molecular and cellular biology of wound repair*, 2nd edn (ed. R. A. F. Clark), p. 339. Plenum Press, New York.
5. Jackson, D. A. (1982). In *Collagen in health and disease* (ed. J. B. Weiss and M. Jayson), p. 466. Churchill-Livingstone, Edinburgh.
6. Porter, R. A., Brown, R. A., Eastwood, M., Occleston, N. L., and Khaw, P. T. (1998). *Wound Repair Regen.*, **6**, 157.
7. Putnam, A. J. and Mooney, D. J. (1996). *Nature Med.*, **2**, 824.
8. Yannas, I. V., Burke, J. F., Orgill, D. P., and Skrabut, E. M. (1982). *Science*, **215**, 174.
9. Brown, R. A., Blunn, G. W., and Ejim, O. S. (1994). *Biomaterials*, **15**, 457.
10. Ahmed, Z., Idowu, B. D., and Brown, R. A. (1999). *Biomaterials*, **20**, 201.
11. Hadlock, T., Elisseeff, J., Langer, R., Vacanti, J., and Cheney, M. (1998). *Arch. Otolaryngol. Head Neck Surg.*, **124**, 1081.
12. L'Heureux, N., Pâquet, S., Labbé, R., Germain, L., and Auger, F. A. (1998). *FASEB J.*, **12**, 47.
13. Whitworth, I. H., Brown, R. A., Dore, C., Green, C. J., and Terenghi, G. (1995). *J. Hand Surg.*, **20B**, 429.
14. Sterne, G. D., Coulton, G. R., Brown, R. A., Green, C. J., and Terenghi, G. (1997). *J. Cell Biol.*, **139**, 709.
15. Whitworth, I. H., Brown, R. A., Dore, C., Green, C. J., and Terenghi, G. (1996). *J. Hand Surg.*, **21B**, 514.
16. Yannas, I. V. (1988). In *Collagen* (ed. M. E. Nimni), Vol. III, p. 87. CRC Press, Florida.
17. Barocas, V. H. and Tranquillo, R. T. (1997). *J. Biomech. Eng.*, **119**, 137.
18. Nerem, R. M., Alexander, R. W., Chappell, D. C., Medford, R. M., Varner, S. E., and Taylor, W. R. (1998). *Am. J. Med. Sci.*, **316**, 169.
19. Silver, F. H. (1994). In *Biomaterials, medical devices and tissue engineering*, p. 46. Chapman and Hall, London.

20. Falanga, V., Margolis, D., Alvarez, O., et al. (1998). *Arch. Dermatol.*, **134**, 293.
21. Rheinwald, J. G. and Green, H. (1975). *Cell*, **6**, 331.
22. van Wachem, P. B., van Luyen, M. J., and da Costa, M. L. (1996). *J. Biomed. Mater. Res.*, **30**, 353.
23. Andreassi, L., Casini, L., Trabucchi, E., et al. (1991). *Wounds*, **3**, 116.
24. Chamberlain, L. J. and Yannas, I. V. (1999). In *Tissue engineering methods and protocols* (ed. J. R. Morgan and M. L. Yarmush), p. 3. Humana Press, New Jersey.
25. Noughton, G., Mansbridge, J., and Gentzkow, G. (1997). *Artif. Organs*, **12**, 1203.
26. Young, J. H., Teumer, J., Kemp, P., and Parenteau, N. (1997). In *Principles of tissue engineering* (ed. R. Lanza, R. Langer, and W. Chick), p. 297. R. G. Landes, Austin, Texas.
27. Petzoldt, J. L., Leigh, I. M., Duffy, P. G., and Masters, J. R. W. (1994). *Urol. Res.*, **22**, 676.
28. Atala, A., Freeman, M. R., Vacanti, J. P., Shepard, J., and Retik, A. B. (1993). *J. Urol.*, **150**, 608.
29. Atala, A. (1998). In *Frontiers in tissue engineering* (ed. C. W. Patrick, A. G. Mikos, and L. V. McIntire), p. 649. Pergamon Press, Oxford.
30. Sabbagh, W., Masters, J. R., Duffy, P. G., Herbage, D., and Brown, R. A. (1998). *Br. J. Urol.*, **82**, 888.
31. Scott, R., Gorham, S. D., Aitcheson, M., Bramwell, P., Speakman, J., and Meddings, R. N. (1991). *Br. J. Urol.*, **68**, 421.
32. Atala, A., Vacanti, J. P., Peters, C. A., Mandell, J., Retik, A. B., and Freeman, M. R. (1992). *J. Urol.*, **148**, 658.
33. Oberpenning, F., Meng, J., Yoo, J. J., and Atala, A. (1999). *Nature Biotechnol.*, **17**, 149.
34. Lundborg, G. (1988). *Nerve injury and repair*, p. 149. Churchill Livingstone, Edinburgh.
35. Doolabh, V. B., Hertl, M. C., and Mackinnon, S. E. (1996). *Rev. Neurol.*, **7**, 47.
36. Molander, H., Olsson, Y., Engkvist, O., Bowald, S., and Eriksson, I. (1982). *Muscle Nerve*, **5**, 54.
37. Favaro, G., Bortolami, M. C., Cereser, S., Pastorello, D. A., Callegrao, L., and Fioi, M. G. (1990). *Trans. Am. Soc. Artif. Intern. Organs*, **36**, 291.
38. Williams, L. R., Danielsen, N., Muller, H., and Varon, S. (1987). *J. Comp. Neurol.*, **264**, 284.
39. Ansselin, A. D., Fink, T., and Davies, D. F. (1997). *Neuropathol. Appl. Neurobiol.*, **23**, 387.
40. Brown, R. E., Erdmann, D., Lyons, S. F., and Suchy, H. (1996). *J. Reconstr. Microsurg.*, **12**, 149.
41. Furnish, E. J. and Schmidt, C. E. (1998). In *Frontiers in tissue engineering* (ed. C. W. Patrick, A. G. Mikos, and L. V. McIntire), p. 514. Pergamon Press, Oxford.
42. Hobson, M. I., Brown, R. A., Green, C. J., and Terenghi, G. (1997). *Br. J. Plastic Surg.*, **50**, 125.
43. Ahmed, Z. A. and Brown, R. A. (1999). *Cell Motil. Cytoskel.*, **42**, 331.
44. Spray, T. L. and Roberts, W. C. (1977). *Am. J. Cardiol.*, **44**, 319.
45. Cohn, L. H. (1984). *Chest*, **85**, 387.
46. Nimni, M. E., Cheung, D. T., Strates, B., Kodama, M., and Sheikh, K. (1988). In *Collagen* (ed. M. E. Nimni), Vol. III, p. 1. CRC Press, Boca Ratan.
47. Ishihara, T., Ferrans, V. J., Jones, M., Boyce, S. W., and Roberts, W. C. (1981). *Am. J. Cardiol.*, **48**, 443.
48. Widmer, M. S. and Mikos, A. G. (1998). In *Frontiers in tissue engineering* (ed. C. W. Patrick, A. G. Mikos, and L. McIntire), p. 107. Pergamon Press, Oxford.
49. Silver, F. H. (1994). *Biomaterials, medical devices and tissue engineering*. Chapman-Hall, London.
50. Freed, L. E., Vunjak-Novakovic, G., Biron, R. J., et al. (1994). *Biotechnology*, **12**, 689.
51. Mooney, D. J., Mazzoni, C. L., Breuer, C., et al. (1996). *Biomaterials*, **17**, 115.
52. Engelberg, I. and Kohn, J. (1991). *Biomaterials*, **12**, 292.

53. Laitinen, O., Tormala, P., Taurio, P., et al. (1992). *Biomaterials*, **13**, 1012.
54. Vacanti, C. A., Langer, R., Schloo, B., and Vacanti, J. P. (1991). *Plast. Reconstr. Surg.*, **88**, 753.
55. Hansbrough, J. F., Cooper, M. L., Cohen, R., et al. (1992). *Surgery*, **111**, 438.
56. Ma, P. X. and Langer, R. (1999). In *Tissue engineering methods and protocols* (ed. J. R. Morgan and M. L. Yarmush), p. 47. Humana Press, Totowa, NJ.
57. Benedetti, L., Cortivo, R., Berti, T., Berti, A., Pea, F., Mazzo, M., et al. (1993). *Biomaterials*, **14**, 1154.
58. Mast, B. A., Flood, L. C., Haynes, J. H., et al. (1991). *Matrix*, **11**, 63.
59. Noble, P. M., Lake, F. R., Henson, P. M., and Riches, D. W. (1993). *J. Clin. Invest.*, **91**, 2368.
60. West, D. C., Hampson, I. N., Arnold, F., et al. (1985). *Science*, **228**, 1324.
61. Campoccia, D., Hunt, J. A., Dohetery, P. J., Zhong, S. P., O'Regan, M., Benedetti, L., et al. (1996). *Biomaterials*, **17**, 963.
62. Chvapil, M. (1977). *J. Biomed. Mater. Res.*, **11**, 721.
63. Doillon, C. J., Silver, F. H., and Berg, R. A. (1987). *Biomaterials*, **8**, 195.
64. van Wachem, P. B., van Luyen, M. J., Olde Damink, L. H., Dijkstra, P. J., Feijen, J., and Nieuwenhis, P. (1994). *Int. J. Artif. Organs*, **17**, 230.
65. Olde Damink, L. H., Dijkstra, P. J., van Luyen, M. J., van Wachem, P. B., Nieuwenhis, P., and Feijen, J. (1996). *Biomaterials*, **17**, 765.
66. Heijmen, F. H., du Pont, J. S., Middlekoop, E., Kreis, R. W., and Hoekstra, M. J. (1997). *Biomaterials*, **18**, 749.
67. Rault, I., Frei, V., Herbage, D., Abdul-Malak, N., and Huc, A. (1996). *J. Mater. Sci. Mater. Med.*, **7**, 215.
68. Frei, V., Huc, A., and Herbage, D. (1994). *J. Biomed. Mater. Res.*, **28**, 159.
69. Tinois, E., Toillier, J., Gaucherand, M., et al. (1991). *Exp. Cell Res.*, **193**, 310.
70. Laquerriere, A., Yun, J., Toillier, J., Hemet, J., and Tadie, M. (1993). *J. Neurosurg.*, **78**, 487.
71. Morgon, A., Disant, F., and Truy, E. (1989). *Acta Otolaryngol.*, **107**, 450.
72. Yannas, I. V. and Tobolsky, A. V. (1967). *Nature*, **215**, 509.
73. Yannas, I. V. and Burke, J. F. (1980). *J. Biomed. Mater. Res.*, **14**, 65.
74. Yannas, I. V. (1990). *Angew. Chem. Int. Ed. Engl.*, **29**, 20.
75. Yannas, I. V., Lee, E., Orgill, D. P., Skrabut, E. M., and Murphy, G. F. (1989). *Proc. Natl. Acad. Sci. USA*, **86**, 933.
76. Heimbach, D., Luterman, A., Burke, J., et al. (1988). *Ann. Surg.*, **208**, 313.
77. Koide, M., Osaki, K., Konishi, J., et al. (1993). *J. Biomed. Mater. Res.*, **27**, 79.
78. De Vries, H. J. C., Middlekoop, E., Mekkes, J. R., Dutrieux, R. P., Wildevuur, C. H. R., and Westerhof, W. (1994). *Wound Repair Regen.*, **2**, 37.
79. Murashita, T., Nakayama, Y., Hirano, T., and Ohashi, S. (1996). *Br. J. Plast. Surg.*, **49**, 58.
80. Ejim, O. S., Blunn, G. W., and Brown, R. A. (1993). *Biomaterials*, **14**, 743.
81. Underwood, S., Afoke, A., Brown, R. A., McLeod, A. J., and Dunnill, P. (1999). *Bioprocess Eng.*, **20**, 239.
82. Ahmed, Z. A. (1999). PhD thesis, University College London.
83. Ahmed, Z., Brinden, A., Hall, S., and Brown, R. A. Stabilisation of large fibronectin cables with micromolar concentrations of copper: cell substrate properties (in preparation).
84. Taravel, M. N. and Damard, A. (1993). *Biomaterials*, **14**, 930.
85. Larsen, N. E., Pollak, C. T., Reiner, K., Leshiner, E., and Balazs, E. A. (1993). *J. Biomed. Mater. Res.*, **27**, 1129.
86. Lee, D. A. and Bader, D. L. (1995). *In Vitro Cell Dev. Biol. Anim.*, **11**, 828.
87. Friman, R., Giaccone, G., Kanemoto, T., Martin, G. R., Gazdar, A. F., and Mulshine, J. L. (1990). *Proc. Natl. Acad. Sci. USA*, **87**, 6698.

88. Silver, F. H., Wang, M. C., and Pins, G. D. (1995). *Biomaterials*, **16**, 891.
89. San-Galli, F., Deminiere, C., Guerin, J., and Rabaud, M. (1996). *Biomaterials*, **17**, 1081.
90. Bonzon, N., Carrat, X., Deminiere, C., Daculsi, G., Lefebvre, F., and Rabaud, M. (1995). *Biomaterials*, **16**, 881.
91. Pandit, A. S., Feldman, D. S., Caulfield, J., and Thompson, A. (1998). *Growth Factors*, **15**, 113.
92. Fournier, N. and Doillon, C. J. (1996). *Biomaterials*, **17**, 1659.
93. Roy, F., DeBlois, C., and Doillon, C. J. (1993). *J. Biomed. Mater. Res.*, **27**, 389.
94. Elsdale, T. and Bard, J. (1972). *J. Cell Biol.*, **54**, 626.
95. Porter, R. A., Brown, R. A., Eastwood, M., Occleston, N. L., and Khaw, P. T. (1998). *Wound Repair Regen.*, **6**, 157.
96. Bell, E., Iverson, B., and Merrill, C. (1979). *Proc. Natl. Acad. Sci USA*, **76**, 1274.
97. Eastwood, M., Porter, R., Kahn, U., McGrouther, G., and Brown, R. (1996). *J. Cell. Physiol.*, **166**, 33.
98. Brown, R. A., Prajapati, R., McGrouther, D. A., Yannas, I. V., and Eastwood, M. (1998). *J. Cell. Physiol.*, **175**, 323.
99. Hanthamrongwit, M., Reid, W. H., and Grant, M. H. (1996). *Biomaterials*, **17**, 775.
100. Kolodney, M. S. and Wysolmerski, R. B. (1992). *J. Cell Biol.*, **117**, 73.
101. Bell, E. (1995). *J. Cell. Eng.*, **1**, 28.
102. Bell, E., Sher, S., and Hull, B. (1984). *Scanning Elect. Microsc.*, **4**, 1957.
103. Grinnell, F. (1994). *J. Cell Biol.*, **124**, 401.
104. Guido, S. and Tranquillo, R. T. (1993). *J. Cell Sci.*, **105**, 317.
105. Dickinson, R. B., Guido, S., and Tranquillo, R. T. (1994). *Ann. Biomed. Eng.*, **22**, 342.
106. Shortkroff, S. and Spector, M. (1999). In *Tissue engineering methods and protocols* (ed. J. R. Morgan and M. L. Yarmush), p. 195. Humana Press, New Jersey.
107. Compton, C. C., Nadire, K. B., Regauer, S., *et al.* (1998). *Differentiation*, **64**, 45.
108. Geppert, T. D. and Lipsky, P. E. (1985). *J. Immunol.*, **135**, 3750.
109. Pober, J. S., Collins, T., Gimbrone, M. A., *et al.* (1993). *Nature*, **305**, 726.
110. Theobald, V. A., Lauer, J. D., Kaplan, F. A., *et al.* (1993). *Transplantation*, **55**, 128.
111. Eaglstein, W. H., Alvarez, O. M., Auletta, M., *et al.* (1999). *Dermatol. Surg.*, **25**, 195.
112. Sun, A. M. (1999). In *Tissue engineering methods and protocols* (ed. J. R. Morgan and M. L. Yarmush), p. 469. Humana Press, New Jersey.
113. Petersen, K. P., Petersen, C. M., and Pope, E. J. (1998). *Proc. Soc. Exp. Biol. Med.*, **218**, 365.
114. Lanza, R. P., Solomon, B. A., Monaco, A. P., *et al.* (1994). In *Pancreatic islet transplantation*, Vol. III (ed. R. P. Lanza and W. L. Chick), p. 154. R. G. Landes Co., Austin.
115. Huber, M., Rettler, T., Bernasloni, K., *et al.* (1995). *Science*, **267**, 525.
116. Morgan, J. R., Barrandon, Y., Green, H., and Mulligan, R. C. (1987). *Science*, **237**, 1476.
117. Hoeben, R. C., Fallaux, F. J., van Tillburg, N. H., *et al.* (1993). *Hum. Gene Ther.*, **4**, 179.
118. Petersen, M. J., Kaplan, J., Jorgenson, C. M., *et al.* (1995). *J. Invest. Dermatol.*, **104**, 171.
119. Wilson, J. M., Birinyi, L. K., Salomon, R. N., Libby, P. L., Callow, A. D., and Mulligan, R. C. (1989). *Science*, **244**, 1344.
120. Prajapati, R. T., Al-Ani, S., Smith, P. J., and Brown, R. A. (1996). *Cell. Eng.*, **1**, 143.
121. Brown, R. A., Talas, G., Porter, R. A., McGrouther, D. A., and Eastwood, M. (1996). *J. Cell. Physiol.*, **169**, 439.
122. Brown, R. A., Terenghi, G., and McFarland, C. D. In *The new angiotherapy* (ed. T.-P. D. Fan and R. Auerbach). Humana Press Inc., Totowa NJ., USA. (in press)
123. Sittinger, M., Schultz, O., Keyszer, G., Minuth, W. W., and Burmester, G. R. (1997). *Int. J. Artif. Organs*, **20**, 57.

124. Sterpetti, A. V., Cucina, A., Fragale, A., Lepidi, S., Cavallaro, A., and Santoro, L. (1994). *Eur. J. Vasc. Surg.*, **8**, 138.
125. Brown, L. F., Yeo, K. T., Berse, B., et al. (1992). *J. Exp. Med.*, **176**, 1375.
126. Stokes, C. L., Rupnick, M. A., Williams, S. K., and Lauffenburger, D. A (1990). *Lab. Invest.*, **63**, 657.
127. Ignotz, R. A. and Massague, J. (1986). *J. Biol. Chem.*, **261**, 4337.
128. Curtis, A. S. G., Wilkinson, C. D., and Wojciak-Stothard, B. (1995). *J. Cell. Eng.*, **1**, 35.
129. Curtis, A. S. G. and Wilkinson, C. D. (1997). *Biomaterials*, **18**, 1573.
130. Curtis, A. S. G. and Varde, M. (1964). *J. Natl. Cancer Inst.*, **33**, 15.
131. Clark, P., Connolly, P., Curtis A. S., Dow, J. A., and Wilkinson, C. D. (1991). *J. Cell Sci.*, **99**, 73.
132. Weiss, P. (1945). *J. Exp. Zool.*, **100**, 353.
133. Wojciak-Stothard, B., Crossan, J., Curtis, A. S. G., and Wilkinson, C. D. (1995). *J. Mater. Sci. Mater. Med.*, **6**, 266.
134. Wojciak-Stothard, B., Denyer, M., Mishra, M., and Brown, R. A. (1997). *In Vitro Cell Dev. Biol.*, **33**, 110.
135. Guido, S. and Tranquillo, R. T. (1993). *J. Cell Sci.*, **105**, 317.
136. Baschong, W., Sutterlin, R., and Aebi, U. (1997). *Eur. J. Cell Biol.*, **72**, 189.
137. Eastwood, M., Mudera, V. C., McGrouther, D. A., and Brown, R. A. (1998). *Cell Motil. Cytoskel.*, **49**, 13.
138. Chiquet, M., Eppenberger, H. M., and Turner, D. C. (1981). *Dev. Biol.*, **88**, 230.
139. Dimilla, P. A., Stone, J. A., Quinn, J. A., Albelda, S. M., and Lauffenburger, D. A. (1991). *J. Cell Biol.*, **122**, 729.
140. Lauffenburger, D. A. and Horowitz, A. F. (1996). *Cell*, **84**, 359.
141. Postlethwaite, A. E., Keski-Oja, J., Balain, G., et al. (1981). *J. Exp. Med.*, **153**, 494.
142. Ungari, S., Katari, R. S., Alessandri, G., and Gullino, P. M. (1985). *Invasion Metast.*, **5**, 193.
143. Postlethwaite, A. E., Seyer, J. M., and Kang, A. H. (1978). *Proc. Natl. Acad. Sci. USA*, **75**, 871.
144. Yannas, I. V. (1988). In *Collagen Vol. III: Biotechnology* (ed. M. E. Nimni), p. 87. CRC Press, Florida.
145. Pilcher, B. K., Kim, D. W., Carney, D. H., and Tomasek, J. J. (1994). *Exp. Cell Res.*, **211**, 368.
146. Ali, S. Y., Evans, L., and Ralphs, J. R. (1986). In *Cell mediated calcification and matrix vesicles* (ed. S. Y. Ali), p. 253. Excerta Medica, Amsterdam.
147. Whitworth, I. H., Terenghi, G., Green, C. J., Brown, R. A., Stevens, E., and Tomlinson, D. R. (1995). *Eur. J. Neurosci.*, **7**, 2220.
148. Ohata, H., Seito, N., Aizawa, H., Nobe, K., and Momose, K. (1995). *Biochem. Biophys. Res. Commun.*, **208**, 19.
149. Ohno, M., Cooke, J. P., Dzau, V. J., and Gibbons, G. H. (1995). *J. Clin. Invest.*, **95**, 1363.
150. Markin, V. S. and Tsong, T. Y. (1991). *J. Biophys.*, **59**, 1317.
151. Wang, N., Butler, J. P., and Ingber, D. E. (1993). *Science*, **206**, 1124.
152. Butt, R. P., Laurent, G. J., and Bishop, J. E. (1995). *Ann. N. Y. Acad. Sci.*, **752**, 387.
153. Levesque, M. J. and Nerem, R. M. (1985). *J. Biomech. Eng.*, **107**, 341.
154. Shirinsky, V. P., Antonov, A. S., Konstantin, G., et al. (1989). *J. Cell Biol.*, **109**, 331.
155. Daifotis, A. G., Weir, F. C., Dreyer, B. F., and Broadus, A. E. (1992). *J. Biol. Chem.*, **267**, 23455.
156. Urban, J. P. (1994). *Br. J. Rheumatol. et al.*, **33**, 901.
157. Burton-Wurster, N., Vernier-Singer, M., Farquhar, T., and Lust, G. (1993). *J. Orthop. Res.*, **11**, 717.
158. Parsons, M., Bishop, J. E., Laurent, G. J., Eastwood, M., and Brown, R. A. Mechanical

loading of collagen gels stimulates dermal fibroblast collagen synthesis and reorganisation. *J. Invest Dermatol.* (in press).

159. Prajapati, R. T., Eastwood, M., and Brown, R. A. Mechanical factors regulating protease production by fibroblasts in 3-dimensional collagen lattices. (submitted).
160. Kain, H. L. and Reuter, U. (1995). *Graefs. Arch. Clin. Exp. Ophthalmol.*, **233**, 236.
161. Davies, P. F. (1995). *Physiol. Rev.*, **75**, 519.
162. Bushmann, M. D., Gluzband, Y. A., Grodzinsky, A. J., and Hunziker, E. B. (1995). *J. Cell Sci.*, **108**, 1497.
163. Helminger, G., Geiger, R., Schreck, S., and Nerem, R. (1991). *ASME J. Biomech. Eng.*, **113**, 123.
164. Oloyede, A., Flaschmann, R., and Broom, N. D. (1992). *Connect. Tissue Res.*, **27**, 211.
165. Malek, A. M. and Izumo, S. (1996). *Cell Sci.*, **109**, 713.
166. Schnittler, H. J., Püschel, B., and Drenckhahn, D. (1997). *Am. J. Physiol.*, **273H**, 2396.
167. Morigi, M., Zoja, C., Figliuzzi, M., *et al.* (1995). *Blood*, **85**, 1696.
168. Papadaki, M. and Eskin, S. G. (1997). *Biotechnol. Prog.*, **13**, 209.
169. Li, S., Piotrowicz, R. S., Levin, E. G., Shyy, Y. J., and Chien, S. (1996). *Am. J. Physiol.*, **271C**, 994.
170. Inoue, N., Ramasamy, S., Fukai, T., Nerem, R. M., and Harrison, D. G. (1996). *Circ. Res.*, **79**, 32.
171. Dimmeler, S., Assmus, B., Hermann, C., Haendeler, J., and Zeiher, A. M. (1998). *Circ. Res.*, **83**, 334.
172. Thoumine, O., Nerem, R. M., and Girard, P. R. (1995). *Lab. Invest.*, **73**, 565.
173. Mudera, V. C., Pleass, R., Eastwood, M., Tarnuzzer, R., Schultz, G., Khaw, P., McGrouther, D. A., and Brown, R. A. Molecular responses of human dermal fibroblasts to dual cues: contact guidance and mechanical load. (2000). *Cell Motility Cytoskel.*, **45**, 1–9.
174. Vandenberg, H. H. (1988). *In Vitro Cell Dev. Biol.*, **24**, 609.
175. Quinn, T. M., Grodzinsky, A. J., Bushmann, M. D., Kim, Y. J., and Hunziker, E. B. (1998). *J. Cell Sci.*, **111**, 573.
176. Kim, Y. J., Bonassar, L. J., and Grodzinsky, A. J. (1995). *J. Biomech.*, **28**, 1055.
177. Wong, M., Wuethrich, P., Bushmann, M. D., Eggli, P., and Hunziker, E. B. (1997). *J. Orthop. Res.*, **15**, 189.
178. Shinagl, R. M., Gurskis, D., Chen, A. C., and Sah, R. L. (1997). *J. Orthop. Res.*, **15**, 499.
179. Shinagl, R. M., Ting, M. K., Price, J. H., and Sah, R. L. (1996). *Ann. Biomed. Eng.*, **24**, 500.
180. Brown, R. A. and Jones, K. L. (1992). *Eur. J. Exp. Musculoskel. Res.*, **1**, 25.
181. Talas, G., Brown, R. A., and McGrouther, D. A. (1999). *Biochem. Pharmacol.*, **85**, 1094.
182. Brouha, X. D. R., Talas, G., Brown, R. A., and Porter, R. A. Preparation and slow release of phenytoin from an implantable drug depot using a fibronectin mat carrier (in preparation).
183. Romanyshyn, L. A., Wichmann, J. K., Kucharczyk, K., Shumaker, R. C., Ward, D., and Sofia, R. D. (1994). *Ther. Drug Monit.*, **16**, 90.
184. Sterne, G. D., Brown, R. A., Green, C. J., and Terenghi, G. (1997). *Eur. J. Neurosci.*, **9**, 1388.

Chapter 7
Cytotoxicity and viability assays

Anne P. Wilson

Oncology Research Laboratory, Derby City Hospital, Uttoxeter Road, Derby DE3 3NE, U.K.

1 Introduction

Drug development programmes involve pre-clinical screening of vast numbers of chemicals for specific and non-specific cytotoxicity against many types of cells. Both are important for indicating the potential therapeutic target and safety evaluation.

Animal models have always played an important role in both contexts, and although cell culture systems have figured largely in the field of cancer chemotherapy, where the potential value of such systems for cytotoxicity and viability testing is now widely accepted, there is increasing pressure for a more comprehensive adoption of *in vitro* testing in safety evaluation. The impetus for change originates partly from financial considerations, since *in vitro* testing has considerable economic advantages over *in vivo* testing. There is also an increasing realization of the limitations of animal models in relation to human metabolism, as more and more metabolic differences between species come to be identified. Finally and as importantly, there is the moral pressure to reduce animal experimentation.

The safety evaluation of chemicals involves a range of studies on mutagenicity, carcinogenicity, and chronic toxicity. The former two aspects have been covered in other books (1-3) and are outside the scope of this text. Whilst the major application of *in vitro* cultures is the analysis of acute toxicity, the existence of adequate culture systems would also improve the prospects of chronic toxicity testing. Toxicology and cancer chemotherapy therefore share the aim of determining the acute toxicity of a range of chemicals against a variety of cell types. In both areas there may be several ultimate goals:

- identification of potentially active compounds
- identification of the mechanism by which the compound exerts its toxic effect
- prediction of anticancer activity
- identification of a potential target cell population

- identification of the toxic concentration range
- relationship of concentration to exposure time (C × T)

The fundamental requirements for both toxicology and screening for anti-cancer activity are similar. They are first that the assay system should give a reproducible dose–response curve with low inherent variability over a concentration range which includes the exposure dose *in vivo*. Secondly, the selected response criterion should show a linear relationship with cell number, and thirdly, the information obtained from the dose–response curve should relate predictively to the *in vivo* effect of the same drug.

2 Background

Use of *in vitro* assay systems for the screening of potential anticancer agents has been common practice almost since the beginnings of cancer chemotherapy in 1946, following the discovery of the antineoplastic activity of nitrogen mustard. A number of articles describe the historical development of these techniques and their application (4-9).

Accumulated experience, both clinical and experimental, demonstrated the heterogeneity of chemosensitivity between tumours, even those of identical histology. The successful development of the *in vitro* agar plate assay for antibiotic screening precipitated interest in the development of an analogous technique for 'tailoring' chemotherapy to suit the individual tumour and patient, thus removing the undesirable combination of ineffective chemotherapy in the presence of non-specific toxicity (10-13).

The correlation between *in vitro* drug sensitivity exhibited by primary cultures of human tumours and their *in vivo* counterparts argued for their use in drug evaluation programmes, since they provided a closer approximation to the human clinical situation than did the limited number of cell lines which were used. The role of the 'human tumour stem cell assay' was investigated in this context (14, 15).

The National Cancer Institute now routinely measures the growth inhibitory properties of every compound under test against a panel of 60 human tumour cell lines which are representative of major human tumour types. For each compound tested, the GI_{50} (concentration of drug needed to inhibit cell growth by 50%) is generated from the dose–response curves for each cell line. The pattern of all the cell lines as a group is then used to rank compounds according to whether a new compound is likely to share a common mechanism of action with compounds already present in the database (COMPARE). In recent years, the COMPARE computer algorithm has been applied to correlate patterns of growth inhibition with particular molecular properties of the cell lines (16).

The developments in the field of safety evaluation of drugs have been reviewed by several authors (17, 18). There are a number of advantages of *in vitro* testing for safety evaluation which include analysis of species specificity, feasibility of using only small amounts of test substances, and the facility to do mechanistic studies. Ref. 18 also highlights the progress which has been made in developing suitable models for different target organs.

3 Specific techniques

The choice of assay depends on the context in which the assay is to be used, the origin of the target cells, and the nature of the test compounds. Parameters which vary between assays include:

- culture method (Section 3.1)
- duration of drug exposure and drug concentration (Section 3.2)
- duration of recovery period after drug exposure (Section 3.3)
- end-point used to quantitate drug effect (Section 4)

3.1 Culture methods

The choice of culture method depends on the origin of the target cells and the duration of the assay and, to some extent, dictates the end-point.

3.1.1 Organ culture

The advantages of organ culture relate to the maintenance of tissue integrity and cell–cell relationships *in vitro*, giving a model that is more representative of the *in vivo* situation than the majority of culture methods available. However, reliable quantitation of drug effect is impaired by variation in size and cellular heterogeneity between replicates. Although the method has been used extensively to study the hormone sensitivity of potentially responsive target tissues, the number of studies relating to drug sensitivity is limited (19).

3.1.2 Spheroids

Spheroids result from the spontaneous aggregation of cells into small spherical masses which grow by proliferation of the component cells. Their structure is analogous to that of a small tumour nodule or micrometastasis, and the use of spheroids for drug sensitivity testing therefore permits an *in vitro* analysis of the effects of three-dimensional relationships on drug sensitivity, without the disadvantages previously mentioned for organ culture. Specific parameters which can be studied are:

- drug penetration barriers in avascular areas
- the effects of metabolic gradients (e.g. nutrients, metabolites, pO_2, pCO_2)
- the effects of proliferation gradients

The majority of studies have been carried out using spheroids derived from cell lines, but primary human tumours also have the capacity to form spheroids in approximately 50% of cases. The spheroid-forming capacity of normal cells is limited in comparison with tumour cells, so stromal elements may be excluded during reaggregation of human tumour biopsy material. Culture times in excess of two weeks are usually necessary for drug sensitivity testing, and the method is therefore not suitable in a situation where results are required quickly.

3.1.3 Suspension cultures

i. Short-term cultures (4–24 hours)

The short-term maintenance of cells in suspension for assay of drug sensitivity is applicable to all cell sources. When the cells are derived from human tumour biopsy material, the assay system has several theoretical advantages in that stromal cell overgrowth and clonal selection are minimized, and results can be obtained rapidly. The method has been used extensively in Germany for chemosensitivity studies on a variety of tumour types (20, 21). A modified method using either tissue fragments or cells has been described (13), again with a variety of tumour types using the incorporation of tritiated nucleotides into DNA or RNA as an end-point. Limitations of the method relate mainly to the short time period of the assay, which precludes long drug exposures over one or more cell cycles and also takes no account of either the reversibility of the drug's effect or of delayed cytotoxicity.

ii. Intermediate duration cultures (4–7 days)

Suspension cultures of intermediate duration are particularly suited to chemosensitivity studies on haematological malignancies, and have been described in several reports (22–24).

3.1.4 Monolayer culture

The technique of growing cells as a monolayer has been most frequently applied to the cytotoxicity testing of cancer cell lines, but the method has also been used with some success for studies on the chemosensitivities of biopsies from a variety of different tumour types. In the case of human biopsy material the greatest problems associated with the method are, first, that the success rate is limited because adherence and proliferation of tumour cells is not always obtained and, secondly, that contamination of tumour cell cultures by stromal cells may occur. Some method of cell identification is therefore an essential part of such assays. The technique has also found wide acceptance with toxicologists since appropriate cell lines may be available for the development of models for specific organ systems (18).

Monolayers offer great flexibility in terms of drug exposure and recovery conditions, and also in methods of quantitation of drug effect. The range of microtitration plates now available together with developments in automation has in turn led to microscale methodologies which are economical in terms of cell numbers required and reagents. There have been major developments in this area in recent years facilitating high throughput drug screening using cell lines. When cell numbers from biopsies are limited it may be feasible to culture the cells until sufficient numbers are available for assay, although reports on changes in chemosensitivity after subculture are conflicting. Two to three subcultures are probably acceptable and, indeed, subculture has been recommended because variability between replicates is reduced (25). When stromal cell contamination is unacceptably high, subculturing also offers the possibility of 'purifying' tumour cell cultures by differential enzyme treatment or physical cell separation.

3.1.5 Clonogenic growth in soft agar

Although monolayer cloning can be applied to cells cultured directly from the tumour, the majority of reports in recent years have used suspension cloning to minimize growth of anchorage-dependent stromal cells. Clonogenic assays have the theoretical advantage that the response is measured selectively in cells with a high capacity for self-renewal (stem cells). However, this is only true if colonies are clones (i.e. were initiated from one cell and not from a clump) and are scored after many cell generations in clonal growth. Cloning efficiencies of 0.01–0.1% may be obtained, but it is difficult to exclude the possibility of clumps; whilst ten generations (about 1000 cells) can be readily obtained in monolayer cloning, suspension colonies are often scored after four to six cell generations (16–64 cells). Given that some of these colonies started out as clumps of three or four cells or larger, the generation number could be as low as two and their capacity for self-renewal in doubt.

Technical problems have been encountered using solid tumours and effusions from patients, which unfortunately influence interpretation of results. These include difficulties in obtaining a pure single cell suspension from epithelial tumours, very low plating efficiencies ($<<$ 1%), the formation of colonies from anchorage-dependent cells under certain growth conditions, requirements for large cell numbers, and finally the somewhat subjective nature of colony quantitation. Critical assessment of the technical difficulties which have been encountered with the 'human tumour stem cell assay' can be found in several reports (26–29). Results obtained using the double agar method developed by Hamburger *et al.* (30) have been described in a review publication (31). An alternative methodology, developed by Courtenay and others (32), gives higher plating efficiencies, and in a comparison with the 'Hamburger–Salmon' system by Tveit *et al.* (33) it was apparent that the methodology used influenced the chemosensitivity profile obtained (*Figure 1*).

In recent years comparatively few publications have dealt with the applied use of the assay, but there have been several reports describing methods of improving plating efficiencies in soft agar (34–36).

3.2 Duration of drug exposure and drug concentrations

The choice of drug concentrations should be dictated by consideration of the therapeutic levels which can be achieved with clinically used drug dosages. When the compound is undergoing pre-clinical screening for potential antitumour activity this is not possible. Accumulated evidence on effective *in vitro* levels of

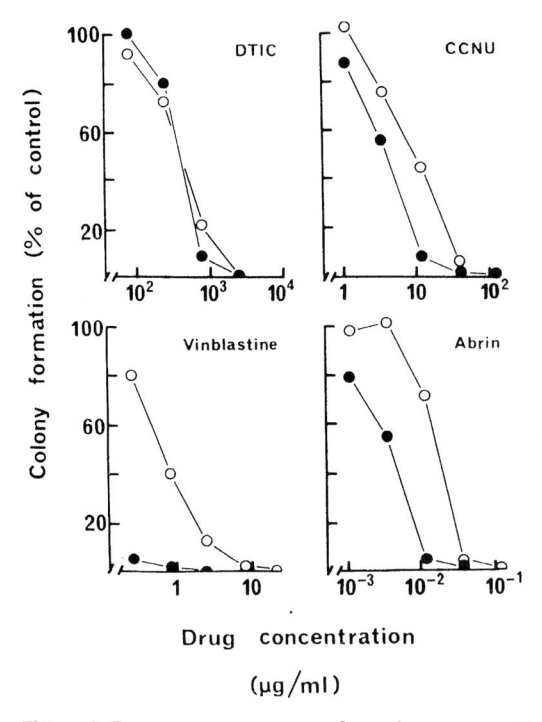

Figure 1. Dose–response curves of a melanoma xenograft cultured in soft agar using the 'Courtenay' method (●) and the 'Hamburger–Salmon' method (○). Cells were exposed to the drugs for 1 h, plated at 3×10^4 cells per tube or dish, and scored for colonies after 14 days incubation. Control cultures produced about 400 colonies using both methods. (Reproduced with permission of the publisher from ref. 33.)

compounds with known *in vivo* activity suggests an upper limit of 100 µg/ml. Pharmacokinetic data are available for many of the clinically used anticancer drugs, and parameters which are relevant to *in vitro* assays include the peak plasma concentration and the plasma clearance curves (*Figure 2*). A detailed description of pharmacological considerations may be found elsewhere (37). Pharmacokinetic data on most cancer chemotherapeutic agents, which includes peak plasma concentration, the C × T parameter (where C is concentration and T is time in hours), and the terminal half-life ($t_{1/2}$) of the drug in plasma, have been summarized (38). When no pharmacokinetic data are available, an approximation of the plasma levels can be obtained by calculation of the theoretical concentration obtained when the administered dose is evenly distributed throughout the total body fluid compartment. It is axiomatic that the concentration range adopted should give a dose–response curve. The same provisos apply when deciding on appropriate drug concentrations for the assessment of acute toxicity, though upper concentration levels may be in excess of 100 µg/ml.

Pharmacokinetic data show that maximum exposure to drug occurs in the first hour after intravenous injection and, for this reason, an exposure period of 1 h has been chosen by many investigators. Whilst this may be adequate for

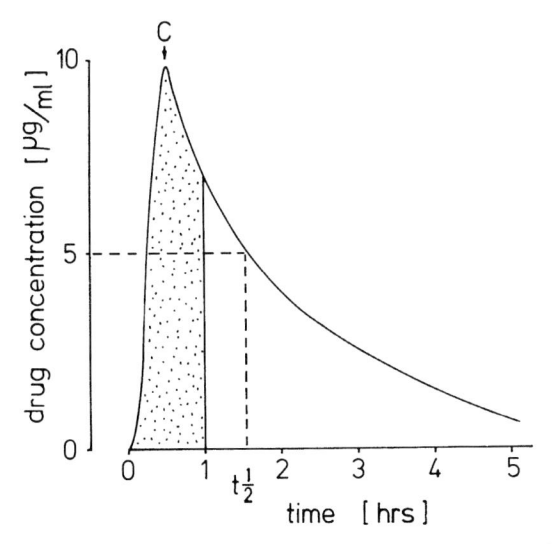

Figure 2. Typical plasma clearance curve for intravenously administered drug. C, peak plasma concentration; $t_{1/2}$, terminal half-life of drug in plasma. The hatched area is the area under the curve for T = 1 h.

cycle-specific drugs, such as the alkylating agents, longer exposure times over several cell cycles may be necessary for phase-specific drugs. Prolonged drug exposure using a variety of cancer chemotherapeutic agents has been shown to result in gradually decreasing ID_{50} values, as exposure times increase (39) (*Figure 3*). In some cases these reach a stable minimum plateau value. Rate of penetration of the drug may also be a limiting factor when short exposure times are used. Ultimately the question of duration of drug exposure becomes one of practicality. If a significant effect is achievable in 1 h, then this should be used. Many drugs may bind irreversibly to intracellular constituents and the actual exposure may therefore be in excess of 1 h due to drug retention. Others, principally the antimetabolites and antitubulins, are more likely to be reversible if not present at the sensitive phase of the cell cycle, and prolonged exposure spanning one or more cell cycles may be required. Resistance of the surviving fraction when short exposures are used may be due to the phase of the cell cycle during drug exposure, but genetically resistant cells (i.e. with resistance even at the appropriate phase of the cell cycle) can only be demonstrated unequivocally after a prolonged exposure.

When prolonged drug exposure times are used, it should be remembered that the theoretical C × T value is only equal to the actual C × T value when the drug retains full activity at 37°C over the entire exposure period and the response is linear with time. Data on the stability of some anticancer drug solutions at 37°C have been reported (38). The effective concentration of drug may also be reduced by binding of drug to the surface of the incubation vessel and medium components. Considerations such as those described above are also relevant for assessment of the acute toxicity of a compound.

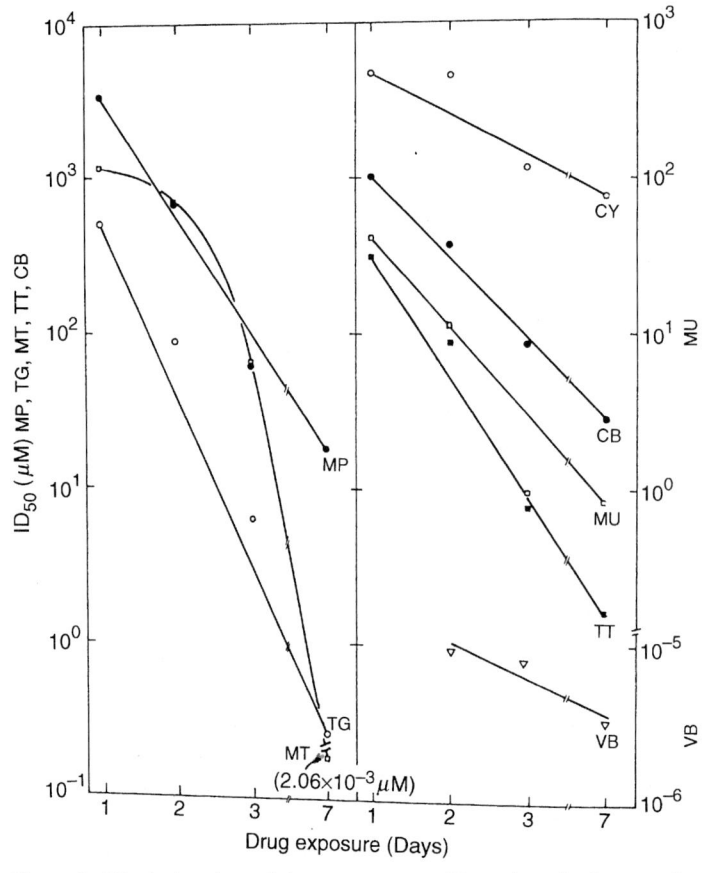

Figure 3. Effect of prolonged drug exposure on ID_{50} values for 6-mercaptopurine (MP), thioguanine (TG), methotrexate (MT), cyclophosphamide (CY), chlorambucil (CB), mustine (MU), thiotepa (TT), and vinblastine (VB), tested against HeLa cells using a microtitration plate assay. Drugs were replaced at 24 h intervals for the 48 h and 72 h exposures, and at 24, 48, and 72 h for the 7 day exposure. A break in the time axis is therefore shown between 3 and 7 days. (Reproduced with permission of the publishers from ref. 39.)

3.3 Recovery period

The inclusion of a recovery period following drug exposure may be important for three reasons:

(a) When metabolic inhibition is used as the end-point, it allows the cells to recover from metabolic perturbations which are unrelated to cell death.

(b) Sublethal damage can be repaired and therefore not interfere with the assay result.

(c) Delayed cytotoxicity, perhaps not expressed until one to two cell cycles after drug treatment, can be measured.

Depending on the drug and the end-point, absence of a recovery period can either underestimate or overestimate the level of cell kill achieved. However, it

CYTOTOXICITY AND VIABILITY ASSAYS

is equally important that the recovery period is not too long, because cell kill can then be masked by overgrowth of a resistant population. In monolayer assays which monitor cell counts or precursor incorporation, the cells must remain in the log phase of growth throughout the exposure and recovery period.

4 End-points

4.1 Cytotoxicity, viability, and survival

Cytotoxicity assays measure drug-induced alterations in metabolic pathways or structural integrity which may or may not be related directly to cell death, whereas survival assays measure the end-result of such metabolic perturbations, which may be either cell recovery or cell death. Theoretically, the only reliable index of survival in proliferating cells is the demonstration of reproductive integrity, as evidenced by clonogenicity. Metabolic parameters may also be used as a measure of survival when the cell population has been allowed time for metabolic recovery following drug exposure.

4.2 Cytotoxicity and viability

Some cytotoxicity assays offer instantaneous interpretation, such as the uptake of a dye by dead cells, or release of ^{51}Chromium or fluorescein from pre-labelled cells. These have been termed tests of viability, and are intended to predict survival rather than measure it directly. On the whole these tests are good at identifying dead cells but may overestimate long-term survival. Most imply a breakdown in membrane integrity and irreversible cell death.

Other aspects of cytotoxicity, measuring metabolic events, may be more accurately quantified and are very sensitive, but prediction of survival is less certain as many forms of metabolic inhibition may be reversible. In these cases impairment of survival can only be inferred if depressed rates of precursor incorporation into DNA, RNA, or protein are maintained after the equivalent of several cell population doubling times has elapsed.

4.2.1 Membrane integrity

This is the commonest measurement of cell viability at the time of assay. It will give an estimate of instantaneous damage (e.g. by cell freezing and thawing), or progressive damage over a few hours. Beyond this, quantitation may be difficult due to loss of dead cells by detachment and autolysis. These assays are of particular importance for toxic agents which exert their primary effect on membrane integrity.

i. ^{51}Chromium release

Labelling cells with ^{51}Cr results in covalent binding of chromate to basic amino acids of intracellular proteins. These labelled proteins leak out of the cell when the membrane is damaged, at a rate which is proportional to the amount of damage. The method is used in immunological studies for determining cytotoxic

T cell activity against target cells. Natural leakage of ^{51}Cr from undamaged cells may be high, and therefore the time period over which the assay can be used is restricted to approximately 4 h. Additionally, the target cells must be pre-labelled prior to incubation with drug or effector cells, and the final preparation for counting as well as counting itself is time-consuming. In one comparative study which evaluated ^{51}Cr release as an end-point for anticancer drug cytotoxicity testing the method was found to be of no value (40).

ii. Enzyme release assays

Enzyme release assays for measuring membrane integrity have been developed which overcome some of the problems of the ^{51}Cr release assay. Different enzymes have been used, though LDH has been found to be generally useful, since it is released by a range of cell types (41). This paper describes the assay for determining cytotoxicity of antibody-dependent and T cell-mediated reactions. The assay also has application in the wider context of toxicity testing, and has been used to investigate hepatotoxicity (42).

iii. Dye exclusion

Viability dyes used to determine membrane integrity include trypan blue, eosin Y, naphthalene black, nigrosin (green), erythrosin B, and fast green. Staining for

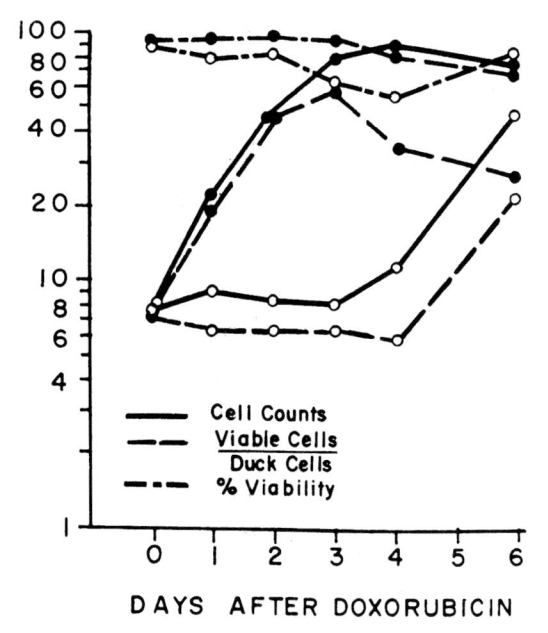

Figure 4. Effect of doxorubicin (0.12 μg ml^{-1}.h) on MDAY-D2 cells as assessed by three different techniques: Coulter counter particle counts (–) (y axis units are cells/ml × 10^{-4}); ratio of living tumour cells to duck red blood cells, normalized to the same scale as the Coulter counts (– –), and percentage viability (– - –) y axis units are percentage viability (living cells/living and dead cells × 100). ●, control cultures; ○, doxorubicin-treated cultures. (Reproduced with permission of the publishers from ref. 43.)

viability assessment is more suited to suspension cultures than to monolayers, because dead cells detach from the monolayer and are therefore lost from the assay. A major disadvantage may be the failure of reproductively dead cells to take up dye, as demonstrated when cells with impaired clonogenicity showed 100% viability according to dye exclusion (40). The method has been renovated, however, and technical innovations introduced which attempt to circumvent some of the problems commonly associated with such assays. In the methodology developed by Weisenthal *et al.* (43) a four day assay period is used to permit reproductively dead cells to lose their membrane integrity and the inaccuracies produced by either overgrowth of viable cells or lysis of dead cells is compensated for by incorporation of fixed duck erythrocytes as an internal standard. Comparison of cell counts versus percentage viability versus viable cell/duck cell ratio demonstrated the increased sensitivity of the latter method (*Figure 4*). The method has been applied with equal success to solid tumours, effusions, and haematological malignancies.

iv. Fluorescent dyes

There are a number of fluorescent probes now available for measuring membrane integrity. These are shown in *Table 1*. Many more are detailed in the *Handbook of fluorescent probes and research chemicals* (44).

Table 1. Commercially available kits for cytotoxicity assays

Company	Kit	Method
Lumitech	Vialight	Bioluminescent detection of ATP.
Lumitech	Apoglow	Measures changes in ADP/ATP ratio.
Stratagene	Quantos Cell proliferation assay kit	Measures fluorescence of a dye when bound to DNA.
Sigma	MTT-based assay kit	Metabolism of tetrazolium salts to insoluble formazan products by mitochondrial enzymes.
Sigma	XTT-based assay kit	Metabolism of tetrazolium salts to soluble (XTT) formazan products by mitochondrial enzymes.
Sigma	Acid phosphatase-based assay kit	Uses *p*-nitrophenol phosphate as substrate.
Sigma	Neutral red-based assay kit	Neutral red taken up by lysosomes and Golgi bodies.
Sigma	Kenacid blue-based assay kit	Binds to protein.
Sigma	Sulforhodamine B-based assay kit	Binds to protein.
Sigma	Lactate dehydrogenase-based assay kit	Measures membrane integrity.
Sigma	Cell Census Plus™ system for cell proliferation analysis using flow cytometry	Cells pre-labelled with fluorescent dye. After drug exposure immunostaining can be used to detect drug effects on specific subsets of cells.

Table 1. Continued

Company	Kit	Method
Sigma	Adenosine 5'-triphosphate (ATP) bioluminescent assay kit	Bioluminescent detection of ATP.
Pharmingen	Annexin-V-FITC apoptosis detection kit	Detection of phosphatidyl serine in outer leaflet of plasma membrane in apoptotic cells using flow cytometry.
Pharmingen	APO-BRDU kit	Two colour staining method for labelling DNA breaks and total cellular DNA using flow cytometry.
Pharmingen	APO-DIRECT kit	Single step staining method for labelling DNA breaks to detect apoptotic cells using flow cytometry.
Pharmingen	APO-ENHANCE kit	Single colour assay for labelling total cellular DNA to detect apoptotic cells using a single parameter DNA histogram.
Pharmingen	Absolute-S™ kit	Two colour staining method for flow cytometry analysis of DNA replication and cellular DNA content/cell cycle position using BrdU and SBIP (strand break-induced photolysis).
Amersham Pharmacia Biotech	Cytostar-T™ Scintillating microplates	Microplates have scintillant incorporated into a transparent base. It uses scintillation proximity assay (SPA) principle. Radiolabelled whole cells can be counted with minimal processing using appropriate liquid scintillation counter.
Amersham Pharmacia Biotech	Cell proliferation ELISA	Measures cell proliferation by quantifying incorporation of BrdU into DNA.
DCS, Innovative Diagnostik Systeme	TCA-100 assay kit	Bioluminescent detection of ATP.
Molecular Probes	CyQUANT™ cell proliferation assay kit	Uses proprietary CyQUANT GR dye, which shows strong fluorescent enhancement when bound to nucleic acids.
Molecular Probes	Fluoreporter fluorimetric and colorimetric cell protein assay kits	Uses either sulforhodamine B which binds to basic amino acids, or sulforhodamine 101.
Molecular Probes	Live/dead viability kits (includes cell-mediated cytotoxicity, viability/cytotoxicity, sperm viability)	Uses calcein-AM and ethidium homodimer to differentially stain live and dead cells, or different fluorescent probes for different contexts.
Oncogene Research Products	Annexin V–biotin kit Annexin V–FITC kit	Detection of apoptosis using fluorescent cytometry/microscopy.

4.2.2 Respiration and glycolysis

Drug-induced changes in respiration (oxygen utilization) and glycolysis (carbon dioxide production) have been measured using Warburg manometry (45, 46). Other authors have determined dehydrogenase activity by incorporating methylene blue into agar containing drug-treated cells, cell death being indicated by non-reduction of the dye (47). The latter method has the disadvantage of being non-quantitative, whilst the former, although quantitative, has not been widely adopted because the technical manipulations involved are extensive and unsuited to multiple screening. The more direct approach of monitoring pH changes in cultures containing an appropriate pH indicator has also been described (48).

4.2.3 Radioisotope incorporation

Measurement of the incorporation of radiolabelled metabolites is a frequently used end-point for cytotoxicity assays of intermediate and short-term duration.

i. Nucleotides

Measurement of [^{3}H]thymidine incorporation into DNA and [^{3}H]uridine incorporation into RNA are commonly used methods of quantitation of drug cytotoxicity (13, 20, 21, 23). In short-term assays, which do not include a recovery period, there are a number of disadvantages, all of which relate to a failure of [^{3}H]thymidine incorporation to reflect the true DNA synthetic capacity of the cell. These are:

(a) Changes may relate to changes in size of the intracellular nucleotide pools rather than changes in DNA synthesis.

(b) Some drugs such as 5-fluorouracil and methotrexate which inhibit pyrimidine biosynthesis (*de novo* pathway) cause increased uptake of exogenous [^{3}H]thymidine due to a transfer to the 'salvage' pathway, which utilizes preformed pyrimidines.

(c) Continuation of DNA synthesis in the absence of [^{3}H]thymidine incorporation can occur (49).

The low labelling index of human tumours with resultant low levels of nucleotide incorporation in short-term assays necessitates the use of high cell densities, which can restrict the number of drugs and range of concentrations tested when cell numbers are limited. Two 'hybrid' techniques have been reported, which combine the advantages of the soft agar culture system with the facilitated quantitation offered by the use of radioisotopes. Both assays are of intermediate duration (about four days) and use [^{3}H]thymidine incorporation into DNA as an end-point. In one method the cells are grown in liquid suspension over soft agar (50), whilst in the other the cells are incorporated in the soft agar (51). Given that a homogeneous cell population is available, [^{3}H]nucleotide incorporation can be used after an appropriate recovery period to measure survival or, in the presence of drug, to measure an antimetabolic effect, but with the reservations expressed above.

ii. [^{125}I]Iododeoxyuridine ([^{125}I]Udr)

[^{125}I]Udr is a specific, stable label for newly synthesized DNA which is minimally reutilized and can therefore be used over a 24 h period to measure the rate of DNA synthesis (52); quantitation is facilitated because the isotope is a gamma emitter. Disadvantages include its variable toxicity to different cell populations, which therefore means that more cells are required because [^{125}I]Udr must be used at low concentrations.

iii. [^{32}P]Phosphate (^{32}P)

The rate of release of ^{32}P into the medium from pre-labelled cells is a function of the cell type and is increased in damaged cells. This has been used as a measure of drug efficacy (53). The incorporation of ^{32}P into nucleotides has also been used as an index of drug cytotoxicity (54). Neither method has been routinely adopted.

iv. [^{14}C]Glucose

Glucose incorporation is used as a cytotoxicity end-point because it is a precursor which is common to a number of biochemical pathways (55). The method has not been widely used.

v. [^{3}H]Amino acids

Protein synthesis is an essential metabolic process without which the cell will not survive, and incorporation of amino acids into proteins has been used as an index of cytotoxicity. The most extensive studies have utilized monolayers of cells growing in microtitration plates, using either incorporation of [^{3}H]leucine (39) measured by liquid scintillation counting, or [^{35}S]methionine incorporation, measured using autofluorography (56).

vi. ^{45}Calcium (^{45}Ca)

Unrelated compounds may produce alterations in the permeability of cell membranes to calcium, such that increased calcium uptake results. Measurements of ^{45}Ca uptake can therefore be used (57).

4.2.4 Total protein content

Protein content determination is a relatively simple method for estimating cell number. It is particularly suited to monolayer cultures, and has the advantage that washed, fixed samples can be stored refrigerated for some time before analysis without impairment of results, facilitating large scale screening. Overestimation of cell number may arise with drugs which inhibit replication without inhibiting protein synthesis (e.g. BrdU, methotrexate). Assessment of cytotoxicity requires the demonstration of an alteration in the accumulation of protein per culture over time, preferably at several points, or at one point after prolonged drug exposure and recovery, as described above.

4.2.5 Colorimetric assays

The advent of sophisticated microplate readers which allow rapid quantitation of colorimetric assays has paralleled the development of a variety of assays which use some form of colour development as an end-point for quantitating cell number. These include methods which reflect:

- protein content (methylene blue, Coomassie blue, Kenacid blue, sulforhodamine B, Bichinoninic acid)
- DNA content (Hoechst 33342) or DNA synthesis (BrdU uptake)
- lysosome and Golgi body activity (neutral red)
- enzyme activity (hexosaminidase, mitochondrial succinate dehydrogenase)

Linear relationships between end-point and cell number have been demonstrated for all these methods. Discrimination between live and dead cells in monolayer assays is not relevant, since dead cells will usually detach, given sufficient time. In suspension cultures this aspect is relevant, however, and needs to be considered when choosing an appropriate assay. Methods which make use of fluorescent dyes have increased in the past five years and bioluminescent assays have also been developed for measuring ATP levels.

i. Protein content

Several methods are available for measuring the protein content of cell monolayers. These include the use of the Folin–Ciocalteau reagent according to the method of Lowry (58), and amido black (6). Several new techniques for colorimetric determination of protein content are also available. These include methylene blue (59, 60), sulforhodamine B (61, 62), Kenacid blue (63), Coomassie blue G-250 (64), and Bichinoninic acid (BCA) (65–67).

ii. DNA content

DNA content may be measured in microtitration plates by staining with dyes whose fluorescence is enhanced by intercalation at AT-specific sites on the chromatin, such as Hoechst 33342 and 2-diamidinophenylindole (DAPI) (68). DNA synthesis may also be determined using bromodeoxyuridine (BrdU). The amount of BrdU incorporated is detected immunohistochemically using a monoclonal antibody to BrdU, and the binding may be quantitated using appropriate conjugates and chromogenic substrates (69).

iii. Lysosomal and Golgi body activity

The uptake of neutral red by lysosomes and Golgi bodies has been used to quantitate cell number (70, 71). The stain appears to be specific for viable cells, but the main limitation of the method is the difference in uptake between cell types. Thus, some cell types, such as activated macrophages and fibroblasts, take up large amounts very rapidly whereas others, such as lymphocytes, show negligible staining.

iv. Tetrazolium dye reduction

The most widely used technique involves the use of a tetrazolium salt (e.g. MTT, 3,(4,5-dimethylthiazol-2-yl)-2,5-diphenyl tetrazolium bromide) which is metabolized to an insoluble coloured formazan salt by mitochondrial enzyme activity in living cells. The method was first described in 1983 (72) as a rapid colorimetric method for immunological studies, and modifications for this application have been described (73). The technique is particularly useful for assaying cell suspensions because of its specificity for living cells. One disadvantage is the need to use unfixed cells, which may impose time restraints. The potential of the technique for drug sensitivity testing of human tumours was recognized, and early reports of its use are promising (74–78). More recent publications describe some technical problems which affect interpretation of the assay results and need to be considered when planning protocols. These include the possibility of increased mitochondrial enzyme activity in drug-treated cells and the effect of medium conditioning by cells on formazan production (79). The absorption spectrum of MTT–formazan has also been shown to be dependent upon pH and cell density (80), and a method has been described which overcomes these problems and increases linearity between MTT–formazan and cell density especially at high cell numbers (80). Another derivative, XTT (2,3-bis (2-methoxy-4-nitro-5-sulfophenyl)-5-[(phenylamino)carbonyl]-2H-tetrazolium hydroxide) (81) produces a soluble formazan end-product and is therefore easier to use than MTT because there is no need to carry out the solubilization procedure. The XTT derivative has been used successfully for quantitating colony formation in soft agar (82).

v. Luminescence-based cell viability testing

Levels of ATP in a cell population provide a sensitive indication of cell viability. The luminescence reaction used to determine ATP levels is based on the following reaction.

$$ATP + \text{D-luciferin} + O_2 \xrightarrow[Mg^{2+}]{\text{Luciferase}} AMP + 2P + CO_2 + \text{light}$$

The sensitivity range has been reported as 20 cells/ml to 2×10^7 cells/ml (83). A number of commercial kits are available for measuring ATP by luminescence (see *Table 2*).

vi. Apoptosis

Many anticancer drugs kill cells by apoptosis (84), and measurement of apoptosis is therefore important in the evaluation of cytotoxicity. Apoptosis can be determined in a number of ways (see Chapter 11), including:

- morphological criteria
- DNA laddering
- detection of phosphatidyl serine in the outer plasma membrane using annexin V conjugated to FITC or biotin (see *Tables 1* and *2*)

CYTOTOXICITY AND VIABILITY ASSAYS

Table 2. Fluorescent probes used to investigate membrane integrity[a]

Dye	Characteristics
Calcein-AM	Membrane-permeant esterase substrate; cleaved by esterase in living cells to yield calcein, which fluoresces green in the cytoplasm.
Ethidium-homodimer	High affinity red fluorescent nucleic acid stain which penetrates membrane of dead but not live cells.
Propidium iodide	Red fluorescent nucleic acid stain which penetrates membrane of dead but not live cells.
DiOC$_{18}$	Green fluorescent membrane stain which can be used in combination with propidium iodide for cell-mediated cytotoxicity assays studying membrane integrity.
Chromatide™ BODIPYF®L-14-dUTP (Molecular Probes)	Fluorescent labelled nucleotide used in TUNEL assay.
Annexin V–FITC	Annexin V binds to phosphatidyl serine and fluorescent label allows demonstration of transfer of phosphatidyl serine from inner to outer plasma membrane which is a feature of apoptosis.
CyQUANT™ GR (Molecular Probes)	Proprietary green fluorescent dye which shows enhanced fluorescence when bound to cellular nucleic acids (see *Table 1*).

[a] See *Table 1* and ref. 44 for more specific details about fluorescent probes.

- TUNEL (terminal deoxynucleotidyl transferase-mediated dUTP nick end-labelling) assay

4.3 Survival (reproductive integrity)

Survival assays give a direct measure of reproductive cell death by measuring plating efficiency either in monolayer or in soft agar. The end-points described in the previous section can also be used as an index of reproductive integrity providing the design of the assay incorporates a recovery period.

4.3.1 Cloning in monolayer

Cells generally have a higher cloning efficiency in monolayer than in soft agar. Normal cells and tumour cells will form colonies in monolayer, and the method is not therefore applicable to tumour biopsy material, unless criteria are available to discriminate between tumour and non-tumour cells. Feeder layers of irradiated or mitomycin C-treated cells can be used to increase plating efficiencies, and indeed small drug-resistant fractions are more likely to be detected in the improved culture conditions existing when feeder cells are used (85).

4.3.2 Cloning in soft agar

The advantages and disadvantages of this method have been discussed in Section 3.1.5.

4.3.3 Spheroids

Various methods can be used to quantitate the effect of drugs on spheroid growth. These are:

- relative changes in volume of treated and untreated spheroids
- cloning efficiency in soft agar of disaggregated spheroids
- cell proliferation from spheroids adherent to culture surfaces

The first method is rather insensitive because spheroid growth tends to plateau, and the second may be affected by difficulties with disaggregation to a single cell suspension and low plating efficiencies.

4.3.4 Cell proliferation

Growth curves may be determined and the doubling time during exponential growth derived. An increase in doubling time is taken as an indication of cytotoxicity, but it must be stressed that this is a kinetic measurement averaged over the whole population and cannot distinguish between a reduced growth rate of all cells and an increase in cell loss at each cell generation. Estimates of cytotoxicity based on cell growth in mass culture must utilize the whole growth curve, or they may be open to misinterpretation. If 50% of cells die at the start of the experiment, the growth rate of the residue, determined in log phase, may be the same, but will show a delay. In practice, it is difficult to distinguish between early cell loss and a prolonged lag period. Cell growth rates must be taken as only a rough guide to cytotoxicity, and accurate measurements of cell survival and cell proliferation should be made by colony-forming efficiency (survival) and colony size (proliferation).

5 Assay comparisons

In spite of the diversity of methodologies used for cytotoxicity and viability testing, approximately the same levels of correlation between *in vitro* sensitivity and *in vivo* response have been reported when the methods have been applied to human tumour biopsy material. A number of comparative studies have been undertaken, which generally indicate that appropriately designed cytotoxicity assays give results which are comparable to clonogenic assays (24, 25, 33, 40, 50, 75, 85, 86). The relative merits of clonogenic and non-clonogenic assays have been discussed and reviewed (87).

6 Technical protocols

6.1 Drugs and drug solutions

6.1.1 Drug sources

Pharmaceutical preparations for intravenous administration frequently contain additives which may be cytotoxic. Such preparations are therefore not suitable for aliquoting by weight and, if they are used as a stock solution, the cytotoxicity of the additional components should be determined. This problem can be avoided by obtaining pure compounds from the drug manufacturers.

6.1.2 Storage

It is recommended that dry compounds be stored at $-20\,°C$ to $-70\,°C$ over desiccant; this is especially important with compounds which are unstable in aqueous solution. It is routine practice to make up stock drug solutions which are then aliquoted and stored at $-70\,°C$. Storage at higher temperature is not recommended, and some drugs (nitrosoureas) are unstable even under these conditions (88). As some drugs may bind to conventional cellulose nitrate or acetate filters, sterile filtration must be carried out under controlled conditions to check for binding of drug. Use the maximum drug concentration and maximum volume to saturate the filter within the first few millilitres of filtrate, which may then be discarded. Alternatively, use non-absorbent filters, e.g. nylon. In either case the filtrate should be assayed to make sure no activity has been lost. In practice it has been found that handling of non-pharmaceutical preparations of pure drugs under aseptic conditions (e.g. use of sterile blade for weighing out into sterile container) is sufficient to prevent contamination, even in experiments of prolonged duration.

6.1.3 Diluents

i. Solvents

Solvents commonly used for compounds which are not soluble in aqueous solution include ethanol, methanol, and dimethyl sulfoxide (DMSO). Different cell populations exhibit different sensitivities to these organic solvents, and appropriate solvent controls should therefore always be included. A minimum dilution factor of 1000 in aqueous solution is recommended to avoid solvent toxicity.

ii. Medium components

Certain drugs bind avidly to serum proteins (e.g. cisplatinum), and the presence of serum in the incubation medium may therefore reduce the amount of available drug. Components of some media can protect cells against the cytotoxic effects of antimetabolites (e.g. thymidine and hypoxanthine protect against methotrexate, thymidine protects against 5-bromodeoxyuridine and 5-fluorodeoxyuridine). The presence of amino acids may also affect uptake of drugs that use the amino acid transport system. A comprehensive list has been detailed elsewhere (6).

6.1.4 Drug activation

Many components that are not themselves cytotoxic are converted to cytotoxic metabolites by the P450 mixed oxidase system of the liver. Active metabolites for drugs known to require metabolic conversion may be available, e.g. phosphoramide mustard and 4-hydroperoxycyclophosphamide for cyclophosphamide (89) but for drugs with less clearly characterized pathways of conversion some method of *in vitro* activation is necessary. Methods include the use of S9 microsome fractions, cultured rat or human hepatocytes, or liver biopsy material (90–93). The use of intact cells, in which the level of cofactors resemble those *in vivo* and are high enough to sustain the associated reactions, provides a closer approximation to the *in vivo* situation, and may give results which are different to those obtained using liver homogenates. Species differences exist in the cytochrome P450 system, and for this reason human hepatocytes may be preferred. The major problem with human hepatocyte culture is loss of cytochrome P450 activity with increasing time in culture, though appropriate conditions can reduce this (94).

6.2 Drug incubation

It is common procedure to incubate cells with drug solutions immediately after enzyme disaggregation of solid tissue, or harvesting of cell monolayers by trypsinization. There is evidence to suggest that susceptibility of cells to drug is altered by enzyme treatment and does not return to control levels until approximately 12 h after enzyme exposure (95). It may therefore be expedient to include a pre-incubation recovery period for freshly disaggregated cells to allow for this.

Maintenance of pH at 7.4 is essential during the incubation period since alterations in pH will alter cell growth, and alkaline pH particularly will reduce cell viability.

Comparison of static versus non-static incubations of cell suspensions revealed marked differences in dose–response profiles (96), and cells should therefore be kept in continuous suspension to allow equal drug distribution. If the surface area:depth ratio of the incubation vessel is small (e.g. test-tubes) incubation in a water-shaker bath will not keep the cells in suspension, and intermittent shaking by hand is recommended.

6.3 Assay by survival and proliferative capacity

6.3.1 Clonogenicity

One of the most generally accepted methods of assaying for survival is the measurement of the ability of cells to form colonies in isolation. This is usually achieved by simple dilution of a single cell suspension, and determination of survival by counting the colonies that form. A lower threshold must be set in line with the doubling time of the cells being studied and the total duration of the assay, usually five or six doublings (32 or 64 cells/colony).

As some drugs may have an effect on cell proliferation as well as, or instead of, survival *per se*, it may be necessary to do a colony size analysis as well. This may be done by counting the number of cells per colony (very tedious and only possible in small colonies), by measuring the diameter (prone to error if cell size or degree of piling up changes), or by measuring absorbance of colonies stained with 1% crystal violet.

CYTOTOXICITY AND VIABILITY ASSAYS

Protocol 1
Monolayer cloning

Equipment and reagents
- Colony counting equipment
- Methanol or 2.5% glutaraldehyde (10 ml of 25% glutaraldehyde plus 90 ml PBS)
- 1% crystal violet (1 g crystal violet in 100 ml distilled water)

Method

1. Prepare replicate 50 ml (25 cm^2) flasks, two for each intended concentration of drug, and two for controls.
2. When cultures are at the required stage of growth (usually mid-log phase but, in special circumstances, the plateau phase of the growth cycle) add the drug[a] to the test and solvent to the control for 1 h at 37 °C.
3. Remove the drug, rinse the monolayer 3 × with PBS, and prepare a single cell suspension by conventional trypsinization (see Chapter 1).
4. Count the cells in each suspension and dilute to the appropriate cell concentration to give 100–200 colonies[b] per 5–6 cm diameter Petri dish.
5. Plate the appropriate number of cells per dish and place in a humid incubator at 37 °C, gassed with 5% carbon dioxide in air.
6. Grow until the colonies are visible and colony size in control plates exceeds the threshold value.[c]
7. Rinse the dishes with phosphate-buffered saline (PBS), then fix the cells in methanol or 2.5% glutaraldehyde, and stain with 1% aqueous crystal violet for 10 min. Rinse the dishes in tap-water to remove unbound dye and air dry.
8. Count the number of colonies above the threshold diameter in drug-treated and control dishes. Then calculate the plating efficiency of the drug-treated cells as a fraction of the control plating efficiency. Plot the plating efficiency on a log scale against drug concentration.

[a] Drug treatment is best performed before subculture for plating, where highly toxic substances are being tested. For low-grade toxins, where chronic application is required, they may be applied 24–48 h after plating, and retained throughout the clonal growth period, provided they are stable.

[b] This figure depends on the plating efficiency of the cells and the effect of the drug (e.g. control plates with no drug from a cell culture of a known plating efficiency of 20% will require 500–1000 cells per dish; at the ID$_{50}$ of the drug they will require 1000–2000 cells per dish, and at the ID$_{90}$, 5000–10 000 cells per dish). A trial plating should be done first, to determine the plating efficiency of the cells, and to determine approximately the ID$_{50}$ and ID$_{90}$ of the drug. In practice it is more usual to set up dishes at two cell concentrations, one to give a satisfactory number of colonies at low drug concentrations and controls, and one for higher drug concentrations, with some overlap in the middle range. Experience will usually determine where this is likely to fall.

> **Protocol 1** continued
>
> [c] For rapidly growing cells (15–24 h doubling time) this will take seven to ten days; for slower growing cells (36–48 h doubling time) two to three weeks are required. In general, for a survival assay the colonies should grow to 1000 cells or more on average (ten generations). This will be the threshold value used when counting colonies. As the colonies increase in size the growth rate (particularly of normal cells) will slow down as the colonies tend to grow from the edge, and the slower growing colonies will tend to catch up. Hence if cell proliferation (colony size) is the main parameter, clonal growth should be determined at shorter incubation times giving smaller colonies with a wider size distribution.

Protocol 2

Clonogenicity in soft agar using a double layer agar system

Equipment and reagents

- Petri dishes
- Colony counting equipment
- Agar
- Growth medium
- Saline

Method

1. Prepare 35 mm Petri dishes with a 1 ml base layer of 0.5% agar in growth medium by mixing 1% agar (melted by autoclaving[a] or boiling) at 45 °C with an equal volume of double-strength culture medium at 45 °C and dispensing 1 ml aliquots in a preheated pipette.[b] Allow four dishes per test condition and 3 × 4 sets of controls.

2. Prepare a single cell suspension of the target cell population, adjust the cell concentration in growth medium to give 20 times the final concentration desired at plating, and store at 4 °C until ready to use.

3. Prepare the selected drug dilutions in growth medium and aliquot out 900 μl of each concentration into duplicate tissue culture grade tubes, including control tubes containing growth medium only and also appropriate solvent controls.[c]

4. Check that the stock cell suspension still comprises single cells, and add 100 μl to each of the prepared tubes.

5. Incubate the tubes at 37 °C for 1 h (see Section 8.2).

6. At the end of the incubation period centrifuge the tubes (2 min, 100 g) and wash the cells in 5 ml of saline. Repeat once more, and resuspend the cell pellet in 2 ml of growth medium. Keep the cells on ice to maintain cell viability whilst replicates are plated out.

7. Centrifuge the duplicate set of tubes for one drug concentration; remove medium and add 2 ml of warmed growth medium containing 0.3% agar to each tube. Disperse cells evenly and plate out 1 ml aliquots on to each of four prepared bases.[d]

Protocol 2 continued

8. Put the dishes to solidify on a cooled, horizontal surface.
9. Repeat for all test conditions, including controls at appropriate intervals throughout.
10. Incubate the plates in an humidified atmosphere of 95% air/5% carbon dioxide at 37 °C.
11. Score the plates for colonies when control colonies have reached a predetermined size. This is usually more than 50 cells for cell lines, but a size of more than 20 or 30 cells has been used when the population has a slow growth rate and a low plating efficiency, as with human tumour biopsies.

[a] Autoclaved agar may be toxic (97).

[b] Pre-warm the pipette by taking up 5 ml of warm culture medium and ejecting it before taking up the agar solution. Add 2 ml of melted agar into 18 ml of warmed medium and mix well by drawing a few millilitres up into the pipette and then releasing it. Avoid introducing air bubbles. Pipette 5 × 1 ml into five dishes and immediately swirl the dishes gently by hand to ensure that the agar covers the entire bottom of the dish before it sets. Again, avoid introducing air bubbles. Repeat, doing five dishes at a time. This volume is sufficient for about 18 dishes. If more are needed, repeat the procedure with more 18 ml aliquots. If larger volumes are used, the agar will set before all of it has been plated out.

[c] Because of the time-span involved when plating out multiple drugs and concentrations, additional controls are recommended for plating out at the beginning, middle, and end of setting up the experiment. The controls thus incorporate variability due to the time involved in setting up the assay.

[d] The final plating out is most easily accomplished using a 1 ml micropipette. If about 2 mm is cut off the end of the tips this prevents problems due to blockage of the small aperture by solidified agar.

6.3.2 Modifications

i. The 'Courtenay' method

The 'Courtenay' method for suspension cloning (96) utilizes rat red blood cells as feeder cells, and a 5% oxygen tension. The procedure outlined above may be used, with appropriate modification of final plating conditions.

ii. Feeder cells

A linear relationship between plating efficiency and plated cell numbers may not occur from low to high cell numbers. This problem can be circumvented by incorporating homologous feeder cells in the assay which have either been lethally irradiated with a ^{60}Co or ^{137}Cs source, or treated for approx. 12 h with 10 μg/ml cells of mitomycin C. The radiation dose needs to be established for each cell type, e.g. 20–30 Gy for lymphocytes, 60 Gy for lymphoid cell lines.

iii. Use of 2-(p-iodopheny)-3-(p-nitropheny)-5 phenyltetrozolium chloride (INT)

Viable cells reduce colourless tetrazolium salts to a water insoluble coloured formazan product, and the reaction has been used to distinguish viable colonies from degenerate clumps when scoring. The stain is made by dissolving INT violet in buffered saline to a final concentration of 0.5 mg/ml; dissolution is slow and the stain needs to be prepared 24 h prior to use. Add 0.5–1 ml to each 35 mm Petri dish and incubate overnight at 37 °C. Viable colonies then stain a reddish-brown colour.

iv. 20% oxygen versus 5% oxygen

The lower oxygen tension, which more closely resembles physiological levels of oxygen, is recommended for clonogenic assays because it results in higher cloning efficiencies. There is evidence to suggest that it also modifies the chemosensitivity profile of the cell population (98), producing enhanced cytotoxicity.

6.3.3 Spheroids

The experimental protocols outlined below are based on techniques described elsewhere (99). Three end-points may be used to determine the cytotoxic effect of drugs on spheroids; these are:

- volume growth delay
- clonogenic growth
- outgrowth as a cell monolayer

Pre-selection of similarly sized spheroids and drug incubation is a common starting point.

Protocol 3
Spheroid incubation

1. Pre-select spheroids in the chosen size range.[a]
2. Incubate the spheroids with the chosen drug concentrations, solvent controls, and medium controls.[b]
3. At the end of the incubation period rinse the spheroids in two to three 5 min washes of drug-free medium and proceed to the appropriate protocol for the selected end-point.[c]

[a] Sizes in the range 150–250 μm are just visible to the naked eye and can be selected using a Pasteur pipette.

[b] The method of incubation with drug depends to some extent on the exposure period to be used. For 1 h exposures, which have been most commonly utilized, incubate spheroids in agar-coated (0.5–1%) Petri dishes or in glass Universal containers. For longer exposure times, incubate the spheroids in a spinner vessel which will keep them in continuous suspension and prevent adherence to the vessel walls.

[c] Treated spheroids can be grown as a mass culture and the volume of individual randomly chosen spheroids determined at set time intervals. However, from a statistical viewpoint it is recommended that successive measurements are made on individually isolated spheroids placed in either 24-well or 96-well multidishes with agar-coated bases. The smaller wells can be used for spheroids up to 600 μm in diameter.

CYTOTOXICITY AND VIABILITY ASSAYS

Protocol 4
Volume growth delay

Equipment

- Inverted microscope with eyepiece graticule

Method

1. Plate out 12–24 spheroids per drug concentration and 24 controls into individual wells over a base layer containing 1% agar in culture medium.[a]

2. At two or three day intervals measure two diameters (x, y) at right angles to each other, using an eyepiece graticule in an inverted microscope.[b]

[a] For a 24-well plate use 0.5 ml of 1% agar for the base and add the spheroids in 1 ml of growth medium.

[b] Calculate the size of the spheroid as 'mean diameter' ($\sqrt{x,y}$) or as 'volume'

$$\frac{4}{3}\pi\left(\frac{\sqrt{xy}}{2}\right)^3$$

and use to construct growth curves for drug-treated and control spheroids. Results are usually normalized to pre-treatment spheroid size and expressed as V_t/V_0 where V_t is the volume at time t and V_0 is the initial volume. An example is shown in *Figure 5* for spheroids derived from the xenograft of an adenocarcinoma of the lung (100). The method assumes that the spheroid is symmetrical, but flattening of spheroids to a dome shape has been observed for some cell lines, and this will lead to overestimation of volume.

Protocol 5
Clonogenic growth of disaggregated spheroids

Equipment and reagents

- 0.25% trypsin, or 0.125% trypsin plus 1000 U/ml type I CLS grade collagenase (Worthington)
- See *Protocol 1*

Method

1. Disaggregate the spheroids to a single cell suspension[a] using for example 0.25% trypsin, or 0.125% trypsin plus 1000 U/ml type I CLS grade collagenase.

2. Adjust the cell number to 20–200/ml for cell lines, or up to 10^3/ml for primary cloning. If cloning in suspension use 10^4/ml for cell lines and up to 5×10^5/ml for primary cloning.

3. Follow *Protocol 1* for cloning in monolayer.

[a] Obtaining a pure single cell suspension may be virtually impossible, and plating efficiencies may be very low. The method described in *Protocol 6* circumvents these problems.

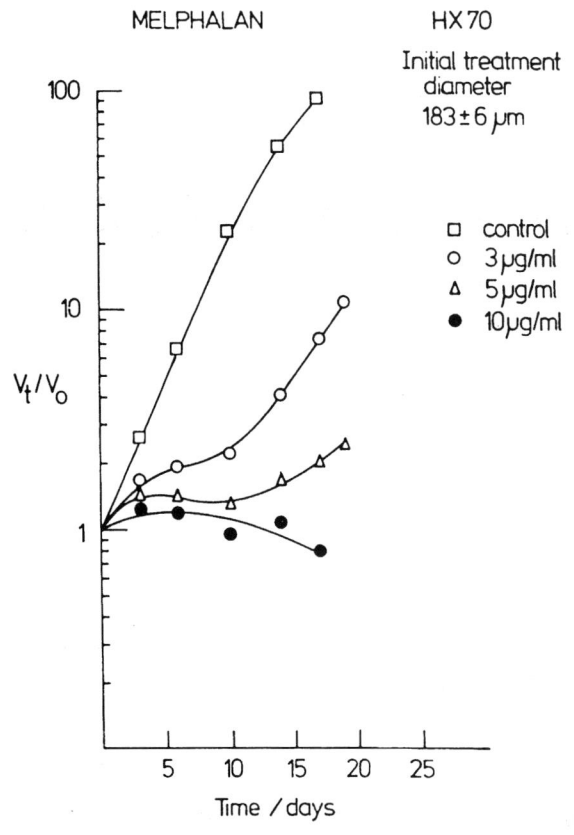

Figure 5. The effect of increasing concentrations of a 1 h exposure of melphalan on the growth of adenocarcinoma lung-derived spheroids. V_t, volume at time interval shown; V_0, volume at start of experiment. Significant differences ($p < 0.001$; Student's t test) were observed at day 14 between control and 3.0 µg/ml, and 3.0 µg/ml and 10.0 µg/ml. The diameter shown on the figure is the mean diameter ± s.e. of the plated spheroids. (Reproduced with permission of the publishers from ref. 100.)

Protocol 6
Outgrowth as cell monolayer from intact spheroids

1. Allow the individual spheroids to attach to individual wells of tissue culture plastic microtitration plates.
2. After two to three weeks remove the central spheroid and determine the number of cells in the monolayer.[a]

[a] This may be done directly by trypsinizing and counting the detached cell population. Alternatively, the methodology described for quantitation of drug effect on cell monolayers in *Protocols 8* and *9* could be adopted.

6.4 Cytotoxicity assays

The following methods have all been developed as automated cytotoxicity assays in microtitration plates. The same procedure for drug exposure of cells can be used as a starting point for all the methods described.

Protocol 7
Preparation of microtitration plates

Equipment and reagents
- Microtitration plates
- Growth medium
- Drugs
- Saline

Method
1. Add 200 µl of cells in growth medium to each well of a microtitration plate.[a]
2. When the cells are in exponential growth,[a] remove the medium and add drugs serially across the plate using replicates of three to six wells per concentration and including appropriate solvent controls. Add fresh growth medium to control wells.
3. Incubate the plates for selected drug exposure times at 37°C in 5% carbon dioxide/ 95% air, re-feeding with fresh medium or drugs at appropriate intervals.
4. Remove the medium and wash wells three times with 200 µl of saline to remove drugs.[b] Add 200 µl of fresh growth medium to each well.
5. Incubate plates for selected recovery period, changing the medium if necessary.
6. Proceed to the selected end-point.

[a] Evaporation from outer wells may be excessive (edge effect) resulting in poor cell growth, and it is therefore recommended that a 10 × 6 matrix is used for the cytotoxicity assay. Edge effects due to evaporation can be reduced by using plate sealers (Mylar, Flow Labs). Some drugs are volatile, or give off volatile metabolites (e.g. formamides release formaldehyde) and this can cause variable non-specific cytotoxicity in adjacent wells. The use of sealers minimizes this effect. Outer wells may be filled with either saline or cell suspension. The appropriate cell concentration depends on the cells used, and varies as a function of growth at low density, cell size, rate of growth, and saturation density. The plating conditions need to be optimized to ensure that cells are in exponential growth phase over the period of the assay. That is, cells are exposed to drugs after lag phase, and the assay is terminated before, or as, controls reach confluence. This needs to be optimized for each cell line used.

[b] Note that it is important to avoid damaging the cell monolayer during wash procedures. This is best done by tilting the plate to 45° and inserting the pipette tip in the angle between the base and side of the well.

The previous methodology using liquid scintillation counting is labour-intensive during processing, and can cause problems when many plates are processed, particularly when β-counting facilities are restricted. Minivials tend to lose scintillant through their walls after some days storage and, although this

Protocol 8

Protein determination of cell monolayers using methylene blue (59, 60)

Equipment and reagents

- Plate shaker (Dynatech)
- Plate washer (optional but easier than washing by hand and more reproducible)
- Plate reader with filter at 660 nm
- 1% (w/v) methylene blue in borate buffer
- 0.01 M borate buffer pH 8.4 (3.8 g/litre in distilled water)
- 0.1 M HCl
- 10% (v/v) formalin in PBS

Method

1. Follow *Protocol 7* for drug incubation and ensure that there is a cell-free row or column of wells for use as a reagent blank.
2. Remove medium from wells and wash twice with 200 µl of PBS including wells without cells for reagent blank.
3. Add 200 µl of 10% (v/v) formalin in PBS and fix cells at room temperature for 10 min.
4. Remove formalin, and wash cells twice with 200 µl of borate buffer.
5. Add 100 µl of methylene blue per well and stain for 10 min at room temperature.
6. Remove methylene blue and wash extensively with borate buffer, blotting the plate surface between washes.[a]
7. Leave washed plates to air dry at room temperature for 2-3 h.
8. Add 200 µl of 0.1 M HCl per well and shake the plate for 15 min at room temperature using a plate shaker to solubilize the stain.
9. Read optical density in microplate reader (Dynatech, Molecular Devices) at 660 nm (maximum absorbance).

[a] A minimum of five washes is needed to remove excess dye.

Protocol 9

Protein determination of cell monolayers using sulforhodamine B (SRB) (61, 62)

Equipment and reagents

- Microplate reader with a dual wavelength facility and filters at 540 nm and 630 nm
- 10% trichloroacetic acid (TCA): 10 g trichloroacetic acid in 100 ml of distilled water
- 0.05–0.1% SRB[a] in 1% acetic acid
- 10 mM unbuffered Tris
- 1% acetic acid

Protocol 9 continued

Method

1. Remove medium from the cells and fix in 10% TCA for 30 min at 4°C.
2. Wash cells five times with tap-water.
3. Add 100 µl of SRB[a] (0.05-0.1%) in 1% acetic acid and stain for 15 min at room temperature.
4. Wash cells four times in 1% acetic acid.
5. Air dry plates and dissolve stain in 200 µl 10 mM unbuffered Tris.
6. Measure the absorbance at 540 nm, using a reference filter of 630 nm in a microplate reader with a dual wavelength facility.

[a] The concentration of SRB needs to be optimized for each batch of dye.

Protocol 10

Protein determination of cell monolayers using Kenacid blue (63, 101)

Equipment and reagents

- Microplate reader with a dual wavelength facility and filters at 404 nm and 577 nm
- Kenacid blue solution: 0.4 g Kenacid blue (BDH) in 250 ml ethanol and 630 ml water (stock); immediately before use add 12 ml of glacial acetic acid to 88 ml of stock solution and filter
- Glacial acetic acid/ethanol/water (1:50:49, by vol.)
- Ethanol/glacial acetic acid/water (10:5:85, by vol.)
- Destain solution: 1 M potassium acetate in 70% ethanol

Method

1. Wash cells three times with PBS and fix in 150 µl glacial acetic acid/ethanol/water for 20 min at room temperature.
2. Remove fixative and stain with 150 µl Kenacid blue solution. Stain for 20 min at room temperature with gentle agitation.
3. Remove unbound stain using three washes with ethanol/glacial acetic acid/water. The final wash should be done for 20 min with agitation.
4. Add 150 µl of destain solution per well and agitate rapidly for 20 min until homogeneous colour distribution is obtained.
5. Measure absorbance at 577 nm against a reference filter of 404 nm.

can be reduced by storing at 4°C prior to counting, it does mean that samples should be counted as soon as possible after preparation. An alternative method, which is well suited to automation, has been developed by Freshney et al. (56).

Protocol 11

Determination of amino acid incorporation into cell monolayers (39, 86)

Equipment and reagents
- Minivials
- Beta counter
- L-4,5 [^{3}H]leucine (Amersham plc)
- 1 M NaOH
- PBS
- Methanol
- 10% trichloroacetic acid (TCA): 10 g in 100 ml distilled water
- 1.1 M HCl
- Scintillant (e.g. Fisofluor, Fisons)

Method

1. At the end of the assay period add 50 μl of 5–20 μCi (185–740 kBq)/ml [^{3}H]leucine in growth medium to each well and incubate for 3 h at 37°C.
2. Remove the isotope and wash the cell monolayer three times in PBS.[a]
3. Add 100 μl of PBS and 100 μl of methanol to each well, remove, and add 100 μl of methanol per well. Fix the cells for 30 min at room temperature.[b]
4. Remove methanol and air dry the plates.[c]
5. Put plates on to ice and wash the monolayers in three 5 min washes of ice-cold 10% trichloroacetic acid (TCA).[d]
6. Wash off TCA with methanol, using three washes from a wash bottle, and air dry the monolayer.
7. Add 100 μl of 1 M NaOH to each well; leave overnight at room temperature to solubilize protein.
8. Transfer NaOH to minivials and add 2.4 ml of scintillant (e.g. Fisofluor, Fisons) to each vial followed by 100 μl of 1.1 M HCl to acidify the contents.
9. Cap the vials, mix well to homogenize and clarify the contents, and count for 5–10 min on a β-counter.

[a] The method can also be used for non-adherent cell suspensions provided the plates are centrifuged at 1000 g for 3–5 min prior to medium removal.
[b] Fixation prevents detachment of the cells during subsequent processing.
[c] Fixed plates can be stored for at least 48 h at 4°C before further processing.
[d] Following fixation, wash solutions can be removed by inversion of the plate and shaking sharply.

Problems associated with the use of the counting systems as described in these protocols have been detailed in a technical review (102).

A modified procedure has been described for measuring [^{3}H]thymidine incorporation into DNA in cells growing in liquid suspension over soft agar in Petri dishes (50).

Protocol 12
Incorporation of [³H]nucleotides into DNA/RNA

Equipment and reagents

- Cell harvester (e.g. Titertek, Dynatech, Inotech)
- 10% TCA
- Emulsifier cocktail scintillant (e.g. Beckman Ready Solv HP)
- 6-[³H]uridine (22 Ci (814 GBq)/mmol); methyl-[³H]thymidine (5 Ci (185 GBq)/mmol); 6-[³H]deoxyuridine (25 Ci (925 GBq)/mmol)

Method

1. Add isotopes to each well to give a final concentration of 2.5 µCi (92.5 kBq)/ml, e.g. 6-[³H]uridine (22 Ci (814 GBq)/mmol)); methyl-[³H]thymidine (5 Ci (185 GBq)/mmol); 6-[³H]deoxyuridine (25 Ci (925 GBq)/mmol) and incubate for 1–3 h at 37°C.
2. Remove the isotope by pipetting into a suitable vessel for disposal of radioactive waste and wash cells in three washes of PBS to remove the unincorporated radio-activity.[a]
3. For cell monolayers, add 100 µl of appropriate enzyme solution and incubate cells at 37°C to detach the cells. Shake plate on a plate shaker for a few minutes to ensure that cell detachment is complete.
4. Harvest the cells on to filter paper using a cell harvester, and wash the filters three times with 10% TCA.
5. Wash filters with methanol to remove water, and air dry.
6. Transfer filters to minivials and add about 3 ml of an emulsifier cocktail scintillant (e.g. Beckman Ready Solv HP). A system for counting dry filters is now available (Inotech).
7. Leave the samples for at least 1 h in the dark before counting.[b]
8. Count the samples for 5–10 min or as long as is required to reduce the counting error to ± 5%.

[a] When using cell suspensions, plates must be centrifuged prior to medium removal as in Protocol 11.

[b] This removes chemiluminescence, which produces artefactually high counts.

Protocol 13
Total DNA synthesis measured by [¹²⁵I]Udr uptake (52)

Equipment and reagents

- Gamma counter
- (2.22 kBq)/ml [¹²⁵I]Udr

Protocol 13 continued

Method

1. Add 0.06 µCi (2.22 kBq)/ml [^{125}I]Udr to the growth medium of monolayer cultures for 24 h.[a]
2. Wash the culture three times in saline and add fresh growth medium.
3. After 6 h, repeat the wash procedure to remove unbound label and label released from lysed dead cells, and count bound ^{125}I activity in a gamma counter.

[a] As for the microtitration assay, confluence should be avoided in control tubes because overcrowding leads to reduced [^{125}I]Udr incorporation, and therefore underestimation of cell kill. When cells were grown as monolayers in inclined test-tubes, the upper limit of the assay was found to be about 2×10^5 cells, and the lower limit about 2×10^4 cells.

Protocol 14

Measurement of viable cell number using neutral red

This protocol is based on a procedure described by Fiennes *et al.* (71).

Equipment and reagents

- Elisa plate reader with a 540 nm filter
- 0.1% (w/v) neutral red: 0.1 g neutral red in 100 ml distilled water, acidify with two drops glacial acetic acid/100 ml. Store in the dark at 4°C for up to three weeks. Immediately before use dilute the stock 1 in 10 ml using warm PBS and check that precipitation of neutral red does not occur by adding a drop to a cold glass slide.
- Glacial acetic acid
- 0.1 M citrate buffer pH 4.2: 21.01 g citric acid, 200 ml of 1 M NaOH per litre (A); 60 ml A + 40 ml of 0.1 M HCl
- Absolute ethanol/0.1 M citrate buffer, mixed 1:1 (v/v)

Method

1. Add 200 µl of neutral red solution to each well of the microtitration plate and incubate for 90 min at 37°C.
2. Remove the dye and wash the cells with two washes of warm PBS.
3. Extract the dye by adding 100 µl of absolute ethanol/0.1 M citrate buffer to each well of the microtitration plate.
4. Agitate the plate for 20 min at room temperature and read optical density at 540 nm.

Protocol 15

Measurement of viable cell number using MTT (80)

Equipment and reagents

- Elisa plate reader with a 570 nm filter
- MTT (Sigma) in PBS at a concentration of 2–5 mg/ml
- Culture medium containing 10 mM Hepes pH 7.4
- Dimethyl sulfoxide (DMSO)
- Sorensen's glycine buffer: 0.1 M glycine, 0.1 M NaCl equilibrated to pH 10.5 with 0.1 M NaOH

Method

1. Remove medium from the wells of the microtitration plate and add 200 μl of fresh medium per well, containing 10 mM Hepes pH 7.4.
2. Immediately add 50 μl MTT in PBS at a concentration of 2–5 mg/ml.[a]
3. Wrap the plates in foil and incubate for 4 h at 37°C.
4. Remove the medium and MTT and immediately add 200 μl of DMSO per well, followed by 25 μl of Sorensen's glycine buffer.
5. Read the optical density of the plates at 570 nm.

[a] The optimal concentration may need determining for individual cell lines.

Protocol 16

Measurement of viable fresh human tumour and normal cell cytotoxicity by differential staining cytotoxicity (DiSC) assay[a]

Equipment and reagents

- Cytospin centrifuge (Shandon)
- Eyepiece counting graticule
- Medium for density gradient centrifugation (e.g. J prep, Lymphoprep)
- Enzymes for disaggregation: 0.1 mg/ml DNase Type IV, 1.5 mg/ml collagenase Type I, 0.33 mg/ml hyaluronidase Type IS (Sigma)
- 2% fast green and 1% nigrosin in PBS (BDH)
- Duck red blood cells (DRBC) fixed in 2% glutaraldehyde in PBS[b]
- Romanowsky stain (e.g. Diff Quick)

Method

1. Isolate mononuclear cells from blood or bone marrow using density gradient centrifugation. Disaggregate solid tumour tissue by mincing, chopping, or enzymatic disaggregation according to the method which is optimal.

Protocol 16 continued

2. Centrifuge the separated tumour cells at 100 g for 5 min, discard the supernatant, and resuspend the cells in 10–20 ml of culture medium.

3. Repeat step 2 twice more to wash the cells free of disaggregation medium or density gradient medium.

4. Make up the drug solutions at 10 × the final concentrations which have been chosen, using PBS as diluent.

5. Count the total number of cells in the final cell suspension and resuspend them at 4×10^6/ml in culture medium.

6. Add 90 μl of cell suspension and 10 μl of culture medium or drug dilutions into polystyrene microtitration plates.[c]

7. Incubate the plates for four days at 37°C in 95% air/5% carbon dioxide.

8. Add 10 μl of 2% fast green/1% nigrosin containing 5×10^6/ml of fixed DRBC to each 100 μl of cells.

9. Cytocentrifuge the cells onto microscope slides.[d]

10. Air dry the slides at room temperature, fix in methanol, and stain with a Romanowsky stain.

11. Identify tumour cells and normal cells.[e]

12. With the aid of an eyepiece counting graticule, count the number of live (purple stained) tumour cells per unit area on control and drug-treated slides. Ignore the dead (blue-black) cells.[f,g]

13. Repeat step 12, counting the number of live normal cells per unit area on control and drug-treated slides.

14. Plot dose–response curves for normal cells and tumour cells using live cells per unit area against drug concentration.

15. Calculate the LC_{90} values for normal cells and tumour cells.

16. Calculate the therapeutic index for each drug (LC_{90} normal/LC_{90} tumour).

[a] Based on the method of Bosanquet and Bell (103) and personal communication.

[b] Duck red blood cells are not available commercially. The help of a local duck is required. 10 ml of blood removed from a wing vein is sufficient for a large number of assays.

[c] Polystyrene rather than polypropylene should be used to avoid cell adherence.

[d] It is anticipated that the assay will be scaled down further, using Octaspot (Bath Analytical). This is a system which allows eight spots of cells to be cytocentrifuged onto one slide from an 8-well strip. 12 strips can be centrifuged in one run.

[e] The help of a cytologist is essential.

[f] DRBC are not counted. They are included as an internal control to ensure that full recovery of cells has been achieved during cytocentrifugation.

[g] Once experience has been gained in obtaining quantitative counts the LC_{90} evaluation may be used, with appropriate validation (103). This is a quicker way of evaluating slides which is more subjective but has been statistically validated.

Protocol 17

Measurement of intracellular ATP using luminescence (83 and personal communication)

Equipment and reagents

- Luminometer (EG & G; Berthold-Wallac)[a]
- White 96-well polystyrene microplates
- ATP releasing agent (e.g. Sigma FL-SAR)[b]
- Luciferin–luciferase reagent (e.g. Sigma)[b]
- Dilution buffer[b]
- ATP standards (e.g. Sigma FL-AAS)
- Maximum inhibition reagent (MI)[c]

Method

1. Perform the drug assay using routine methodology.[d]
2. At the end of the assay period, extract ATP from the cells using 50–75 μl of ATP releasing agent/200 μl of medium.[e,f]
3. Transfer 50 μl aliquots from each well of the drug assay plate into a white microplate.
4. Make up an ATP standard with six to ten different dilutions, covering the range of ATP values expected in the test samples. Add the standards to another microplate.[g]
5. Use a multichannel pipette to add the luciferin–luciferase reagent to the plates.[h]
6. Read the ATP standard plate, then the drug assay plates, then the ATP standard plate again.[g]
7. Analyse the data using % inhibition = $1 - \frac{(\text{Test} - \text{MI})}{(\text{MO} - \text{MI})}$ %

   where MO = ATP value without drug, MI = maximum inhibition produced by non-volatile maximum inhibitor of cell growth.

[a] The luciferin–luciferase reagent should ideally produce a glow which is stable over ~ 20 min rather than a transient flash. If it does **not** have glow kinetics an injection facility is needed on the luminometer.

[b] The luciferin–luciferase reagent, the ATP extraction agent, and the dilution buffer should be purchased from one supplier to ensure that their contents are compatible with regard to pH and maintenance of ATPase activity.

[c] The MI reagent should be non-volatile and checked to ensure that it does not inhibit ATPase. Sodium azide is not suitable, for this reason.

[d] The cytotoxic drugs under test should also be checked for non-interference with the assay. If a drug does interfere, then an alternative end-point should be chosen.

[e] Extracted plates can be stored for up to one month at −20 °C before measuring ATP levels. Thaw them out at room temperature for ~ 30 min. Loss of activity will occur if they are left for too long.

[f] The amount added needs to be optimized for maximum ATP extraction.

> **Protocol 17 continued**
>
> [g] The ATP standards need to be read before and after the drug assay plate to ensure that luciferase activity has been retained for the full assay period. This is particularly important if the luciferin–luciferase reagent shows flash kinetics and the plate is being read using a luminometer with an injector. The ATP standard curves also need to be generated whilst standardizing the assay and as an internal quality control for assay runs.
>
> [h] The amount of luciferin–luciferase reagent to be added is dependent upon its activity. Follow the manufacturer's instructions for optimization.

> **Protocol 18**
>
> **Measurement of viable cell number using XTT (81, 82)**
>
> **Equipment and reagents**
> - XTT working solution: dissolve XTT at 1 mg/ml in warmed Hanks' balanced salt solution and immediately add 5 μl/ml of phenazine methosulfate from a stock solution of 5 mM
> - Microplate reader with dual wavelength facility and filters at 450 nm and 650 nm
>
> **Method**
> 1. Follow a routine procedure for drug assay in microtitration plates[a] and ensure that a cell-free row is available to use as a blank.
> 2. Add 50 μl of XTT working solution to the medium in each well containing controls and drug-treated cells. Add 50 μl of XTT to a cell-free row containing the same volume of culture medium as that used in the assay.
> 3. Incubate the plate at 37°C in an atmosphere of 95% air/5% carbon dioxide for 6–18 h.[b]
> 4. Measure the optical density of each well in a plate reader using a test wavelength of 450 nm and a reference wavelength of 650 nm.[c] Use the cell-free wells containing XTT as blanks.
>
> [a] The method can be used for monolayers, liquid suspension cultures, and for cells growing in soft agar (82).
>
> [b] The length of incubation is dependent on the cell density and the metabolic activity, and the optimal incubation period should be determined for each cell line.
>
> [c] The final colour is an intense orange. Whilst the dual wavelength helps to prevent interference by pH-induced colour changes in the phenol red, this potential problem can be eliminated by using phenol red-free culture medium.

7 Interpretation of results

7.1 Relationship between cell number and cytotoxicity index

The validity of a cytotoxicity index is dependent upon the degree of linearity between cell number and the chosen end-point, and this should be confirmed

CYTOTOXICITY AND VIABILITY ASSAYS

for any cytotoxicity assay. In clonogenic assays a linear relationship may not occur at low cell numbers due to the dependence of clonogenic growth on conditioning factors, whilst at high cell densities linearity is lost due to nutritional deficiencies. In cytotoxicity assays linearity may be lost at the upper end due to density-dependent inhibition of the relevant metabolic pathway, whilst the sensitivity limit of the assay may affect linearity at the lower end. This would cause apparent stimulation at low drug levels and an overestimation of cell kill at higher concentrations. Control cell numbers at the end of the assay must therefore fall on the linear portion of the curve. The accuracy at higher levels of cell kill is dependent on the range over which linearity extends, and influences the number of decades of cell kill which can be measured. If results are plotted on a log scale this may imply that the assay is accurate down to three or four decades of cell kill, which should be confirmed before expressing results in this way. This is particularly important in *in vitro* drug combination studies when synergism or additivity is often observed beyond the second decade of cell kill.

7.2 Dose–response curves

Results are commonly plotted as dose–response curves using a linear scale for percentage inhibition (of isotope incorporation, for example) and a log scale for surviving fraction on clonogenic tests. Assay variation for replicate points is routinely depicted as mean ± standard deviation; a minimum of three replicates is therefore required for each test point. Some method is required for defining the sensitivity of a cell population in relation to other cell populations, or different test conditions; several parameters are available and are shown in *Figure 6*.

7.2.1 Area under curve method

The use of this method acknowledges the probability that the shape of the dose–response curve may be instrumental in influencing the outcome of drug exposure, rather than cell kill at any one concentration. It is calculated using the trapezoidal method which adds the area of rectangles and triangles under the survival curve. The method has been applied most extensively in the 'human tumour stem cell assay' (104). Some drugs do not produce curves with an easily defined IC_{50} value and AUC may provide a more easily quantifiable result.

7.2.2 Cut-off points for definition of sensitivity and resistance

If plots of dose–response curves from comparisons of cell lines show that they maintain their relative sensitivity rankings at different concentrations (i.e. crossing-over of dose–response curves is minimal), then information on the relative sensitivities of different cell lines can be obtained by defining sensitivity at one concentration, and this is the most commonly used method for *in vitro* predictive testing. When retrospective correlations between *in vitro* data are made for defining these cut-off points, an intermediate zone is found where

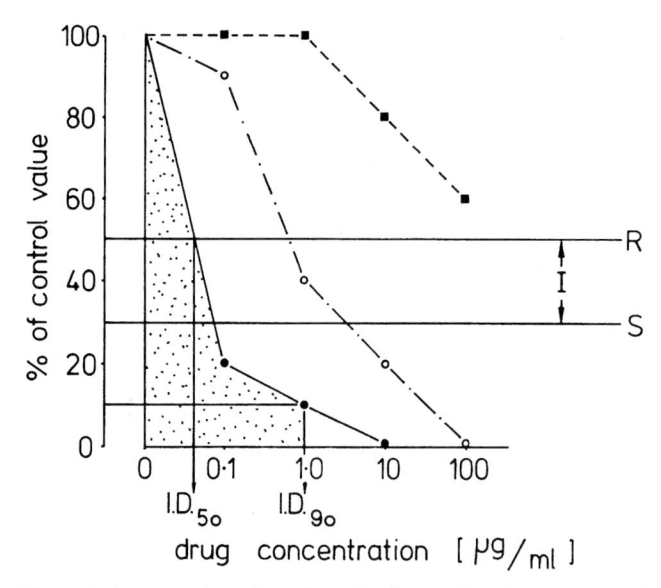

Figure 6. Interpretation of results using linear plot of response against drug concentration. ID_{50} and ID_{90} are the concentrations required to reduce cytotoxicity index to 50% or 90% of control values. The hatched region is the area under curve between 0–1 µg/ml (for peak plasma concentration of 10 µg/ml). S, cut-off boundary for sensitivity; R cut-off boundary for resistance; I, intermediate zone. ●, sensitive population; ○-○, intermediate population; ■, resistant population.

tumours cannot be defined as sensitive or resistant, and there is no clear-cut correlation between *in vitro* results and clinical response. The size of the intermediate zone will be at least partly related to the inherent variability of the assay, larger zones being associated with higher standard deviations. Although sensitivity may be defined at one concentration it is recommended that more than one concentration is tested, particularly in the developmental stage of the assay.

7.2.3 ID_{50} and ID_{90} values

Cell sensitivity may also be defined by the ID_{50} and ID_{90} values (i.e. drug concentration required to inhibit viability by 50% or 90%). These values may also be termed IC (inhibitory concentration), LC (lethal concentration), or GI (growth inhibition), but they are all determined in the same way.

7.2.4 Correlation between *in vitro* and *in vivo* results

Criteria for defining tumours as sensitive or resistant are based on retrospective correlations between *in vitro* results and clinical responses, using a training set of data. Even when a laboratory is using an established method for tumour sensitivity testing, 'own laboratory' sets of training data should be obtained to allow for inter-laboratory variation. The response of patients with tumours of intermediate sensitivity may be influenced by prognostic factors other than tumour sensitivity (e.g. tumour burden at onset of chemotherapy, stage of disease, histology, tumour cell doubling time, previous chemotherapy, and performance status). When analysing results for correlations, some attempt to stratify patients according to these parameters may assist in providing more meaningful data. Quantitative assessment of tumour response is also of paramount importance. It is pointed out that *in vitro* chemosensitivity can be expected only to indicate that some degree of cell kill will be achieved *in vivo*, not that the patient will achieve a complete response to treatment. The true positive correlation rate of an assay is defined as:

$$\frac{S^t/S^p + S^t/R^p}{S^t/S^p}$$

where the numerator of each fraction is the *in vitro* test response (t) and the denominator is the *in vivo* patient response (p). S^t and R^t are respectively the number of tumours showing *in vitro* sensitivity or resistance according to the selected cut-off points. S^p and R^p are respectively the number of patients achieving a clinical response or no clinical response to the drug tested *in vitro*.

The true negative rate is defined as:

$$\frac{R^t/R^p + R^t/S^p}{R^t/R^p}$$

In assessing the significance of the correlation rates obtained, these should be compared with the correlation rates which would be obtained were the *in vitro* results randomly distributed (105). For example, a drug gives a 50% response rate *in vivo*, and 50% of tumours show *in vitro* sensitivity to this drug. If the *in vitro* results are randomly distributed between sensitivity and resistance, then the chances of obtaining a positive correlation between *in vitro* sensitivity and *in vivo* response are 50% of 50% (i.e. 25%), and also of obtaining a positive correlation between *in vitro* resistance and *in vivo* resistance. The overall apparent positive correlation rate is therefore 50%. *In vitro* versus *in vivo* correlations are also complicated by the use of combination regimes to treat patients. Strictly speaking, correlations should be made only when *in vitro* data is available for all drugs used. Whether or not they are tested in combination depends on the treatment protocol, since some drugs are administered sequentially. Also, if the assay can only measure two decades of cell kill it may be too insensitive to detect additive or synergistic effects.

7.2.5 Combinations of drugs

A 96-well plate clonogenic assay based on limiting dilutions has been used to look at additive and supra-additive (synergistic) effects of drug combinations (106–108). The data is analysed using the formula plating efficiency (PE) = −ln (number of negative wells)/(total number of wells)/number of cells plated per well (109). Fractional survival data are then fitted to the linear quadratic model, $F = \exp[-(\alpha D + \beta D^2)]$ and AUC analysed by numerical integration using appropriate software.

8 Pitfalls and troubleshooting

Problems which may be encountered with these assays include:

- large standard deviations
- variability between assays done on the same cell population
- stimulation to above control levels

8.1 Large standard deviations

Possible reasons for large standard deviations include:

(a) Faults in aliquoting cell suspension, which are most likely to be made due to inadequate mixing of cell suspension during dispensing leading to uneven distribution of cells between replicates.

(b) The presence of large cell aggregates in the original cell suspension, leading to uneven distribution of cells between replicates.

(c) Non-specificity of cytotoxicity end-point (e.g. due to measurement of non-specific binding of radioactivity—see *Protocol 10*).

8.2 Variation between assays

Replicate assays on different days cannot be performed on human tumour biopsy material to check day-to-day reproducibility, but this can be evaluated using cell lines. It is a recognized problem that cell lines which show consistent sensitivity profiles may show 'deviant' results occasionally, for reasons which cannot be identified. Specific reasons for failure to obtain reproducible results may include:

(a) Failure to harvest the cell population at an identical time point (e.g. exponential growth versus early confluence versus late confluence).

(b) Deterioration of stock drug solutions (see Section 5).

(c) When drug solutions have a short half-life they must be used immediately after diluting to ensure consistency in the drug levels available to cells in each assay.

(d) Failure to standardize incubation conditions (Section 6.2).

The assay system must be checked for reproducibility before applying it to human biopsy material.

8.3 Stimulation to above control levels

Stimulation can be a true measure of cellular events but may be due to technical artefacts. These include:

(a) Non-specific binding of radioactivity.

(b) Density-dependent inhibition of metabolic pathways in controls which is not evident in test situations where some cell kill has been achieved.

(c) Stimulation of uptake of metabolic precursors by antimetabolites (e.g. thymidine by 5-fluorouracil and methotrexate).

CYTOTOXICITY AND VIABILITY ASSAYS

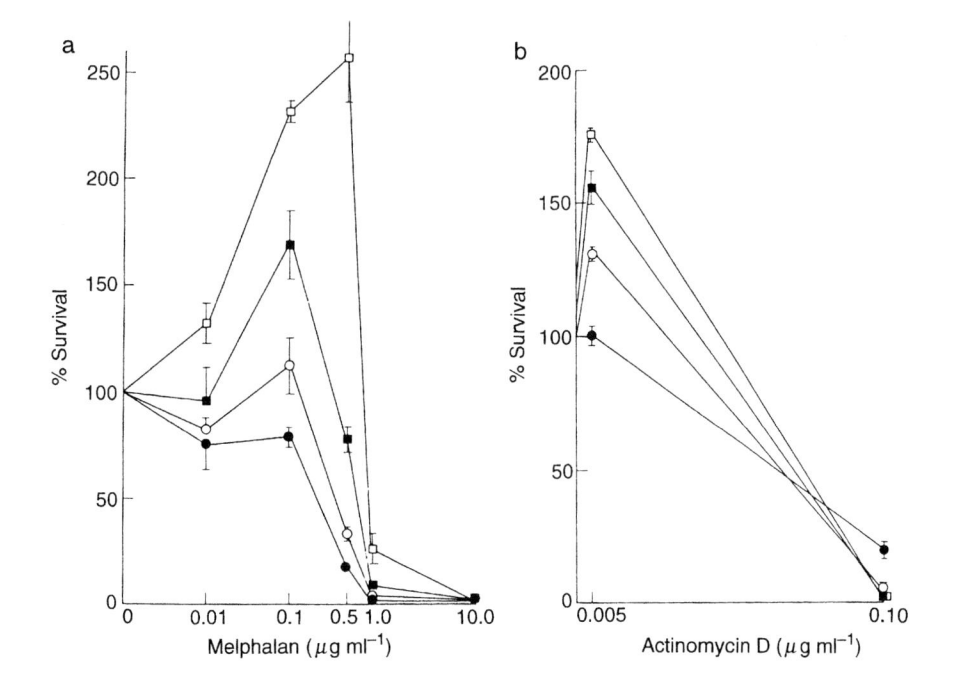

Figure 7. The effect of growth unit size on the survival curves of a murine melanoma cell line (CCL) to melphalan (a) and of a human melanoma biopsy to actinomycin D (b). Growth unit size and frequency was measured using FAS II automated image analysis system. ●, ≥ 60 μm; ○, ≥ 104 μm; ■, ≥ 124 μm; □, ≥ 140 μm. Mean ± s.e. shown. (Reproduced with permission of the publishers from ref. 105.)

Plated cell density influences the distribution in size of growth units in clonogenic assays, with large units decreasing as plated cell numbers increase (110). The effect of this on drug sensitivity profiles was examined and, as expected, the dose–response curve was strongly influenced by the size criterion used for colony scoring, with stimulation to above control levels occurring when large colonies were scored (*Figure 7*).

References

1. Balmain, A. and Brown, K. (1988). *Adv. Cancer Res.*, **51**, 147.
2. Kilbey, B. J., Legator, M., Nichols, W., and Ranel, C. (ed.) (1984). *Handbook of mutagenicity testing procedures*. Elsevier, Amsterdam.
3. Veritt, S. and Parry, J. M. (ed.) (1984). *Mutagenicity testing: a practical approach*. IRL, Oxford.
4. Gellhorn, A. and Hirschberg, E. (1955). *Cancer Res.*, **15**, Suppl. **3**, 1.
5. Foley, G. E. and Epstein, S. S. (1964). *Adv. Chemother.*, **1**, 175.
6. Hakala, M. I. and Rustrum, Y. M. (1979). In *Methods in cancer research: cancer drug development*. Part A (ed. V. T. DeVita and H. Busch), p. 247. Academic Press, New York.
7. Dendy, P. P. (ed.) (1976). *Human tumours in short-term culture: techniques and clinical application*. Academic Press, New York.

8. Dendy, P. P. and Hill, B. T. (ed.) (1983). *Human tumour drug sensitivity testing in vitro: techniques and clinical applications.* Academic Press, New York.
9. Eagle, H. and Foley, G. E. (1956). *Am. J. Med.*, **21**, 739.
10. Wright, J. C., Cobb, J. P., Gumport, S. L., Golomb, F. M., and Safadi, D. (1957). *N. Engl. J. Med.*, **257**, 1207.
11. Von Hoff, D. D., et al. (1983). *Cancer Res.*, **43**, 1926.
12. Wilson, A. P. and Neal, F. E. (1981). *Br. J. Cancer*, **44**, 189.
13. Silvestrini, R., Sanfilippo, O., and Daidone, M. G. (1983). In *Human tumour drug sensitivity testing in vitro* (ed. P. P. Dendy and B. T. Hill), p. 281. Academic Press, New York.
14. Salmon, S. E. (1980). In *Cloning of human tumour cells. Progress in clinical and biological research*, Vol. 48 (ed. S. E. Salmon), p. 281. Alan R. Liss, New York.
15. Von Hoff, D. D., et al. (1986). *J. Clin. Oncol.*, **4**, 1827.
16. Monks, A., Scudiero, D. A., Johnson, G. S., Pauull, K. D., and Sausville, E. A. (1997). *Anticancer Drug Des.*, **12**, 533.
17. Turner, P. (ed.) (1983). *Animals in scientific research: an effective substitute for man?* Macmillan, London.
18. Atterwill, C. K. and Steele, C. E. (ed.) (1987). *In vitro methods in toxicology.* Cambridge University Press.
19. Masters, J. R. W. (1983). In *Human tumour drug sensitivity testing in vitro* (ed. P. Dendy and B. T. Hill), p. 163. Academic Press, New York.
20. Volm, M., Wayss, K., Kaufmann, M., and Mattern, J. (1979). *Eur. J. Cancer*, **15**, 983.
21. KSST. Group for Sensitivity Testing of Tumours (1981). *Cancer*, **48**, 2127.
22. Durkin, W. J., Ghanta, V. K., Balch, C. M., Davis, D. W., and Hiramoto, R. N. (1979). *Cancer Res.*, **39**, 402.
23. Raich, P. C. (1978). *Lancet*, **i**, 74.
24. Weisenthal, L. M. and Marsden, J. (1981). *Proc. Am. Assoc. Cancer Res.*, **22**, 155.
25. Morgan, D., Freshney, R. I., Darling, J. L., Thomas, D. G. T., and Celik, F. (1983). *Br. J. Cancer*, **47**, 205.
26. Hill, B. T. (1983). In *Human tumour drug sensitivity testing in vitro* (ed. P. P. Dendy and B. T. Hill), p. 91. Academic Press, New York.
27. Agrez, M. W., Kovach, J. S., and Lieber, M. M. (1982). *Br. J. Cancer*, **46**, 88.
28. Bertoncello, I., et al. (1982). *Br. J. Cancer*, **45**, 803.
29. Rupniak, H. T. and Hill, B. T. (1980). *Cell Biol. Int. Rep.*, **4**, 479.
30. Hamburger, A. W., Salmon, S. E., Kim, M. B., Trent, J. M., Soehnlen, B., Alberts, D. S., et al. (1978). *Cancer Res.*, **38**, 3438.
31. Salmon, S. E. (ed.) (1980). *Cloning of human tumour stem cells. Progress in clinical and biological research*, Vol. 48. Alan R. Liss, New York.
32. Courtenay, V. D. and Mills, J. (1978). *Br. J. Cancer*, **37**, 261.
33. Tveit, K. M., Endersen, L., Rugstad, H. E., Fodstad, O., and Pihl, A. (1981). *Br. J. Cancer*, **44**, 539.
34. Von Hoff, D. D., Forseth, B. J., Huong, M., Buchok, J. B., and Lathan, B. (1986). *Cancer Res.*, **46**, 4012.
35. Hanauske, A.-R., Hanauske, U., and Von Hoff, D. D. (1987). *Eur. J. Cancer, Clin. Oncol.*, **23**, 603.
36. Baker, F. L., Ajani, J., Spitzer, G., Tomasovic, B. J., Williams, M., Finders, M., et al. (1988). *Int. J. Cell Cloning*, **6**, 95.
37. Alberts, D. S., George Chen, H.-S., and Salmon, S. E. (1980). In *Cloning of human tumour stem cells* (ed. S. E. Salmon), p. 197. Alan R. Liss, New York.
38. Alberts, D. S. and Chen, H.-S. G. (1980). In *Cloning of human tumour stem cells* (ed. S. E. Salmon), Appendix 4. Alan R. Liss, New York.

39. Freshney, R. I., Paul, J., and Kane, L. M. (1975). *Br. J. Cancer*, **31**, 89.
40. Roper, P. R. and Drewinko, B. (1976). *Cancer Res.*, **36**, 2182.
41. Korzeniewski, C. and Callewaert, D. M. (1983). *J. Immunol. Methods*, **64**, 313.
42. Chenery, R. J. (1987). In *In vitro methods in toxicology* (ed. C. K. Atterwill and C. E. Steele), p. 211. Cambridge University Press.
43. Weisenthal, L. M., Dill, P. L., Kurnick, N. B., and Lippman, M. E. (1983). *Cancer Res.*, **43**, 258.
44. Haugland, R. P. (1996). *Handbook of fluorescent probes and research chemicals*, 6th edn. Molecular Probes.
45. Bickis, I. J., Henderson, I. W. D., and Quastel, J. H. (1966). *Cancer*, **19**, 103.
46. Dickson, J. A. and Suzanger, M. (1976). In *Human tumours in short-term cultures: techniques and clinical Aapplications* (ed. P. P. Dendy), p. 107. Academic Press, New York.
47. Buskirk, H. H., Crim, J. A., Van Giessen, G. J., and Petering, H. G. (1973). *J. Natl. Cancer Inst.*, **51**, 135.
48. Cosma, G. N. and Wenzel, D. G. (1984). *J. Tissue Culture Methods*, **9**, 29.
49. Grunicke, H., Hirsch, F., Wolf, H., Bauer, V., and Kiefer, G. (1975). *Exp. Cell Res.*, **90**, 357.
50. Friedman, H. M. and Glaubiger, D. L. (1983). *Cancer Res.*, **42**, 4683.
51. Sondak, V. K., *et al.* (1984). *Cancer Res.*, **44**, 1725.
52. Dendy, P. P., Dawson, M. P. A., Warner, D. M. A., and Honess, D. J. (1976). In *Human tumours in short-term culture: techniques and clinical applications* (ed. P. P. Dendy), p. 139. Academic Press, New York.
53. Forbes, L. J. (1963). *Aust. J. Exp. Biol.*, **41**, 255.
54. Izsak, F. Ch., Gotlieb-Stematsky, T., Eylan, E., and Gazith, A. (1968). *Eur. J. Cancer*, **4**, 375.
55. Edwards, A. J. and Rowlands, G. F. (1968). *Br. J. Surg.*, **55**, 687.
56. Freshney, R. I. and Morgan, D. (1978). *Cell Biol. Int. Rep.*, **2**, 375.
57. Ramos, K. and Acosta, D. (1984). *J. Tissue Culture Methods*, **9**, 3.
58. Oyama, V. I. and Eagle, H. (1956). *Proc. Soc. Exp. Biol. Med.*, **91**, 305.
59. Pelletier, B., Dhainaut, F., Pauly, A., and Zahnd, J.-P. (1988). *J. Biochem. Biophys. Methods*, **16**, 63.
60. Dent, M. F., Hubbold, L., Radford, H., and Wilson, A. P. (1995). *Cytotechnology*, **14**, 1.
61. Skehan, P., Storeng, R., Scudicro, D., Monks, A., McMahon, J., Vistica, D., *et al.* (1989). *Proc. Am. Assoc. Cancer Res.*, **30**, 612.
62. Fricker, S. (1990). BACR Meeting 1990, Poster 30.
63. Knox, P., Uphill, P. F., Fry, J. R., Berford, J., and Balls, M. (1986). *Food Chem. Toxic*, **24**, 457.
64. Bradford, M. M. (1976). *Anal. Biochem.*, **72**, 248.
65. Laughton, C. (1984). *Anal. Biochem.*, **140**, 417.
66. Smith, P. K., *et al.* (1985). *Anal. Biochem.*, **150**, 76.
67. Lane, R. D., Federman, D., Flora, J. L., and Beck, B. L. (1986). *J. Immunol. Methods*, **92**, 261.
68. McCaffrey, T. A., Agarwal, L. A., and Weksler, B. B. (1988). *In Vitro Cell Dev. Biol.*, **24**, 247.
69. In *Amersham research news*. (1989). **3**, 13.
70. Parish, C. R. and Muilbacher, A. (1983). *J. Immunol. Methods*, **58**, 225.
71. Fiennes, A. G. T. W., Walton, J., Winterbourne, D., McGIashar, D., and Hermon-Taylor, J. (1987). *Cell Biol. Int. Rep.*, **11**, 373.
72. Landegren, U. (1984). *J. Immunol. Methods*, **67**, 379.
73. Mosmann, T. (1983). *J. Immunol. Methods*, **65**, 55.
74. Denizot, F. and Lang, R. (1986). *J. Immunol. Methods*, **89**, 271.

75. Cole, S. P. C. (1986). *Cancer Chemother. Pharmacol.*, **17**, 259.
76. Carmichael, J., DeGraff, W. G., Gazdar, A. F., Minna, J. D., and Mitchell, J. B. (1987). *Cancer Res.*, **47**, 936.
77. Park, J.-G., Kramer, B. S., Steinberg, S. M., Carmichael, J., Collins, J. M., Minna, J. D., et al. (1987). *Cancer Res.*, **47**, 5875.
78. Pieters, R., Huismans, D. R., Leyva, A., and Veerman, A. J. (1989). *Br. J. Cancer*, **59**, 217.
79. Jabbar, S. A. B., Twentyman, P. R., and Watson, J. V. (1989). *Br. J. Cancer*, **60**, 523.
80. Plumb, J. A., Milroy, R., and Kaye, S. B. (1989). *Cancer Res.*, **49**, 4435.
81. Paull, K. D., Shoemaker, R. H., and Boyd, M. R. (1988). *J. Heterocyclic Chem.*, **25**, 911.
82. Dent, M. F., Hubbold, L., Radford, H., and Wilson, A. P. (1996). *Cytotechnology*, **18**, 219.
83. Cree, I. A. (1998). *Methods Mol. Biol.*, **102**, 169.
84. Kerr, J. F. R., Wyllie, A. H., and Currie, A. R. (1972). *Br. J. Cancer*, **26**, 239.
85. Freshney, R. I., Celik, F., and Morgan, D. (1982). In *The control of tumour growth and its biological base* (ed. W. Davis, C. Maltoni, and St. Tanneberger). Fortschritte in der Onkologie, Band 10, Berlin, Akademie-Verlag.
86. Wilson, A. P., Ford, C. H. J., Newman, C. H., and Howell, A. (1984). *Br. J. Cancer*, **49**, 57.
87. Weisenthal, L. M. and Lippman, M. E. (1985). *Cancer Treat. Rep.*, **69**, 615.
88. Bosanquet, A. G. (1984). *Br. J. Cancer*, **49**, 385.
89. Powers, J. F. and Sladek, N. E. (1983). *Cancer Res.*, **43**, 1101.
90. Fry, J. R. (1983). In *Animals in scientific research: an effective substitute for man?* (ed. P. Turner), p. 69. Macmillan, London.
91. Begue, J. M., Le Bigot, J. F., Guguen-Guillouzo, C., Kiechel, J. R., and Guillouzo, A. (1983). *Biochem. Pharmacol.*, **32**, 1643.
92. Davies, D. S. and Boobis, A. R. (1983). In *Animals in scientific research: an effective substitute for man?* (ed. P. Turner), p. 69. Macmillan, London.
93. Guillouzo, A. and Guguen-Guillouzo, C. (ed.) (1986). *Research in isolated and cultured hepatocytes*. John Libbey, Eurotext Ltd./INSERM, London.
94. Guillouzo, A., Beaune, P., Gascoin, M.-N., Begue, J. M., Campion, J.-P., Guengerich, P. F., et al. (1985). *Biochem. Pharmacol.*, **34**, 2991.
95. Barranco, S. C., Bolton, W. E., and Novak, J. K. (1980). *J. Natl. Cancer Inst.*, **64**, 913.
96. Courtenay, V. D. and Mills, J. (1981). *Br. J. Cancer*, **44**, 306.
97. Dixon, R. A., Linch, D., Baines, P., and Rosendaal, M. (1981). *Exp. Cell Res.*, **131**, 478.
98. Gupta, V. and Krishnan, A. (1982). *Cancer Res.*, **42**, 1005.
99. Nederman, T. and Twentyman, P. (1984). In *Spheroids in cancer research. Methods and perspectives* (ed. H. Acker, J. Carlsson, R. Durand, and R. M. Sutherland), Ch. 5. Springer–Verlag, Berlin.
100. Jones, A. C., Stratford, I., Wilson, P. A., and Peckham, M. J. (1982). *Br. J. Cancer*, **46**, 870.
101. INVITTOX (1990). *The FRAME cytotoxicity test*. INVITTOX Protocol number 3. FRAME, Nottingham.
102. Kolb, A. I. (1981). *Lab. Equip. Dig.*, **19**, 87.
103. Bosanquet, A. G. and Bell, P. B. (1996). *Leukemia Res.*, **20**, 143.
104. Moon, T. E. (1980). In *Cloning of human tumour stem cells* (ed. S. E. Salmon), p. 209. Alan R. Liss, New York.
105. Berenbaum, M. C. (1974). *Lancet*, **ii**, 1141.
106. Grenman, R., Burk, D., Virolainen, E., Buick, R. N., Church, J., Schwartz, D. R., et al. (1989). *Int. J. Cancer*, **44**, 131.
107. Rantanen, V., Grenman, S., Kulmala, J., and Grenman, R. (1994). *Br. J. Cancer*, **69**, 482.
108. Engblom, P., Rantanen, V., Kulmala, J., Helenius, H., and Grenman, R. (1999). *Br. J. Cancer*, **79**, 286.

109. Thilly, W. G., DeLuca, J. G., Furth, E. E., Hoppe, H., Kaden, D. A., Krolenski, J. J., *et al.* (1980). In *Chemical mutagens* (ed. F. J. de Serpes and A. Hollander), p. 31. Plenum Press, NY.
110. Meyskens, F. L. Jr., Thomson, S. P., Hickie, R. A., and Sipes, N. J. (1983). *Br. J. Cancer*, **46**, 863.

Chapter 8
Fluorescence *in situ* hybridization (FISH)

W. Nicol Keith
CRC Department of Medical Oncology, Garscube Estate, Switchback Road, Glasgow G61 1BD, UK.

1 Introduction

Classical karyotyping uses dyes that differentially stain the chromosomes. Thus, each chromosome is identified by its banding pattern. However, traditional banding techniques are limited to the identification of large and relatively simple changes.

In situ hybridization is the most sensitive and specific means of identifying the location of genes on chromosomes. In addition, using chromosome- and gene-specific probes, numerical and structural aberrations can be analysed within individual cells. Thus, *in situ* hybridization has applications in the diagnosis of genetic disease and in the identification of gene deletions, translocations, and amplification (1–7).

New karyotyping methods based on chromosome painting techniques, called spectral karyotyping (SKY) and multicolour fluorescence *in situ* hybridization (M-FISH), have been developed. These techniques allow the simultaneous visualization of all 24 human chromosomes in different colours (8, 9). Computerized analysis of the fluorescence patterns makes these techniques more sensitive than the human eye (3, 10). Thus FISH techniques are proving highly successful in the identification of chromosomal alterations which could not be resolved by traditional approaches.

FISH requires nucleic acid probes that are labelled by incorporation of nucleotides modified with molecules such as biotin or digoxigenin. After hybridization of the labelled probes to the chromosomes, detection of the hybridized sequences is achieved by forming antibody complexes that recognize the biotin or digoxigenin within the probe. The visualization is achieved by using antibodies conjugated to fluorochromes. The fluorescent signal can be detected in a number of ways. If the signal is strong enough, standard fluorescence microscopy can be used. However, data analysis and storage can be improved considerably by the use of digital imaging systems such as confocal laser scanning microscopy or cooled CCD camera (3).

The major advantages of fluorescence *in situ* hybridization (FISH) are that radioactivity is not required, it is rapid, amenable to computerized storage and manipulation, sensitive, gives accurate signal localization, allows simultaneous analysis of two or more fluorochromes, and provides a quantitative and spatial distribution of the signal.

The best way to get started is to visit a laboratory where FISH is done routinely. This need not take long, as the basics can be learnt in one week. A number of companies market the reagents and equipment and offer practical training courses. You will find that many hospital and university genetics, clinical genetics, or cytogenetics units will have all the expertise you need.

2 Probes

For the quantitative assessment of gene copy number, high efficiency probes are required. We find that P1 or cosmid clones containing the gene of interest work best. cDNA clones are difficult to work with and virtually useless for interphase cytogenetics due to their low sensitivity. If your laboratory is inexperienced in FISH, do not be tempted to start FISH with cDNA clones, use a P1 or cosmid clone. Repetitive sequence probes such as those that recognize the centromeric sequences of specific chromosomes are the easiest probes to work with and a good place to start. Some probe properties are shown in *Table 1*.

Test all probes on normal chromosomes prepared from lymphocytes. This is essential no matter what the final application is. Pay particular attention to the

Table 1. Probe properties

Probe type	Insert size	Target	Application	Characteristics	Sensitivity
Unique sequences		Genes			
P1 clones Genomic sequences	~90 kb		Mapping Interphase cytogenetics	Usually make good FISH probes	High
Cosmid clones Genomic sequences	~45 kb		Mapping Interphase cytogenetics	Usually make good FISH probes	High
Lambda Genomic sequences	~12 kb		Mapping Interphase cytogenetics	Better to use P1 or cosmid clones	Medium
cDNA clones	Wide range: 1–4 kb plus		Mapping	Not much use for interphase cytogenetics	Low
Repetitive sequences		Satellite DNA	Chromosome enumeration	Easy to use	Very high
		Telomeres	Telomere length	Very specialized	Tricky!
Chromosome paints		Whole chromosomes	Translocations Chromosome identification	Easy to use	High

FLUORESCENCE IN SITU HYBRIDIZATION (FISH)

specificity of hybridization. Some probes, such as chromosome-specific repeat sequence probes, can bind to several chromosomes if the hybridization conditions are not correct. Use commercial probes to test your reagents.

There are many companies offering a wide range of probes (see Section 5). Commercial probes are invaluable for testing your reagents and for troubleshooting.

With the vast numbers of FISH publications, you can also ask other researchers for probes (4, 11). There are probe databanks on the Web and some are given in the list of Web sites (see Section 5). You can also develop them (12), or purchase the aid of a commercial company like Genome Systems Inc. (1, 3). This can be expensive but very efficient.

Protocol 1
Probe labelling

Reagents

- Biotin-Nick Translation Mix (Boehringer Mannheim, Cat. No. 1745 824)
- DIG-Nick Translation Mix (Boehringer Mannheim, Cat. No. 1745 816)

A. Incorporation of biotin-16-dUTP

1. Mix the following:
   - DNA (1 µg)   X µl
   - dH$_2$O   (16 µl − X µl)
   - Biotin-Nick Mix   4 µl
   - Final volume   20 µl

2. Mix and centrifuge briefly and incubate at 16°C for 1–4 h (usually about 90 min).

3. Then precipitate probe.

B. Incorporation of digoxigenin-11-dUTP

1. Follow part A, but use Dig-Nick Mix instead of Biotin.

Protocol 2
Probe precipitation

Reagents

- See *Protocol 1*
- Human cot-1 DNA
- 3 M NaAc pH 8
- Ethanol
- 50% formamide hybridization mix

Protocol 2 continued

Method

1. Nick translate 1 μg DNA (Bio Nick or Dig Nick as *Protocol 1*). Volume of reaction is 20 μl.
2. To precipitate probe add to the 20 μl nick translation mix:
   - 3 M NaAc pH 8   5 μl
   - Human cot-1 DNA   25–100 μl
   - Glycogen   1 μl
   - 100% ethanol   300 μl

   Mix well.
3. Place on dry ice for 30 min, or −20 °C overnight.
4. Spin down at 20 000 g for 15 min in microcentrifuge.
5. Pour off ethanol carefully.
6. Wash pellet in 100 μl of 70% ethanol and spin at 13 K for 5 min.
7. Carefully pour off ethanol and dry pellet (air dry or rotary evaporate).
8. Resuspend in 60 μl of 50% formamide hybridization mix (for single copy probes).

Protocol 3

Chromosome preparation from lymphocytes

Reagents

- Chromosome medium (lymphocyte culture medium from Gibco BRL)
- 5 ml of whole blood, collected in a heparinized tube
- Hanks' balanced salt solution (HBSS)
- 6 mg/ml thymidine stock (use at final concentration of 0.35 mg/ml): dissolve in water, filter, and store at −20 °C
- Hypotonic solution: 0.075 M potassium chloride
- 3 mg/ml bromodeoxyuridine stock (use at final concentration of 0.03 mg/ml): dissolve in water, filter, and store at −20 °C
- 10 μg/ml colcemid (100 × stock, Gibco BRL): add 50 μl per tube containing 5 ml of chromosome medium, giving a final concentration of 0.1 μg/ml
- Methanol/acetic acid (3:1), make fresh

Method

1. Add 200 μl of whole blood to 5 ml chromosome medium.
2. Incubate for 72 h, mixing every day, at 37 °C in 5% CO_2.
3. Add 294 μl of 6 mg/ml thymidine and incubate at 37 °C for 15–17 h in 5% CO_2.
4. Wash three times in HBSS, centrifuge at 500 g for 7 min for each wash.
5. Resuspend in 5 ml of fresh chromosome medium containing 50 μl of 3 mg/ml BrdU.

Protocol 3 continued

6. Incubate for 7-8 h at 37°C in 5% CO_2 in the dark.
7. Add 50 μl colcemid per tube and incubate 1-3 h.
8. Spin at 500 g for 5 min.
9. Remove supernatant and resuspend in 10 ml of hypotonic solution.
10. Incubate at 37°C for 10-15 min.
11. Add 2-3 ml of methanol/acetic acid (3:1).
12. Centrifuge at 500 g for 5 min.
13. Remove supernatant and resuspend in 10 ml methanol/acetic acid.
14. Leave at room temperature for 10-15 min.
15. Repeat steps 12-14 at least three times.
16. Finally spin at 500 g, resuspend in a small volume (3-5 ml) of methanol/acetic acid, and store at −20°C.

Protocol 4
Chromosome preparation from cell lines

Reagents
- See *Protocol 3*

Method

1. Add colcemid to each flask at 0.1 μg/ml final concentration.
2. Incubate at 37°C for 2-3 h.
3. Trypsinize cells as normal and transfer to a 50 ml Falcon tube.
4. Spin at 500 g for 5 min.
5. Remove supernatant and resuspend in 10 ml of hypotonic solution.
6. Incubate at 37°C for 10-15 min.
7. Add 2-3 ml of methanol/acetic acid (3:1).
8. Centrifuge at 500 g for 5 min.
9. Remove supernatant and resuspend in 10 ml methanol/acetic acid.
10. Leave at room temperature for 10-15 min.
11. Repeat steps 8 and 9 at least three times.
12. Finally spin at 500 g for 5 min, resuspend in a small volume (3-5 ml) of methanol/acetic acid, and store at −20°C.

Protocol 5

Fluorescence *in situ* hybridization: chromosome preparation

Reagents

- Methanol/acetic acid (3:1)
- 100 μg/ml RNase
- Pepsin solution (0.01% in 10 mM HCl), make from frozen pepsin stock
- SSC
- Fixative
- Ethanol

Method

1. Drop a volume of chromosome preparation onto a slide from a height.
2. Mark an area of metaphase spreads on the slide using a diamond pen.
3. Fix for 1 h in methanol/acetic acid (3:1) at room temperature.
4. Air dry.
5. Incubate for 1 h in 100 μg/ml RNase in 2 × SSC at 37 °C. Make from frozen RNase stock.
6. Rinse in 2 × SSC.
7. Digest in pepsin solution for 10 min at 37 °C.
8. Rinse in water.
9. Fix for 10 min in STF (Streck tissue fixative, Alpha Labs) at room temperature. Alternatively use 1% formaldehyde (add 4 ml of 37% formaldehyde—this is the concentration usually supplied—to 146 ml PBS, 50 mM $MgCl_2$).
10. Dehydrate 2 × 2 min 70% EtOH, 2 × 2 min 100% EtOH, and leave to air dry.
11. Proceed to *Protocols* 6 and 7.

Protocol 6

Target denaturation

Equipment and reagents

- Coplin jars
- Ethanol
- 70% formamide: 35 ml formamide, 15 ml of 2 × SSC

Method

1. Warm 70% formamide to temperature required. Immerse the slide with chromosomes in 70% formamide for time required. Temperature and time variables — try:
   (a) 75 °C for 3 min.

Protocol 6 continued

      (b) 75 °C for 5 min.
      (c) 80 °C for 3 min.
      (d) 80 °C for 5 min.

    Carry out in fume hood using plastic Coplin jars.

2. Rinse in large volume (500 ml) 70% EtOH.
3. Dehydrate in 70% EtOH for 2 min, then 100% EtOH for 2 min.
4. Air dry.

Protocol 7

Probe denaturation

Equipment and reagents

- Water-bath
- 22 × 22 coverslip
- Cow Gum

A. For satellite probes

1. Heat probe (12–15 μl per slide) to 70–80 °C for 5 min. Place on ice for 5 min.
2. Apply to slide, use 22 × 22 coverslip and seal edges with Cow Gum.
3. Incubate 37 °C or 42 °C overnight.

B. For cosmid probes

1. Heat probe to 70 °C for 5 min, to 80 °C for 10 min.
2. Place probe at 37 °C for 15–60 min to allow for suppression of repetitive sequences in the probe by cot-1 DNA in hybridization mix.
3. Apply to slide (12–15 μl per slide) and seal with Cow Gum.
4. Incubate at 37 °C overnight.
5. Proceed to *Protocol 8*.

Protocol 8

Wash steps

IMPORTANT! Do not let slides dry out at any stage.

Reagents

- 2 × SSC
- 50% formamide/1 × SSC

> **Protocol 8** continued
>
> **Method**
> 1. Remove coverslips by soaking in 2 × SSC for 2 min, then peel off the Cow Gum.
> 2. Wash in 50% formamide/1 × SSC for 20 min at 42 °C.
> 3. Wash in 2 × SSC for 20 min at 42 °C.
> 4. Proceed to probe detection steps (*Protocols 9–12*).

3 Probe detection

There are a number of protocols for the detection of hybridized probe and these are summarized in *Table 2*. The choice of detection system will be dictated by your choice of probe label (biotin or digoxigenin), and whether you are carrying out single or double hybridizations. For single hybridizations with biotin labelled probes we favour the avidin detection protocol, and for double hybridizations with biotin and dig labelled probes simultaneously, the avidin/anti-digoxigenin protocol.

Table 3 lists the properties of a number of fluorochromes. The combination of fluorochromes you use will depend on your specific application. Remember, you must match the fluorochromes you use with the capabilities of your microscope.

Table 2. Probe detection

Hybridization	Probe label	Immunological detection
Single probe	Biotin	Avidin
	Biotin	Antibody: anti-biotin
	Digoxigenin	Antibody: anti-digoxigenin
Two probes simultaneously	Biotin/digoxigenin	Avidin/anti-digoxigenin
	Biotin/digoxigenin	Anti-biotin/anti-digoxigenin

Table 3. Properties of fluorochromes

Fluorochrome	Max. excitation wavelength (nm)	Max. emission wavelength (nm)	Colour of fluorescence
Signal generating systems			
Coumarin AMCA	350	450	Blue
Fluoroscein FITC	495	515	Green
Cy3	550	570	Red
Rhodamine	550	575	Red
Rhodamine TRITC	575	600	Red
Texas Red	595	615	Red
Cy5	650	680	Far red
DNA counterstains			
Chromomycin A3	430	570	Yellow
DAPI	355	450	Blue
Hoechst 33258	356	465	Blue
Propidium iodide (PI)	340, 530	615	Red

Choose your microscope and filter system very carefully. There are two common DNA counterstains, DAPI and PI. PI fluoresces red and cannot be used in conjunction with red emitting signal generating systems like Texas Red but can be used with the green emitting signal generating system, FITC. DAPI is a blue emitting DNA counterstain and is compatible with both Texas Red and FITC.

Protocol 9
Avidin detection for biotinylated probes (FITC)

All detection steps are carried out in humidity chambers in the dark and under Parafilm coverslips. Use 100 µl of detection reagent per slide.

Reagents

- 4 × SSC-T: 4 × SSC, 0.05% Tween
- 4 × SSC-TB: 4 × SSC-T/0.5% block
- FITC–avidin DCS
- Biotinylated anti-avidin D
- FITC–avidin DCS
- Antifade with PI and DAPI

Method
1. Rinse in 4 × SSC-T for 3 min.
2. Block slides in 4 × SSC-TB for 10 min at room temperature under Parafilm coverslips in an humidity chamber.
3. First layer detection: add 100 µl to each slide of FITC–avidin DCS at 1:200 dilution in 4 × SSC-TB for 45 min at room temperature.
4. Wash in 4 × SSC-T for 10 min at room temperature.
5. Second layer detection: biotinylated anti-avidin D at 1:100 dilution in 4 × SSC-TB for 45 min at room temperature.
6. Wash in 4 × SSC-T for 10 min at room temperature.
7. Third layer detection: FITC–avidin DCS at 1:200 dilution in 4 × SSC-TB for 45 min at room temperature.
8. Wash in 4 × SSC-T for 20 min at room temperature.
9. Dehydrate and mount slides in Antifade with PI and DAPI.

Protocol 10
Digoxigenin antibody detection (FITC)

All detection steps are carried out in an humidity chamber in the dark and under Parafilm coverslips. Use 100 µl of detection reagent per slide.

Reagents

- 4 × SSC-T: 4 × SSC, 0.05% Tween
- 4 × SSC-TB: 4 × SSC-T/0.5% block
- Sheep anti-digoxigenin
- Donkey anti-sheep FITC
- Antifade with PI and DAPI

Protocol 10 continued

Method
1. Rinse in 4 × SSC-T for 3 min.
2. Block slides in 4 × SSC-TB for 10 min at room temperature under Parafilm coverslips in an humidity chamber.
3. First layer detection: add 100 μl to each slide of sheep anti-digoxigenin at 1:200 dilution in 4 × SSC-TB for 1 h at room temperature.
4. Wash in 4 × SSC-T for 10 min at room temperature.
5. Second layer detection: donkey anti-sheep FITC 1:300 in 4 × SSC-TB for 1 h at room temperature.
6. Wash in 4 × SSC-T for 20 min at room temperature.
7. Dehydrate and mount slides in Antifade with PI and DAPI.

Protocol 11

Probe detection using avidin and antibodies for two colours (FITC/Texas Red)

All detection steps are carried out in humidity chambers in the dark and under Parafilm coverslips. Use 100 μl of detection reagent per slide.

Reagents
- 4 × SSC-T: 4 × SSC, 0.05% Tween
- 4 × SSC-TB: 4 × SSC-T/0.5% block
- FITC–avidin DCS
- Biotinylated anti-avidin D
- Sheep anti-digoxigenin
- Donkey anti-sheep Texas Red
- Antifade with DAPI

Method
1. Rinse in 4 × SSC-T for 3 min.
2. Block slides in 4 × SSC-TB for 10 min at room temperature under Parafilm coverslips in an humidity chamber.
3. First layer detection: add 100 μl to each slide of FITC–avidin DCS at 1:200 dilution in 4 × SSC-TB for 45 min at room temperature.
4. Wash in 4 × SSC-T for 10 min at room temperature.
5. Second layer detection: biotinylated anti-avidin D 1:100 and sheep anti-digoxigenin 1:200 in 4 × SSC-TB for 1 h at room temperature.
6. Wash in 4 × SSC-T for 10 min at room temperature.
7. Third layer detection: FITC–avidin DCS at 1:200 and donkey anti-sheep Texas Red 1:300 in 4 × SSC-TB for 1 h at room temperature.
8. Wash in 4 × SSC-T for 20 min at room temperature.
9. Dehydrate and mount slides in DAPI only Antifade.

FLUORESCENCE *IN SITU* HYBRIDIZATION (FISH)

Protocol 12
Two colour antibody detection

All detection steps are carried out in humidity chambers in the dark and under Parafilm coverslips. Use 100 µl of detection reagent per slide.

Reagents

- 4 × SSC-T: 4 × SSC, 0.05% Tween
- 4 × SSC-TB: 4 × SSC-T/0.5% block
- Mouse anti-biotin
- Sheep anti-digoxigenin
- Goat anti-mouse Texas Red
- Donkey anti-sheep FITC
- Antifade with DAPI

Method

1. Rinse in 4 × SSC-T for 3 min.
2. Block slides in 4 × SSC-TB for 10 min at room temperature under Parafilm coverslips in an humidity chamber.
3. First layer detection: add 100 µl to each slide of mouse anti-biotin and sheep anti-digoxigenin at 1:200 dilution in 4 × SSC-TB for 1 h at room temperature.
4. Wash in 4 × SSC-T for 10 min at room temperature.
5. Second layer detection: goat anti-mouse Texas Red and donkey anti-sheep FITC 1:300 in 4 × SSC-TB for 1 h at room temperature.[a]
6. Wash in 4 × SSC-T for 20 min at room temperature.
7. Dehydrate and mount slides in DAPI only Antifade.

[a] The second antibody step could also be goat anti-mouse FITC and donkey anti-sheep Texas Red 1:300 dilution in 4 × SSC-T.

4 A final word on the tricky bits

If we assume the probe is perfect then the critical points in the procedure are specimen preparation and denaturation of the target DNA. In order for efficient hybridization to occur, the DNA in the target specimen must be made accessible by denaturation. This is achieved through pepsin digestion followed by immersion of the target in hot formamide. The critical steps are therefore length of time in pepsin, length of time in hot formamide, and the temperature of the formamide. Suggestions have been made concerning these steps, but some in-house experimentation will probably be necessary. Too much denaturation can lead to a loss of target DNA and therefore a drop in signal intensity, and it will lead to a loss of chromosome or tissue morphology, which becomes quite obvious down the microscope. Chromosomes will appear fuzzy and faintly counterstained, nuclei will appear ghost-like, and central areas of DNA may even be removed from nuclei. Too little denaturation will lead to poor signal strength with strong counterstaining of well defined chromosomes and nuclei.

5 FISH resources

5.1 Solutions

(a) Antifade + PI + DAPI: add to Antifade solution 0.3 μg/ml propidium iodide (PI) and 0.1 μg/ml DAPI (Boehringer Mannheim).

(b) 10% block (Boehringer Mannheim, Cat. No. 1096 176): make up in maleic acid buffer (100 mM maleic acid, 150 mM NaCl pH 7.5) and autoclave.

(c) 1 × SSC: 0.15 M NaCl, 0.015 M sodium citrate pH 7.

(d) 4 × SSC-T: 4 × SSC, 0.05% Tween 20.

(e) 4 × SSC-TB: 4 × SSC-T, 0.5% Boehringer block solution.

(f) RNase stock: 10 mg/ml in H_2O; aliquot and keep at $-20\,°C$.

(g) Pepsin: 10% in H_2O; aliquot and keep at $-20\,°C$.

(h) Methanol/acetic acid (3:1, v/v): 75 ml methanol + 25 ml acetic acid; make fresh just before use.

(i) 50% formamide hybridization mix (for single copy probes, such as cosmids): 2 × SSC, 500 μg/ml salmon sperm DNA, 10% dextran sulfate, 50% formamide.

(j) 70% formamide hybridization mix (for repeat sequence probes, such as centromeric probes): 2 × SSC, 500 μg/ml salmon sperm DNA, 10% dextran sulfate, 70% formamide.

5.2 Useful books

(a) *In situ hybridisation: a practical guide* (ed. A. R. Leatch, T. Schwarzacher, D. Jackson, and I. J. Leitch). (1994). Bios Scientific Publishers, Wiley Ltd. BUY THIS BOOK IF NO OTHER.

(b) *Diagnostic molecular pathology: a practical approach*, Vol. I and II (ed. C. S. Herrington and J. O'D McGee). (1992). IRL Press, Oxford University Press, Oxford.

(c) *Human cytogenetics: a practical approach*, Vol. I and II (ed. D. E. Rooney and B. H. Czepulkowski). (1992). IRL Press, Oxford University Press, Oxford.

(d) *Human cytogenetics: essential data series* (ed. D. E. Rooney and B. H. Czepulkowski). (1994). Bios Scientific Publishers Ltd., Wiley Ltd.

(e) *Human chromosomes, principles and techniques* (2nd edn) (ed. R. S. Verma and A. Babin). (1995). McGraw-Hill Inc. VERY GOOD AND VERY EXPENSIVE.

5.3 Useful Web sites

(a) `http://www.vysis.com` Vysis markets an integrated line of advanced genetic imaging workstations, nucleic acid probes, and reagents.

(b) http://www.lifescreen.com Information on apoptosis, FISH, lab equipment.

(c) http://www.cytocell.co.uk Cytocell develops and manufactures products for fluorescence *in situ* hybridization (FISH).

(d) http://www.cambio.co.uk Cambio sell FISH reagents including chromosome paints for mouse chromosomes.

(e) http://cc.ucsf.edu/people/waldman/index.html Waldman Group Home Page, Program in Cancer Genetics, UCSF Cancer Center. Protocols for CGH and FISH.

(f) http://141.42.37.2/cgh/index.html Chromosome and CGH protocols from Institut für Pathologie, Rudolf-Virchow-Haus, University Hospital Charité, Schumannstr. 20/21, 10117 Berlin.

(g) http:www.hybaid.co.uk Hybaid specializes in the manufacture and supply of equipment and reagents for molecular biology.

(h) http://www.genomesystems.com Genome Systems have genomic libraries which they will screen for you. This will give you great probes for FISH.

(i) http://www.resgen.com Research Genetics have assembled genomic libraries and genome resources. They will screen libraries for you to get genomic probes that you can then use for FISH.

(j) http://www.hgmp.mrc.ac.uk UK Human Genome Mapping Project Resource Centre. This is an excellent source of probes and filters that you can screen yourself for sequences of interest.

References

1. Soder, A. I., Hoare, S. F., Muire, S., Balmain, A., Parkinson, E. K., and Keith, W. N. (1997). *Genomics*, **41**, 293.
2. Gause, P. R., LluriaPrevatt, M., Keith, W. N., Balmain, A., Linardopolous, S., Warneke, J., *et al.* (1997). *Mol. Carcinogenesis*, **20**, 78.
3. Soder, A. I., Hoare, S. F., Muir, S., Going, J. J., Parkinson, E. K., and Keith, W. N. (1997). *Oncogene*, **14**, 1013.
4. McLeod, H. L. and Keith, W. N. (1996). *Br. J. Cancer*, **74**, 508.
5. Withoff, S., Keith, W. N., Knol, A. J., Coutts, J. C., Hoare, S. F., Mulder, N. H., *et al.* (1996). *Br. J. Cancer*, **74**, 502.
6. Withoff, S., deVries, E. G. E., Keith, W. N., Nienhuis, E. F., van der Graaf, W. T. A., Uges, D. R. A., *et al.* (1996). *Br. J. Cancer*, **74**, 1869.
7. Murphy, D. S., Hoare, S. F., Going, J. J., Mallon, E. E. A., George, W. D., Kaye, S. B., *et al.* (1995). *J. Natl. Cancer Inst.*, **87**, 1694.
8. Macville, M., Schrock, E., PadillaNash, H., Keck, C., Ghadimi, B. M., Zimonjic, D., *et al.* (1999). *Cancer Res.*, **59**, 141.
9. Dixkens, C., Posseckert, G., Keller, T., and Hameister, H. (1998). *Chromosome Res.*, **6**, 329.
10. AlMulla, F., Keith, W. N., Pickford, I. R., Going, J. J., and Birnie, G. D. (1999). *Genes Chromosomes Cancer*, **24**, 306.
11. Coutts, J., Plumb, J. A., Brown, R., and Keith, W. N. (1993). *Br. J. Cancer*, **68**, 793.
12. Hoare, S. F., Freeman, C. A., Coutts, J. C., Varley, J. M., James, L., and Keith, W. N. (1997). *Br. J. Cancer*, **75**, 275.

Chapter 9
Genetic modification

Majid Hafezparast

Department of Neurogenetics, Imperial College School of Medicine, Norfolk Place, London W2 1PG, UK.

1 Introduction

To explore the role of a gene, we need to analyse its regulation and expression during and after development, the function of its product, and interactions with other proteins. In most cases, it is necessary to be able to transfer the gene, or its manipulated form, into cells. However, since mammalian cells do not take up foreign DNA efficiently, the availability of effective methods for introducing genes into the cells is essential. Many gene transfer methods have been developed that are routinely used by scientists studying mammalian cells. Moreover, with the advancement of recombinant DNA technology, gene transfer techniques have become powerful tools in gene cloning and mapping, construction of transgenic animals, and gene therapy.

In this chapter some of the most commonly used methods for introducing genes into mammalian cells, including transfection by calcium phosphate precipitation and cationic lipids, electroporation, and the techniques of microcell-mediated chromosome transfer and irradiation fusion gene transfer, will be discussed. Other techniques of gene delivery, which are not discussed here, include DEAE dextran neutralization (1), microinjection of DNA directly into the cells (2), and the use of viral vectors (3).

Due to their versatility and efficiency, viral vectors have become attractive tools for the introduction of genes into human cells in gene therapy studies. Their drawbacks are the biological hazards associated with viruses and in some cases the immunogenicity of viral vectors (4). Alternative therapeutic gene delivery systems use cationic lipids as carriers of the exogenous DNA (see Section 2.2).

2 Transfection

Gene transfer methods by transfection can be classified into two major categories: biochemical and physical. The biochemical methods include calcium phosphate precipitation and lipid-mediated transfections, and physical transfection is achieved by electroporation. The choice of method depends on the cell type. For example, haematopoietic cell lines are more efficiently transfected by electroporation than the biochemical based methods (5), whereas some cell types take up DNA efficiently when it is associated with cationic lipids rather than calcium phosphate precipitates. On the other hand, cationic lipids can be highly toxic to some other cell types.

Following transfection, the exogenous DNA remains in the nucleus for a limited time unless it is integrated into the host genome. However, even in the non-integrated form, the transfected DNA is subject to the regulatory mechanisms that control gene expression and therefore its genes can be expressed transiently. This transient gene expression has been exploited in studies such as the analysis of regulatory sequences (6–8) and optimization of transfection conditions (5). Integration of the transfected gene into the host genome, on the other hand, gives rise to stable retention of the exogenous gene in the cells, providing constitutive expression.

In most transfection experiments the cDNA of the gene of interest is inserted into a mammalian expression vector that carries an expression cassette consisting of a strong promoter, such as the LTR promoter of Rous sarcoma virus (for expression of the gene in mammalian cells), a multiple cloning site, and a polyadenylation sequence. The vector also contains a bacterial origin of replication and a bacterial selectable marker (for propagation and selection of the vector in bacterial cells). In order to isolate the few transfected mammalian cells from the majority of non-transfected ones, a dominant mammalian selectable marker is also utilized. This marker may be on a different vector that is co-transfected into the cells along with the vector or DNA sequence that carries the target gene. In this case, although the co-transfected DNAs are not physically linked, a considerable number of them integrate into the genome in large concatemeric structures that can be up to 2000 kb (9). Therefore it is believed that at some stage after entering the cell co-transfected DNA sequences are ligated to each other before integrating into the host genome (10, 11).

Transfection is a powerful tool for cloning genes that give rise to phenotypic effects in the transfected cells. This gene cloning strategy, however, requires the rescue of the exogenous DNA from the genomes of the cells that exhibit the phenotypic effects. Several procedures have been described to achieve this, but they are often cumbersome and involve second rounds of transfections followed by construction of genomic libraries from the secondary transfectants and screening of this library for the cloned gene (12–15). Hence, other types of vectors have been generated that utilize the Epstein–Barr virus (EBV) origin of replication. The EBV origin of replication allows these vectors to replicate in human cells during each cell division while maintained in an episomal state (i.e. non-integrated). Therefore these vectors have two advantages. First, they give rise to greater transfection efficiencies, because they do not need to integrate to be maintained, and secondly, following transfection, they can be shuttled back into an *E. coli* host, thus making their subsequent isolation easier (see *Protocol 5*). The EBV-based shuttle vectors were successfully used in cloning the human cDNAs for Fanconi's anaemia complementation group C (*FACC*) and xeroderma pigmentosum group C (*XPCC*) genes (16, 17).

2.1 Calcium phosphate–DNA co-precipitation

The method of calcium phosphate–DNA co-precipitation is the most widely used technique for the transfection of mammalian cells. This technique was first introduced by Graham and Van der Eb in 1973 (18). Since then there have been many modifications and optimization protocols for obtaining higher transfection frequencies for both stable and transient expressions of exogenous genes. The principle of this technique is the formation of insoluble calcium phosphate–DNA complexes in a supersaturated solution. This is achieved by adding a solution containing DNA and $CaCl_2$ to a buffered saline solution containing phosphate and incubating the mixture for a period of time, to allow the formation of calcium phosphate–DNA co-precipitates. The mixture is added to culture cells in medium followed by incubation at 37°C for a period, usually between 6–24 hours, depending on the cell type. DNA molecules enter the cells by endocytosis of the calcium phosphate–DNA co-precipitates. Several of the factors that influence the efficacy of this method are described below.

2.1.1 DNA

The concentration, topology, and quality of DNA affect the transfection efficiency. It has been shown that suboptimal concentrations of DNA drastically reduce the efficiency of calcium phosphate transfection (19). In agreement with this, studies on the kinetics of the formation of calcium phosphate–DNA complex indicate that high DNA concentrations inhibit this process and lead to reduction of DNA associated with the insoluble complex (20). Moreover, Chen and Okayama (19) have reported that supercoiled plasmid DNA gives rise to several-fold higher numbers of stable transformants than linearized DNA.

2.1.2 pH

Optimum pH range for the formation of calcium phosphate precipitates for most cell lines is 6.95 to 7.05. It has been suggested that the CO_2 level in the incubator during transfection also influences the transfection frequency (19, 20).

2.1.3 Standing and incubation times

Standing time greatly affects the quality of precipitates for transfections. The longer the calcium phosphate–DNA mixture is incubated prior to addition to the cells the coarser the precipitates will become, and since there is an optimal size of the precipitates to be taken up by cells, fine-tuning is needed to obtain the best standing time. It appears that the solubility of calcium phosphate is reduced by increasing the temperature, however, most protocols suggest room temperature and varying standing times of 1–20 minutes. Incubation time for transfection varies between a few hours up to 16 hours, depending on the cell type.

Protocol 1
Transfection by calcium phosphate–DNA co-precipitation

Reagents

- 2.5 mM CaCl$_2$ (filter sterilized and stored at -20°C)
- 1 × PBS (phosphate-buffered saline)
- 2 × Hepes solution: 140 mM NaCl, 1.5 mM Na$_2$HPO$_4$, 50 mM Hepes pH 7.05 at 23°C (filter sterilized and stored at -20°C)

Method

1. Seed appropriate numbers of exponentially growing cells in 10 cm tissue culture Petri dishes.
2. When cells are between 60–80% confluent (within 48 h after seeding), change the medium.
3. Label one 14 ml round-base polypropylene tube for each culture plate to be transfected.
4. Add 25 μg of DNA to each tube, with the exception of the negative control.
5. Make up the volume to 450 μl in each tube by adding appropriate volume of room temperature ddH$_2$O and mix gently.
6. Add 50 μl of 2.5 mM CaCl$_2$ to each tube and mix gently.
7. Add 500 μl of 2 × Hepes solution dropwise while swirling the mix gently.
8. Incubate at room temperature for 10 min.
9. Mix each tube by swirling gently and slowly add to the culture dishes, containing 10 ml of medium,[a] while rocking the dish gently.
10. Incubate the culture dishes at 37°C for 5 h.
11. Aspirate the medium, wash twice with PBS, and add supplemented medium.
12. Allow 24–48 h for expression of transfected gene(s) before the analysis of gene expression (for transient transfections) or adding selective drug (for stable transfections).

[a] Do not use RPMI 1640.

2.2 Lipid-mediated gene transfer (lipofection)

The use of cationic lipids for DNA transfection into mammalian cells has become widespread because several features of these reagents make them attractive vehicles for gene delivery, particularly in gene therapy. For instance, they are safer than viral vectors, can be produced in large quantities, and can deliver large DNA fragments of up to several megabase pairs long into cells.

There are many formulations of lipid reagents for transfection, but they normally contain a positively charged moiety attached to a neutral lipid component (21). The first generation of lipid reagents for transfection contained

Table 1. Some of the commercially available cationic reagents

Supplier	Name	Comments
Promega	Tfx™	Different ratios of cationic lipid and the neutral lipid DOPE[a]
	TransFast™	Cationic lipid with DOPE
	Transfectam™	Cationic lipid with spermine head groups
Boehringer Mannheim	DOSPER[b]	Polycationic lipid with a spermine head group
	DOTAP[c]	Monocationic lipid
Life Technologies	Lipofectin	1:1 formulation of DOTMA[d] and DOPE
	LipofectAmine	3:1 formulation of polycationic lipid DOSPA[e] and DOPE
	CellFectin	1:1.5 molar mixture of TM-TPS[f] and DOPE
	DMRIE-C[g]	1:1 formulation of DMRIE and cholesterol
Invitrogen	PerFect Lipid™	Eight different lipid reagents supplied as either a single cationic lipid, a combination of two cationic lipids, or a cationic lipid and DOPE
Stratagene	LipoTAXI®	A novel liposome reagent; tested on more than 30 cell lines

[a] L-Dioleoyl phosphatidyl ethanolamine.
[b] 1,3-Di-oleoyloxy-2-(6-carboxy-spermyl)-propylamide.
[c] N-[1-(2,3-dioleoyloxy)propyl]-N,N,N-trimethylammonium methylsulfate.
[d] N-[1-(2,3-dioleyloxy)-propyl]-N,N,N-trimethylammonium chloride.
[e] 2,3-dioleoyloxy-N(2(sperminecarboxamido)ethyl)-N,N-dimethyl-1-propanammonium trifluoroacetate.
[f] N,N1,N2,N3 tetra-methyltetrapalmityl spermine.
[g] 1,2-Dimyristyloxypropyl-3-dimethyl-hydroxy ethylammonium bromide.

neutral lipids that relied on methods for capturing DNA molecules within the liposomes, but today's cationic lipids form lipid–DNA complexes that make the transfection efficiency independent of encapsulation of DNA. On mixing of these reagents with DNA, the charged head groups are drawn towards the phosphate backbone of DNA and form lipid–DNA complexes. When the suspension of these complexes is added to the cells, the positively charged head groups of the lipid are attracted to the negatively charged cell membrane. The end-result is that the lipid–DNA complex is either fused to the cell membrane or enters the cell by endocytosis, transferring its DNA load into the cell. It is due to these properties of the cationic liposomes that efficient DNA delivery has been achieved in many cell lines.

Various liposome formulations are commercially available (see *Table 1*) but it is important to bear in mind that the transfection efficiency depends very much on the cell type and also the chemical and physical structure of the cationic liposomes, particularly with respect to the liposome size and cationic head group (22). Thus preparations must be compared to identify the formulation that gives the best transfection efficiency for the cell line of interest. Other significant factors in using these reagents for DNA transfection are lipid:DNA ratio, their concentrations, DNA quality, cell culture density, and the duration of exposure of cells to the lipid. A starting point for optimizing the lipofection conditions is described in *Protocol 2*.

Protocol 2

Optimization of conditions for lipofection

Reagents

- A β-galactosidase enzyme activity assay system
- pSV-β-galactosidase vector DNA (Promega)

Method

1. Seed appropriate number of cells in nine wells of two 6-well plates to give 60–80% confluency within 24 h.
2. Incubate at 37 °C for 24 h.
3. Prepare the following combinations of the cationic lipid and DNA concentrations, following the manufacturer's instructions.[a]

	Set 1 (1 μg DNA)	Set 2 (2 μg DNA)	Set 3 (5 μg DNA)	Ratio (Lipid:DNA)
Cationic lipid	2.5 μg	5 μg	12.5 μg	2.5
	5 μg	10 μg	25 μg	5
	10 μg	20 μg	50 μg	10

4. Remove medium from the wells.
5. Mix each lipid reagent–DNA suspension and overlay onto the cells.
6. Incubate the plates at 37 °C for 1–5 h.
7. Add complete medium and incubate for 24–48 h.
8. Analyse β-galactosidase expression by using an assay system such as β-galactosidase assay kit from Promega (E2000), or X-gal staining (see *Protocol 4*).

[a] Some lipids transfect cells more efficiently in the absence of serum, but in most cases serum does not interfere with the transfection efficiency. This however has to be determined empirically.

2.3 Electroporation

Electroporation involves exposure of cells in suspension to a pulsed electric field that causes transient formation of pores in the cell membrane, allowing exchange of macromolecules between the extracellular environment and the cytoplasm. Removal of the electric field results in spontaneous sealing of the pores, as long as the field strength and duration of the pulse are within the tolerable range for the cells (23). Parameters that affect the transfection efficiency of electroporation are discussed below.

2.3.1 Voltage

One of the most important factors affecting transfection efficiency by electroporation is the electric field strength across the cell which, at a critical point, leads to the localized breakdown of the cell membrane and formation of pores. Since the field strength is described by:

$$E = V/d$$

where E is the field strength, V is the voltage, and d is the distance between electrodes, there is a direct correlation between the field strength and voltage. Moreover, the field strength is also proportional to the product of the capacitor voltage and the cell diameter (24). Therefore the optimum voltage for different cell types varies and has to be determined empirically. Generally, efficient introduction of DNA into cells occurs at field strengths that give rise to 20–80% cell death. It is noteworthy that after electroporation many cells that exhibit trypan blue exclusion and adhere to the plate die and detach within 48 hours post-transfection (25) and therefore counting the cells by trypan blue exclusion may provide an inaccurate measure of the number of viable cells.

2.3.2 Time constant

Most electroporations are performed using exponential decay waves that are generated by an instant surge of voltage followed by its exponential decline over a time period. This is defined as time constant and is a product of capacitance and the electroporation buffer resistance. Time constant is a useful parameter for ensuring reproducibility of the pulse. Some commercial electroporation apparatuses indicate the time constant after each pulse and provide the opportunity to compare different pulses and hence reproducibility of the electroporation.

In order to modify the time constant, the capacitance of the apparatus and/or the buffer resistance can be altered. Higher ionic strength of the buffer and shorter distance between the electrodes result in lower resistance. The resistance is however inversely proportional to the cross-sectional area of the path and therefore it is possible to manipulate the resistance by changing the volume of the electroporation buffer as well.

2.3.3 Buffer

The composition of the electroporation buffer determines its resistance and hence influences the time constant. Several types of buffer have been used in electroporation of mammalian cells as indicated in *Protocol 3*. In some cases growth media have also been used as the suspension medium. It is therefore essential to determine the best buffer/medium for each cell type empirically.

2.3.4 Temperature

There are different opinions with regards to the cell suspension temperature before, during, and after the electric pulse. Some reports indicate higher transfection frequencies at room temperature whereas others suggest incubation of cells in the presence of DNA on ice at least ten minutes before and after applying the pulse (25, 26).

2.3.5 DNA

Increasing DNA concentration results in higher transfection frequency by electroporation. High transfection efficiencies have been reported with up to 80 μg/ml of DNA. Carrier nucleic acids have also been used to increase the transfection efficiency, but their nature appears to play an important role. For instance salmon sperm DNA increases transfection efficiencies of some human cell types whereas yeast tRNA results in lower transfection frequencies (25).

2.3.6 Cells

Like other transfection methods the type of cells and the cell cycle phase at the time of electroporation are important factors in determining the efficiency of transfection by electroporation. In most cases best results are obtained with 1×10^6 to 2×10^7 cells/ml at exponential growth phase.

Protocol 3

Determination of the best capacitance setting for electroporation of a particular mammalian cell type

Equipment and reagents

- Electroporation apparatus (e.g. Genepulser II from Bio-Rad)
- Appropriate electroporation cuvettes
- Electroporation buffer such as:
- 1 × HeBS: 20 mM Hepes pH 7.05, 137 mM NaCl, 5 mM KCl, 0.7 mM Na$_2$HPO$_4$, 6 mM dextrose
- PBS: phosphate-buffered saline
- 1 × Hepes: 70 mM NaCl, 2.5 mM KCl, 0.35 mM Na$_2$HPO$_4$, 2.8 mM dextrose, 21 mM Hepes pH 7.05
- HBS: 1 × Hepes plus 10% sucrose
- DNA (1 μg/μl) of a mammalian expression vector for the reporter gene *Lac-Z*, e.g. pSV-β-galactosidase vector DNA (Promega)

Method

1. Grow the cells to 60–70% confluency.
2. Replace the growth medium 1–2 h before harvesting.
3. Trypsinize, pool, and count the cells.
4. Dispense the cells into three centrifugation tubes at 2×10^7 cells per tube.
5. Centrifuge the tubes at 500 g for 10 min.
6. Discard the supernatant and resuspend each pellet in 10 ml of the electroporation buffer.
7. Centrifuge the cell suspensions as above and discard the supernatant.
8. Resuspend each pellet in 0.75 ml of the electroporation buffer, to give a total volume of 0.8 ml.
9. Add 25 μg of the plasmid DNA to each cell suspension, mix gently, and incubate the cell–DNA mixtures at room temperature for 10 min.

Protocol 3 continued

10. Mix one of the cell–DNA suspensions gently and transfer that to an electroporation cuvette.
11. Place the cuvette in the electroporation chamber and pulse the cells at 200 V and 250 µF capacitance.[a]
12. Incubate the electroporated cells at room temperature for 10 min.
13. Meanwhile electroporate the remaining two cell suspensions by keeping the voltage constant at 200 V and varying the capacitance to 500 µF and 960 µF.
14. After the incubation period, transfer the contents of each cuvette into 50 ml of growth medium, mix, and dispense into five 10 cm tissue culture plates.
15. Incubate the plates at 37 °C for 24–72 h.
16. Stain the cells for β-galactosidase expression (see *Protocol 4*), add 10 ml of PBS to each plate, and count blue cells.
17. Compare the numbers of blue cells obtained from different electroporation settings and identify the setting that gives rise to the highest number of blue cells, i.e. the highest transfection frequency.

[a] Record the time constant and check the actual voltage of the pulse for reproducibility of the electroporation conditions.

2.4 Staining of cells for expression of β-galactosidase

Factors that influence the efficiency of a gene delivery system have to be tested empirically to obtain the best conditions for a particular cell line. For this purpose, use can be made of reporter genes, including the genes that code for β-galactosidase, luciferase, and chloramphenicol acetyl transferase (CAT). The reporter gene can be transfected into cells and its expression analysed by a biochemical assay and/or histochemical staining 24–72 hours post-transfection. Because there tends to be a correlation between transient expression of the reporter gene and stable transformation of cells, this method can be used to determine the optimum conditions for both transient and stable transfections.

Protocol 4
Staining of cells for β-galactosidase expression

Reagents

- 100 mM phosphate buffer: 57.7 ml of 100 mM Na_2HPO_4 pH 7.0, and 42.3 ml of 100 mM NaH_2PO_4 pH 7.0
- 1 M $MgCl_2$
- 100 mM $K_4Fe(CN)_6.3H_2O$ (freshly made)
- 100 mM $K_3Fe(CN)_6$ (freshly made)
- 50% glutaraldehyde

- X-gal stain: 10 mM phosphate buffer, 1 mM $MgCl_2$, 150 mM NaCl, 33 mM $K_4Fe(CN)_6.3H_2O$, 33 mM $K_3Fe(CN)_6$, 0.1% X-gal
- NP-40 buffer: 1 mM $MgCl_2$, 0.02% Nonidet P-40

Protocol 4 continued

Method

1. Wash attached cells three times with 100 mM phosphate buffer.
2. Fix the cells in 100 mM phosphate buffer containing 1 mM $MgCl_2$ and 0.5% glutaraldehyde at 4°C for 10 min.
3. Wash the cells with NP-40 buffer at room temperature.
4. Repeat step 3 twice more.
5. Stain the cells by overlaying them with the X-gal stain and incubate at 37°C, in a humidified atmosphere, for 2–24 h.
6. Wash the cells three times with PBS.
7. Add 10 ml PBS to each plate and visualize the Lac-Z expressing cells, appearing blue, under microscope.

2.5 Rescue of episomal plasmids

This method was originally described by Hirt (27) for the isolation of viral DNA from mouse cells. The episomal Epstein–Barr virus (EBV) vectors such as the pREP series of expression vectors (Invitrogen) can be rescued from transfected cells and used to transform E. coli cells for their propagation (see Section 2 for the features of EBV-based vectors).

Protocol 5

Plasmid rescue from cells transfected with EBV-based shuttle vectors

Reagents

- 0.6% SDS/10 mM EDTA
- 5 M NaCl
- Phenol/chloroform/isoamyl alcohol (25:24:1 parts)
- Chloroform/isoamyl alcohol (24:1 parts)
- 0.25% linear polyacrylamide[a]
- Highly competent E. coli cells (Stratagene)

Method

1. Culture transfected cells that carry the EBV-based vector to 80% confluency in a 10 ml tissue culture plate.
2. Lyse the cells by adding 0.6 ml of 0.6% SDS/10 mM EDTA and rock to ensure complete spreading of the lysis solution over the cells.
3. Incubate at room temperature for 15 min.
4. Scrape the viscous lysate from the flask with a cell scraper and transfer to a screw top 10 ml polypropylene tube.
5. Add 150 µl of 5 M NaCl to the lysate and mix gently by inverting the tube ten times.

Protocol 5 continued

6. Incubate the tube on ice in the cold room for 2 h or overnight.
7. Centrifuge the tube in a 4°C centrifuge at 10 000 g for 30 min.
8. Pipette the supernatant to a fresh 2 ml Eppendorf tube and phenol/chloroform/isoamyl alcohol extract twice followed by one extraction with chloroform/isoamyl alcohol.
9. Add 5 μl of 0.25% linear polyacrylamide (stored at −20°C) to the extracted DNA suspension, mix, and precipitate the plasmid DNA with 2 vol. of absolute ethanol.
10. Incubate the tube at −20°C for 1 h.
11. Pellet the DNA by centrifugation at high speed in a microcentrifuge at 4°C for 15 min.
12. Wash the pellet with 500 μl of 70% ethanol at room temperature.
13. Centrifuge the tube at high speed for 15 min at room temperature, remove the ethanol, dry, and dissolve the pellet in 10 μl of ddH$_2$O.
14. Transform highly competent *E. coli* cells according to the manufacturer's instructions, using 5–10 μl of the DNA suspension.[b]

[a] Gaillard and Strauss (28).

[b] Keep everything cold while doing the transformation, e.g. store tips at −20°C for about 10 min before using them for aliquoting the bacterial cells etc.; the duration and temperature of heat shocking are critical.

3 Microcell-mediated chromosome transfer

Cell–cell fusion of somatic cells is a simple way of transferring genetic material from one cell to another, where all the chromosomes of the participating cells are engulfed within a single cell membrane. A more refined method of transferring genetic information is the method of microcell-mediated chromosome transfer (MMCT), where only one chromosome, or a small group of chromosomes, is transferred to the recipient cell.

The MMCT technique has been a powerful tool in a variety of studies involving the transfer of human chromosomes to rodent cells and the generation of monochromosomal human–rodent cell hybrids. However, since rodent cell hybrids lose the human chromosomes at high frequencies, one of the first considerations is to have the means for selecting the cells that have received and retained the desired chromosome. Naturally occurring markers such as dihydrofolate reductase, thymidine kinase, and hypoxanthine-guanine phosphoribosyl transferase, on chromosomes 5, 17, and X, respectively, can be used to isolate human–rodent hybrids that carry these chromosomes. However, for the majority of the human chromosomes exogenous markers have to be employed to tag the chromosomes prior to their transfer into the recipient rodent cells. The two most widely used markers for this purpose have been the *neo* gene, which confers resistance to neomycin, and the bacterial xanthine-guanine phosphoribosyl transferase (*gpt*) gene, which enables mammalian cells carrying this gene to utilize xanthine rather than hypoxanthine in the salvage pathway of purine biosynthesis. Methods such as calcium phosphate co-precipitation and retroviral transduction have been used to transfer these markers into human cells and to generate monochromosomal hybrids (29–31).

Microcell-mediated chromosome transfer has been particularly useful in the mapping of DNA repair genes. In these studies, complementing DNA repair genes have been assigned to particular chromosomes by individually transferring human chromosomes into DNA repair defective cell lines and identifying the chromosomes that restore proficient DNA repair (32–39). This technique has also been used in other studies such as chromosomal localization of tumour suppressor genes (40–44), gene regulation (45–49), manipulation of chromosomal alleles (50), and introduction of human minichromosomes into mouse embryonal stem cells (51).

The donor cells in MMCT experiments are usually rodent cells carrying a single human chromosome, tagged with a selectable marker such as *gpt*, *neo*, or *hyg*, and are called mouse–human monochromosomal hybrids. Several groups have constructed monochromosomal hybrid panels that can be obtained directly from the laboratories that constructed them, purchased from commercial centres or from public resource institutes, such as Coriell Cell Repositories (http://locus.umdnj.edu/nigms/hybrids/sumintro.html) and the UK MRC Human Genome Mapping Project Resource Centre (http://www.hgmp.mrc.ac.uk/Research/). It must be noted that in some cases in addition to the selected intact human chromosome, a monochromosomal hybrid may have retained a small fragment(s) of another human chromosome and therefore careful characterization of these hybrids is essential before their use in any chromosome transfer experiment.

The procedure of MMCT involves five main stages, which are illustrated in *Figure 1* and described below.

3.1 Formation of micronuclei

Micronucleation describes the process of formation of micronuclei and is induced by treating donor cells with a mitotic inhibitor such as colcemid. Treatment of the cells with sublethal doses of colcemid results in disruption of microtubule formation during cell division followed by the formation of micronuclei when the cells enter interphase. Each micronucleus contains one or a few chromosomes within a membrane.

The appropriate dose of colcemid has to be determined empirically. In general, micronucleation can be induced in rodent cells with lower doses (0.01–0.10 μg/ml) of colcemid than in human cells, which need higher concentrations (10–20 μg/ml) of this drug (52). In the MMCT protocol that is described below, micronucleation is achieved by prolonged incubation (30–40 hours) of mouse–human monochromosomal hybrid cells, as donors of the human chromosome, in medium containing 0.02 μg/ml of colcemid. Another method has been

GENETIC MODIFICATION

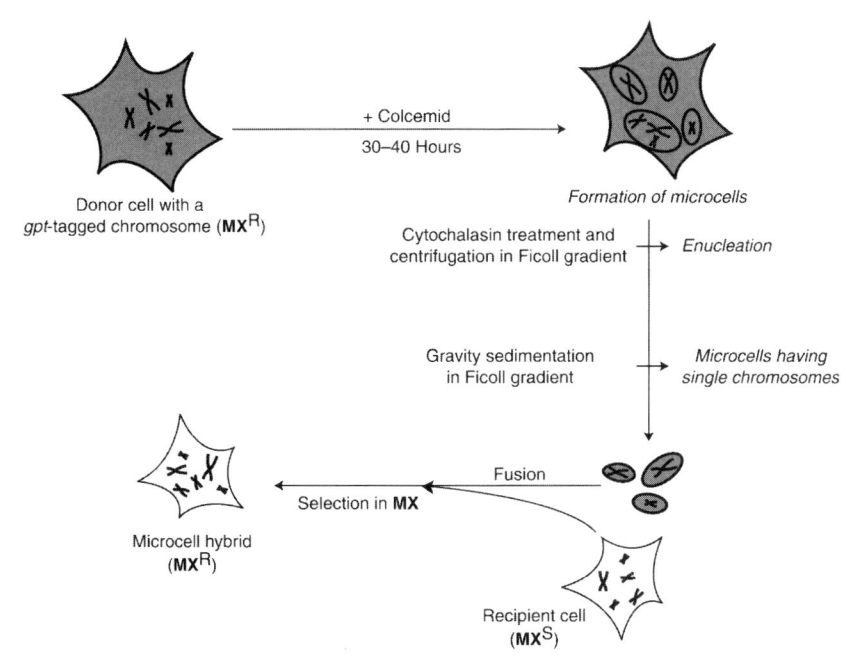

Figure 1. Microcell-mediated chromosome transfer. In this method a single chromosome that is tagged with a dominant marker (such as the bacterial guanine phosphoribosyl transferase (*gpt*) gene) is encapsulated into a microcell by colcemid-induced micronucleation of its donor cell. The microcell is then extruded from the micronucleate cell by centrifugation in a discontinuous Ficoll gradient containing cytochalasin B and then size fractionated in a continuous Ficoll gradient. Fusion of the microcell to a recipient cell and growth in MX medium (medium containing mycophenolic acid and xanthine) gives rise to colonies of microcell hybrids that contain the tagged chromosome from the donor cell. MX = culture medium containing mycophenolic acid and xanthine. MX^R = resistant to MX. MX^S = sensitive to MX.

described in which the donor cells are treated with colcemid for a shorter period of time (8–16 hours) followed by incubation for 4–6 hours at 37 °C in medium containing cytochalasin B (52), which is a mitotic inhibitor and induces nuclear extrusion.

3.2 Enucleation

When micronucleate cells are treated with relatively high concentrations of cytochalasin B (10 µg/ml) the micronuclei can be extruded in two ways: as monolayer cells or as cells in suspension.

3.2.1 Enucleation of monolayer cells

Micronucleate cells can be enucleated while they are still attached to the surface of either a tissue culture flask or to sterile pieces of plastic cut from tissue culture plates in the shape of disks or bullets. If the cells are attached to a tissue culture flask, the flask is filled with serum-free medium containing 10 µg/ml cytochalasin B and then centrifuged. If, on the other hand, micronucleate cells have been generated on disks or bullets, they are enucleated by transferring the disks, cell side down, to 50 ml centrifuge tubes containing serum-free medium supplemented with 10 µg/ml of cytochalasin B and centrifuged at high speed. The g force of the centrifugation, of course, must not exceed the tolerable limit for the tissue culture flask or the pre-cut plastic pieces (52).

3.2.2 Enucleation in suspension

In the method described in *Protocol 6*, enucleation is achieved by first harvesting the micronucleate cells by trypsinization and then centrifuging them in a discontinuous Ficoll gradient in the presence of cytochalasin B.

3.3 Purification of microcells

The end-result of enucleation of micronucleate cells is the release of microcells as compartments containing single, or just a few chromosomes, with a thin rim of cytoplasm encircled by a cell membrane.

Microcells can be purified on the basis of their size either by unit gravity sedimentation (see *Protocol 6*), or by filtration through polycarbonate filters with pore size of 5 µm or 8 µm. Alternatively, the crude microcell preparation is transferred into tissue culture flasks followed by incubation at 37°C for several hours. In this period of incubation, which varies for different donor cell types, the contaminating cells that are intact adhere to the surface of the tissue culture flasks and the microcells, which are non-adhering, remain floating in the medium and can be collected (52).

Protocol 6
Preparation of microcells

Equipment and reagents

- Ultracentrifuge
- PBS/EDTA: 1 mM EDTA in 1 × PBS
- Cytochalasin B: 1 mg/ml in DMSO
- 24% Ficoll: 4 ml of 30% Ficoll, autoclaved, and 1 ml of PBS/EDTA
- 18% Ficoll containing 10 µg/ml cytochalasin B: 6 ml of 30% Ficoll, 4 ml of PBS, and 0.1 ml of 1 mg/ml cytochalasin B
- 15% Ficoll containing 10 µg/ml cytochalasin B: 10 ml of 30% Ficoll, 10 ml of PBS/EDTA, and 0.2 ml of 1 mg/ml cytochalasin B
- 1% and 3% Ficoll in PBS/EDTA
- 0.5% BSA in PBS/EDTA
- 10% sucrose

Method

1. Culture chromosome donor cells to 50% confluency in 10 cm dishes.
2. Replace the growth medium with fresh medium containing 20 ng/ml colcemid and incubate at 37°C for 40 h.

Protocol 6 continued

3. Harvest the cells by trypsinization and transfer to a 50 ml centrifuge tube.
4. Adjust total volume of the cell suspension to 40 ml with PBS/EDTA.
5. Count the cells using a haemocytometer and ensure that at least 2×10^7 cells have been harvested.
6. Centrifuge the cells at 500 g for 10 min.
7. Resuspend the pellet of cells in 10 ml of serum-free medium supplemented with 10 µg/ml cytochalasin B.
8. Incubate the cell suspension in a 37 °C water-bath for 30 min.
9. Add 1.5 ml of 24% Ficoll to two 30 ml centrifuge tubes.
10. Slowly layer 9.5 ml of 15% Ficoll containing 10 µg/ml cytochalasin B over the Ficoll solution in step 9 and keep the tubes at 37 °C.
11. Add 10 ml of 18% Ficoll containing 10 µg/ml cytochalasin B to the cells.
12. Pass the Ficoll–cell suspension six times through a 40 ml Dounce homogenizer.
13. Layer the homogenized cells onto the surface of the Ficoll gradients that were prepared in steps 9 and 10.
14. Slowly add serum-free medium to within approximately one inch of the top of the tubes without disturbing the gradients.
15. Balance the tubes, place them in pre-warmed, 37 °C, centrifuge buckets and centrifuge at 80 000 g for 75 min at approx. 25 °C.
16. Discard the top part of the gradient, leaving some of the 15% and the whole of the 24% phases in the tubes.
17. Transfer the remainder to a 50 ml centrifuge tube, avoiding the pellet, which might contain intact donor cells.
18. Carefully pipette the suspension up and down several times.
19. Make up the total volume to 50 ml with PBS/EDTA and count the microcells using a haemocytometer.
20. Pellet the microcells by centrifugation at 3000 g for 10 min.
21. Resuspend the pellet in 50 ml PBS/EDTA and wash twice.
22. Resuspend the pellet of microcells in 2 ml PBS/EDTA and pass ten times through a 10 ml Dounce homogenizer.
23. Set up a gradient making system as shown in *Figure 2*.
24. Transfer 10 ml of 0.5% BSA to the Universal bottle of the gradient making system.
25. Check that the tap between the gradient maker chambers is closed and the tube that connects the gradient maker to the Universal bottle is sealed with a clamp.
26. Add 30 ml of 1% Ficoll to the gradient maker chamber which is connected to the Universal bottle (i.e. chamber B in *Figure 2*) and 30 ml of 3% Ficoll to the other chamber (i.e. chamber A in *Figure 2*).

Protocol 6 continued

27. Connect the chambers by opening the tap and allow the solution to flow slowly through the system, by loosening the clamp.

28. Stop the flow as soon as the solution reaches the bottom of the reservoir.

29. Layer the microcell suspension very slowly onto the gradient solution in the reservoir.

30. Restart the flow and stop when the volume of the gradient is approx. 46 ml (this process has to take between 10–15 min).

31. Leave the gradient for 2 h to allow size fractionation of the microcells.

32. Add 50 ml of 10% sucrose to the gradient maker chambers while the clamp is tight.

33. Allow the solution to flow through the system and collect two 7 ml fractions from the top of the gradient in the reservoir into two 12 ml centrifuge tubes.

34. Calculate the number of microcells by counting them using a haemocytometer (each fraction should contain at least 1×10^7 microcells).

35. Wash and pellet the microcells twice with PBS/EDTA at 3000 g for 10 min.

36. Resuspend each pellet of microcells in 4 ml of serum-free medium, or combine if few microcells are available (i.e. 4 ml total volume).

3.4 Fusion of microcells to recipient cells

There are two main methods for the fusion of microcells to the recipient cells. One method, described in *Protocol 7*, uses polyethylene glycol (PEG) in combination with phytohaemagglutinin-P (PHA-P). The role of PHA-P in this procedure is to increase agglutination of the microcells to the surface of the recipient cells and hence enhance the fusion efficiency. The optimum duration of PEG treatment has to be determined empirically for each cell type. An alternative method uses inactivated Sendai virus for fusing the microcells to the recipient cells (53), but the former method is more widely used, owing to its efficiency and simplicity.

3.5 Selection of microcell hybrids

In this step the monochromosomal hybrids are selected by growing the fusion products in medium containing an appropriate selective agent. If the human chromosomes are tagged with the *neo* or *hyg* genes, neomycin or hygromycin are used to select the monochromosomal hybrids respectively. If, on the other hand, the human chromosomes are tagged with the *gpt* gene, the selection is performed by growing the cells in MX medium, containing mycophenolic acid and xanthine. The *gpt* gene, as described in Section 3, is the bacterial analogue

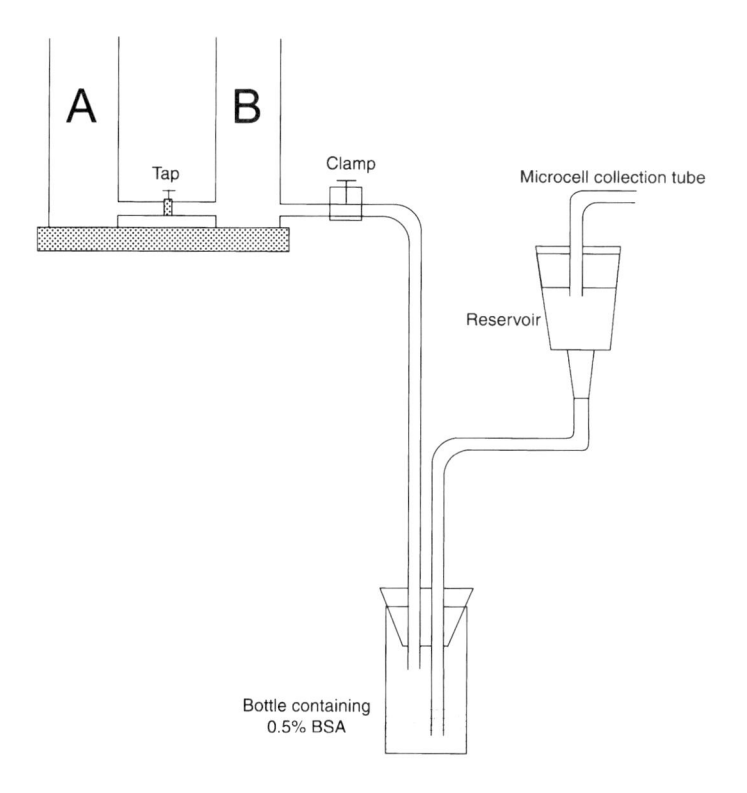

Figure 2. Gradient system for size fractionation of microcells. This is a schematic diagram of a system for purification of microcells by unit gravity sedimentation in a continuous Ficoll gradient. The gradient maker is connected to a 10 ml bottle via a tube that is sealed with a clamp. Thus the clamp can control the flow of solution through the system. The bottle is also connected to the bottom of a reservoir, which can be the barrel of a 50 ml syringe. Two magnetic stirrers, one in chamber B of the gradient maker and one in the bottle, are used to mix the solutions gently. The components of this system have to be sterile and assembled on a non-vibrating bench. In order to prevent contamination, it is necessary to cover the gradient maker chambers and the reservoir with pre-sterilized aluminium foil throughout the process, before and immediately after the addition of solutions. This and supplementing the culture medium with amphotericin B after the fusion step helps prevent yeast and fungal infection of the cells.

of the mammalian *hprt* gene in the purine biosynthesis pathway. The product of this gene utilizes xanthine instead of hypoxanthine, which is the substrate for the *hprt* gene product. Selection in MX medium is therefore based on blocking the endogenous pathway of purine biosynthesis by mycophenolic acid and supplementing the medium with xanthine. Therefore, only the cells containing the *gpt* gene product can utilize the xanthine component of the selection medium to bypass the inhibitory effect of mycophenolic acid and survive.

Protocol 7
Fusion of microcells with recipient cells and selection for microcell hybrid clones

Reagents
- 10 mg/ml phytohaemagglutinin
- 50% (w/v) PEG 1500 (polyethylene glycol 1500), freshly made in serum-free medium and pre-warmed to 37 °C
- Amphotericin B (Gibco)
- MX selection medium: complete medium containing the selective concentrations of mycophenolic acid and xanthine for the recipient cell type

Method
1. Seed an appropriate number of cells, in 10 cm dishes, to have 80% confluent cultures on the day of fusion.
2. Wash the cells twice with 10 ml PBS.
3. Add 4 ml of the microcell preparation from *Protocol 6* to each plate followed by addition of 40 μl of 10 mg/ml phytohaemagglutinin (i.e. 100 μg/ml final concentration).
4. Incubate the plates at 37 °C for 15 min.
5. Remove the medium and add 1.5 ml of pre-warmed 50% (w/v) PEG 1500.
6. Rock the plate to ensure that the PEG has covered the whole surface of the plate and incubate it at room temperature for 2 min (or the length of time that has been empirically determined for the recipient cells).
7. Wash the plates three times with 10 ml of serum-free medium and incubate in 10 ml serum-free medium at 37 °C for 10–15 min.
8. Replace the serum-free medium with 10 ml of complete medium and incubate at 37 °C overnight.
9. Trypsinize the cells and plate them at 5×10^5 cells/dish in MX medium.
10. (Optional) Add amphotericin B (2.5 μg/ml) to avoid yeast and fungal infection.
11. Incubate the cells at 37 °C and media change as required until colonies of microcell fusion hybrid are visible.

4 Irradiation fusion gene transfer

The technique of irradiation fusion gene transfer (IFGT) allows the generation of a varied array of hybrids that can be used in investigations such as gene mapping and/or positional cloning of genes. *Figure 3* is a schematic illustration of the IFGT technique for reducing the amount of human DNA in somatic cell hybrids.

GENETIC MODIFICATION

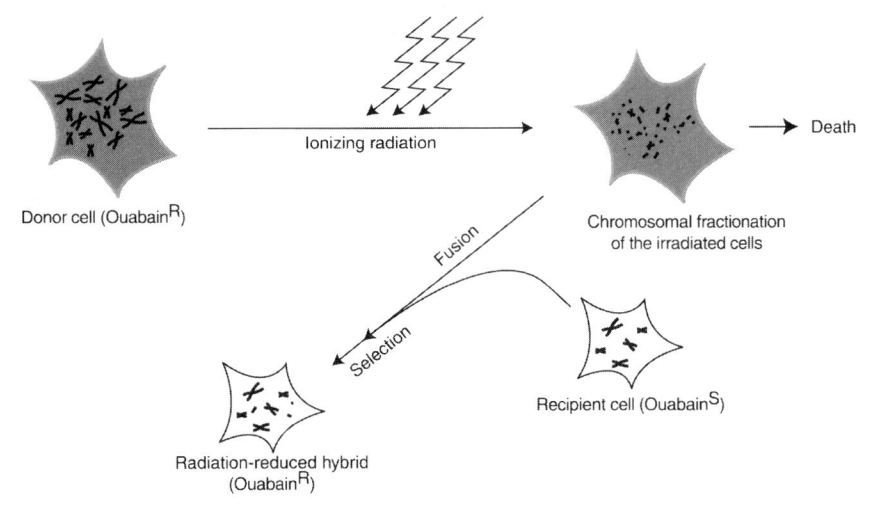

Figure 3. Irradiation fusion gene transfer. Chromosomes of donor cells that carry a selectable marker are fragmented by exposure of the cells to high doses of ionizing radiation. The irradiated cells are then rescued by fusing them to a recipient cell line that is sensitive to the selective marker. The resulting hybrids that retain fragments of the donor cell chromosomes are isolated by selection in medium containing the selective marker. Ouabain^R = resistant to ouabain. Ouabain^S = sensitive to ouabain.

4.1 Use in mapping genes

The IFGT technique was first introduced by Goss and Harris (54). They based their work on two observations:

(a) Ionizing radiation breaks chromosomes into smaller fragments with sizes that are proportional to the radiation dose. Therefore, the closer two genes are together the more likely they are to segregate within a segment of DNA after treatment with ionizing radiation.

(b) Fragments of foreign DNA can, under certain conditions, be introduced into mammalian cells where they can replicate and express their genes.

In their study Goss and Harris determined the linkage between four X chromosome genes and their order. These genes were those coding for the enzymes hypoxanthine guanine phosphoribosyl transferase (HPRT), phosphoglycerate kinase (PGK), glucose-6-phosphate dehydrogenase (G6PD), and α-galactosidase (α-gal). Irradiated lymphocytes were fused to a hamster cell line lacking HPRT activity. Unfused lymphocytes were removed by washing and the hybrids were grown under HAT selection (medium containing hypoxanthine, aminopterin, and thymidine). Therefore, only the recipients of the humant HPRT gene could survive. It was shown that both the gene order and distance between the genes could be determined by this technique.

IFGT was also used to map 14 DNA probes from a region of human chromosome 21q spanning 21 Mb. These experiments started with a Chinese hamster/human somatic cell hybrid containing a single copy of human chromosome 21q

253

and very little other human chromosomal material. The hamster/human hybrid cells were irradiated with 80 Gy X-rays. The lethally irradiated cells were fused to hamster cells lacking the HPRT gene, and the hybrids were selected in HAT medium. In this selection only those hybrids which had received the hamster HPRT gene from the irradiated cells could survive. Therefore, there was no direct selection for human sequences. The presence of the markers in the radiation hybrids was confirmed by Southern blot hybridization analysis and it was found that each of the 14 chromosome 21 markers was retained in 30–60% of the radiation hybrids. Statistical analysis on the results was then used to construct a map of the markers on chromosome 21. These mapping data were in agreement with those obtained from pulse-field gel electrophoresis studies, indicating that it is possible to determine the distance between different markers, and their order on the chromosome, by estimating the frequency of breakage between them (55).

Radiation hybrid panels have become very useful tools in mapping human genes because they can be screened by polymorphic as well as non-polymorphic markers, which are uninformative in linkage mapping. One such panel is the GeneBridge4 radiation hybrid mapping panel (56), which is available from the UK MRC Human Genome Mapping Project Resource Centre, UK MRC HGMP-RC (http://www.hgmp.mrc.ac.uk/Research/). This panel consists of 86 clones of the whole human genome and can be used for ordering genetic markers and estimating their distances. After obtaining the panel, the investigator screens this panel with the marker of interest and submits the data to the UK MRC HGMP-RC, where they are statistically analysed and the results returned.

4.2 Use in positional cloning

The most important factor in the positional cloning of a gene is the identification or isolation of markers that are tightly linked to that gene. One possible way to achieve this is to use the method of IFGT to construct a panel of radiation reduced hybrids as described in Section 4.1. If, for example, the aim is to clone a human gene, for which no tightly linked marker has been identified, a hybrid is generated that contains the smallest possible chromosome fragment carrying the gene of interest. In this case the best donor cell is a human–rodent monochromosomal hybrid or a hybrid that contains only a segment of the human chromosome carrying the gene. The generated radiation reduced hybrids can then be analysed using techniques such as polymerase chain reaction (PCR) or Southern blot hybridization for the presence of known markers in the critical region. This can lead to the identification of a marker that is closely linked to the gene. Moreover, new sequence tagged site (STS) markers can be generated from the hybrid with the least amount of human DNA by the method of inter-Alu-repeat sequence PCR (IRS-PCR), which amplifies sequences between the Alu-repeat elements (57–61). The new STS markers are then analysed for their linkage to the gene and for isolation of yeast artificial chromosomes (YACs) that serve as the primary tools for the next stage of positional cloning.

A point to consider in the generation of radiation reduced hybrids for positional cloning by IFGT is that because the selectable marker on the human

GENETIC MODIFICATION

chromosome is likely to be too far away from the gene, direct selection for the human chromosome cannot be applied. Therefore it is necessary to generate donor monochromosomal hybrid cells with a selectable marker (e.g. ouabain resistance) on the non-human chromosomes, or use recipient cells that are defective in a gene such as HPRT. Thus, selection of the radiation reduced hybrids is achieved by growing the cells in ouabain or HAT medium. This, therefore, limits the use of IFGT in positional cloning of a gene to instances where the gene has phenotypic effects in the resulting hybrids or where flanking markers have been identified. An example of successful use of the IFGT technique in positional cloning of a gene is the identification of ku80 as the mutant gene in the DNA repair deficient xrs cells (36, 62).

4.3 Irradiation doses

The relationship between the dose of irradiation and the size or number of human DNA fragments retained in the recipient cells has been studied by using a hamster/human somatic cell hybrid containing human chromosomes 3 and X as the source of human DNA fragments (63). These cells were irradiated with a lethal dose of 25–250 Gy γ-rays. The irradiated cells were rescued by fusing them to a thymidine kinase$^-$ (TK$^-$) hamster cell line followed by selection of the radiation hybrids in HAT medium, allowing only the survival of hybrids receiving the hamster Tk gene. Alu-PCR analysis showed the presence of unselected human DNA sequences in the isolated hybrids. The presence or absence of markers specific to chromosome 3 or X in the hybrids was also tested. This analysis as well as fluorescence *in situ* hybridization (FISH) studies on the radiation hybrids suggested that the number of human DNA fragments in these hybrids was independent of the irradiation dose. However, the sizes of the fragments were dose-dependent. By increasing the irradiation dose from 50 to 250 Gy a five- to tenfold reduction in the sizes of the largest fragments was observed. Also human marker retention frequencies declined from 0.27 to 0.03. The Xq27–Xq28 region was used as a model to estimate the size range of the retained fragments in the radiation hybrids. 40% of the hybrids generated following 50 Gy γ-rays retained fragments in the range of 3–30 Mb, 10% retained the whole chromosome arm, and the remaining 50% retained fragments of less than 2–3 Mb. The retention of 3 Mb or larger fragments decreased to less than 6% in hybrids generated at the higher dose of 250 Gy.

Protocol 8

Irradiation fusion gene transfer[a]

Equipment and reagents

- 50% (w/v) PEG 1500 (polyethylene glycol 1500), freshly made in serum-free medium and pre-warmed to 37°C
- Ionizing radiation source

Protocol 8 continued

Method

1. Trypsinize and count donor and recipient cells in the log phase growth.
2. Irradiate the donor cells with the desired dose of ionizing radiation (see Section 4.3).
3. Mix 1×10^6 cells each of donor (irradiated) and recipient (non-irradiated) parental cells and plate in a 10 cm dish.
4. Incubate the cells at 37°C for 6 h.
5. Remove the medium from the plate containing the mixed populations of parental cells.
6. Add 1.5 ml of pre-warmed PEG 1500 (50%, w/v) and rock the plate to ensure that the PEG has covered the whole surface of the plate.
7. Incubate the cells at room temperature for 2 min (or the length of time that has been empirically determined for the recipient cells).
8. Wash the plate three times with 10 ml of serum-free medium and incubate in 10 ml serum-free medium at 37°C for 10–15 min.
9. Replace the serum-free medium with 10 ml of complete medium and incubate at 37°C overnight.
10. Trypsinize the cells and plate at 5×10^5 cells/10 cm dish in the appropriate selection medium.
11. Incubate the cells at 37°C and media change as required until colonies of radiation reduced hybrids are visible.

[a] Graw *et al.* (64) with modification.

References

1. McCutchan, J. H. and Pagano, J. S. (1968). *J. Natl. Cancer Inst.*, **41**, 351.
2. Capecchi, M. R. (1980). *Cell*, **22**, 479.
3. Palese, P. and Roizman, B. (1996). *Proc. Natl. Acad. Sci. USA*, **93**, 11287.
4. Crystal, R. G., McElvaney, N. G., Rosenfeld, M. A., Chu, C. S., Mastrangeli, A., Hay, J. G., *et al.* (1994). *Nature Genet.*, **8**, 42.
5. McNally, M. A., Lebkowski, J. S., Okarma, T. B., and Lerch, L. B. (1988). *Biotechniques*, **6**, 882.
6. Starck, J., Doubeikovski, A., Sarrazin, S., Gonnet, C., Rao, G., Skoultchi, A., *et al.* (1999). *Mol. Cell. Biol.*, **19**, 121.
7. Radomska, H. S., Satterthwaite, A. B., Burn, T. C., Oliff, I. A., and Tenen, D. G. (1998). *Gene*, **222**, 305.
8. Li, Q., Hu, N., Daggett, M. A., Chu, W. A., Bittel, D., Johnson, J. A., *et al.* (1998). *Nucleic Acids Res.*, **26**, 5182.
9. Perucho, M., Hanahan, D., and Wigler, M. (1980). *Cell*, **22**, 309.
10. Robins, D. M., Ripley, S., Henderson, A. S., and Axel, R. (1981). *Cell*, **23**, 29.
11. Scangos, G. A., Huttner, K. M., Juricek, D. K., and Ruddle, F. H. (1981). *Mol. Cell. Biol.*, **1**, 111.

12. Perucho, M., Hanahan, D., Lipsich, L., and Wigler, M. (1980). *Nature*, **285**, 207.
13. Lowy, I., Pellicer, A., Jackson, J. F., Sim, G. K., Silverstein, S., and Axel, R. (1980). *Cell*, **22**, 817.
14. Jolly, D. J., Esty, A. C., Bernard, H. U., and Friedmann, T. (1982). *Proc. Natl. Acad. Sci. USA*, **79**, 5038.
15. Tanaka, K., Satokata, I., Ogita, Z., Uchida, T., and Okada, Y. (1989). *Proc. Natl. Acad. Sci. USA*, **86**, 5512.
16. Strathdee, C. A., Gavish, H., Shannon, W. R., and Buchwald, M. (1992). *Nature*, **358**, 434.
17. Legerski, R. and Peterson, C. (1992). *Nature*, **359**, 70.
18. Graham, F. L. and van der Eb, A. J. (1973). *Virology*, **52**, 456.
19. Chen, C. and Okayama, H. (1987). *Mol. Cell. Biol.*, **7**, 2745.
20. Jordan, M., Schallhorn, A., and Wurm, F. M. (1996). *Nucleic Acids Res.*, **24**, 596.
21. Gao, X. and Huang, L. (1995). *Gene Ther.*, **2**, 710.
22. Hug, P. and Sleight, R. G. (1991). *Biochim. Biophys. Acta*, **1097**, 1.
23. Zimmerman, U. (1982). *Biochim. Biophys. Acta*, **694**, 227.
24. Neumann, E., Schaefer-Ridder, M., Wang, Y., and Hofschneider, P. H. (1982). *EMBO J.*, **1**, 841.
25. Chu, G., Hayakawa, H., and Berg, P. (1987). *Nucleic Acids Res.*, **15**, 1311.
26. Potter, H., Weir, L., and Leder, P. (1984). *Proc. Natl. Acad. Sci. USA*, **81**, 7161.
27. Hirt, B. (1967). *J. Mol. Biol.*, **26**, 365.
28. Gaillard, C. and Strauss, F. (1990). *Nucleic Acids Res.*, **18**, 378.
29. Lugo, T. G., Handelin, B., Killary, A. M., Housman, D. E., and Fournier, R. E. (1987). *Mol. Cell. Biol.*, **7**, 2814.
30. Athwal, R. S., Smarsh, M., Searle, B. M., and Deo, S. S. (1985). *Somat. Cell Mol. Genet.*, **11**, 177.
31. Saxon, P. J., Srivatsan, E. S., Leipzig, G. V., Sameshima, J. H., and Stanbridge, E. J. (1985). *Mol. Cell. Biol.*, **5**, 140.
32. Ishizaki, K., Oshimura, M., Sasaki, M. S., Nakamura, Y., and Ikenaga, M. (1990). *Mutat. Res.*, **235**, 209.
33. Kaur, G. P. and Athwal, R. S. (1989). *Proc. Natl. Acad. Sci. USA*, **86**, 8872.
34. Stackhouse, M. A., Ortiz, J. B., Sato, K., and Chen, D. J. (1994). *Mutat. Res.*, **323**, 47.
35. Koi, M., Umar, A., Chauhan, D. P., Cherian, S. P., Carethers, J. M., Kunkel, T. A., *et al.* (1994). *Cancer Res.*, **54**, 4308.
36. Hafezparast, M., Kaur, G. P., Zdzienicka, M., Athwal, R. S., Lehmann, A. R., and Jeggo, P. A. (1993). *Somat. Cell Mol. Genet.*, **19**, 413.
37. Kaur, G. P. and Athwal, R. S. (1993). *Somat. Cell. Mol. Genet.*, **19**, 83.
38. Jeggo, P. A., Hafezparast, M., Thompson, A. F., Broughton, B. C., Kaur, G. P., Zdzienicka, M. Z., *et al.* (1992). *Proc. Natl. Acad. Sci. USA*, **89**, 6423.
39. Lambert, C., Schultz, R. A., Smith, M., Wagner McPherson, C., McDaniel, L. D., Donlon, T., *et al.* (1991). *Proc. Natl. Acad. Sci. USA*, **88**, 5907.
40. Robertson, G. P., Hufford, A., and Lugo, T. G. (1997). *Cytogenet. Cell Genet.*, **79**, 53.
41. Kon, H., Sonoda, Y., Kumabe, T., Yoshimoto, T., Sekiya, T., and Murakami, Y. (1997). *Oncogene*, **16**, 257.
42. O'Briant, K., Jolicoeur, E., Garst, J., Campa, M., Schreiber, G., and Bepler, G. (1997). *Anticancer Res.*, **17**, 3243.
43. Matsuda, T., Sasaki, M., Kato, H., Yamada, H., Cohen, M., Barrett, J. C., *et al.* (1997). *Oncogene*, **15**, 2773.
44. Chekmareva, M. A., Hollowell, C. M., Smith, R. C., Davis, E. M., LeBeau, M. M., and Rinker Schaeffer, C. W. (1997). *Prostate*, **33**, 271.
45. Kanamori, H. and Siegel, J. N. (1997). *Exp. Cell Res.*, **232**, 90.

46. Harrington, R. D. and Geballe, A. P. (1996). *Ann. Clin. Lab. Sci.*, **26**, 522.
47. Meguro, M., Mitsuya, K., Sui, H., Shigenami, K., Kugoh, H., Nakao, M., *et al.* (1997). *Hum. Mol. Genet.*, **6**, 2127.
48. Shapero, M. H., Langston, A. A., and Fournier, R. E. (1994). *Somat. Cell Mol. Genet.*, **20**, 215.
49. Anderson, M. J., Fasching, C. L., Xu, H. J., Benedict, W. F., and Stanbridge, E. J. (1994). *Genes Chromosomes Cancer*, **9**, 251.
50. Dieken, E. S., Epner, E. M., Fiering, S., Fournier, R. E., and Groudine, M. (1996). *Nature Genet.*, **12**, 174.
51. Shen, M. H., Yang, J., Loupart, M. L., Smith, A., and Brown, W. (1997). *Hum. Mol. Genet.*, **6**, 1375.
52. Fournier, R. E. K. (1982). In *Techniques in somatic cell genetics* (ed. J. W. Shay), p. 309. Plenum Press, New York and London.
53. Ringertz, N. R. (1978). *Natl. Cancer Inst. Monogr.*, **48**, 31.
54. Goss, S. J. and Harris, H. (1975). *Nature*, **255**, 680.
55. Cox, D. R., Burmeister, M., Price, E. R., Kim, S., and Myers, R. M. (1990). *Science*, **250**, 245.
56. Gyapay, G., Schmitt, K., Fizames, C., Jones, H., Vega-Czarny, N., Spillett, D., *et al.* (1996). *Hum. Mol. Genet.*, **5**, 339.
57. Lerner, T. J., D'Arigo, K. L., Haines, J. L., Doggett, N. A., Taschner, P. E., de Vos, N., *et al.* (1995). *Am. J. Med. Genet.*, **57**, 320.
58. Oshima, J., Yu, C. E., Boehnke, M., Weber, J. L., Edelhoff, S., Wagner, M. J., *et al.* (1994). *Genomics*, **23**, 100.
59. Nieuwenhuijsen, B. W., Chen, K. L., Chinault, A. C., Wang, S., Valmiki, V. H., Meershoek, E. J., *et al.* (1992). *Hum. Mol. Genet.*, **1**, 605.
60. Desmaze, C., Zucman, J., Delattre, O., Thomas, G., and Aurias, A. (1992). *Hum. Genet.*, **88**, 541.
61. Cotter, F. E., Das, S., Douek, E., Carter, N. P., and Young, B. D. (1991). *Genomics*, **9**, 473.
62. Taccioli, G. E., Gottlieb, T. M., Blunt, T., Priestley, A., Demengeot, J., Mizuta, R., *et al.* (1994). *Science*, **265**, 1442.
63. Siden, T. S., Kumlien, J., Schwartz, C. E., and Rohme, D. (1992). *Somat. Cell Mol. Genet.*, **18**, 33.
64. Graw, S., Davidson, J., Gusella, J., Watkins, P., Tanzi, R., Neve, R., *et al.* (1988). *Somat. Cell Mol. Genet.*, **14**, 233.

Chapter 10
Epithelial stem cell identification, isolation and culture

David Hudson

Institute of Urology, University College London Medical School,
67 Riding House Street, London W1P 7PN, UK.

1 Introduction

This chapter will discuss the ways in which the identification and characterization of stem cells from various epithelial tissues can be approached. Specific protocols are given for techniques that have been shown to be effective in studies involving one of the best-characterized epithelial stem cell models, epidermal keratinocytes. First, methods involved in the separation of an epithelial cell type from other cells will be examined, followed by ways in which the proliferative capacity of such a cell type can be assessed. Secondly, methods used for the maintenance of primary stem cells in culture and ways of characterizing stem cells using immunocytochemistry will be described. Details of Basic Cell Culture have been described in another edition of *The Practical Approach* series (1).

1.1 Clinical application of cultured human stem cells

The study of human stem cells is one of the most rapidly growing areas of cell biology. Recent publications have shown that not only can human embryonic stem (ES) cells be cultured in the laboratory (2), but also that these cells may be manipulated in the laboratory to produce cultures with the characteristics of particular tissues. This opens the possibility of using cultured cells to repair tissues with functions impaired by damage or ageing. Applications may include culture of dopamine-producing neurons for implantation into the brains of patients with Parkinson's disease, or pancreatic insulin-producing cells for diabetes patients (3).

Tissue-specific stem cells have been isolated from human and other animal species and these, unlike ES cells, are less than totipotent. They either lack the ability to form all cell types and have a restricted repertoire of differentiated progeny (multipotency), or can differentiate only into a single cell type and are termed unipotent. Cells from tissues such as brain and bone marrow have a more limited ability to form different cell types, but have an equally useful potential for tissue replacement. One example involves the transplantation of mouse neuronal stem cells which, when placed into mice lacking bone marrow, were shown to develop into blood cells (4). Cultured epithelial cells are already widely used for surgical repair, as in skin grafting (5, 6) and they are also potential host cells for gene therapy (7).

2 Basic principles for identification and purification of stem cells

Epithelial tissues, such as the epidermis and the lining of the gut, share a common feature in that they are constantly shedding cells from their outer surface. This constant cell loss is compensated by continual replacement through cell proliferation in a highly regulated process. In humans the entire outer layer of the skin is shed daily (8), while the entire epithelial lining of the mouse gut is replaced every three to four days (9). Since this process continues throughout life it has been argued that this is proof of the existence of long-lived stem cells. The definition of such a stem cell is that it lacks certain tissue-specific differentiation markers, remains in the tissue throughout life, retaining proliferative capacity, and gives rise to daughters, some of which generate differentiated cells and others which are themselves stem cells. The cells should also be capable of regenerating the tissue after injury (10). It has been demonstrated in the epidermis that the proliferative compartment consists not only of stem cells and post-mitotic cells, but also of a transit amplifying population (11). Transit amplifying cells are produced initially by the division of stem cells and, although they have a high proliferative capacity, their lifespan is limited and eventually all the cells will differentiate and be shed from the tissue.

When approaching the question of whether or not a particular tissue contains a discrete stem cell population it is necessary to establish whether there is true proliferative heterogeneity within the tissue. There are three possibilities: first, all the cells may have equal proliferative potential, secondly, only a subpopulation of non-differentiated cells may divide or thirdly, as in the skin, there are two types of proliferative cell, stem cells and transit amplifying cells. The latter may be represented by a continuous distribution of proliferative capacity, from unlimited to a single division only. Alternatively there may well be two clear proliferative subpopulations that can be distinguished by criteria such as the morphology of colonies produced in cell culture.

3 Assessment of proliferative heterogeneity

In the epidermis it has been shown that there is proliferative heterogeneity within the keratinocyte population. Barrandon and Green (12) showed that

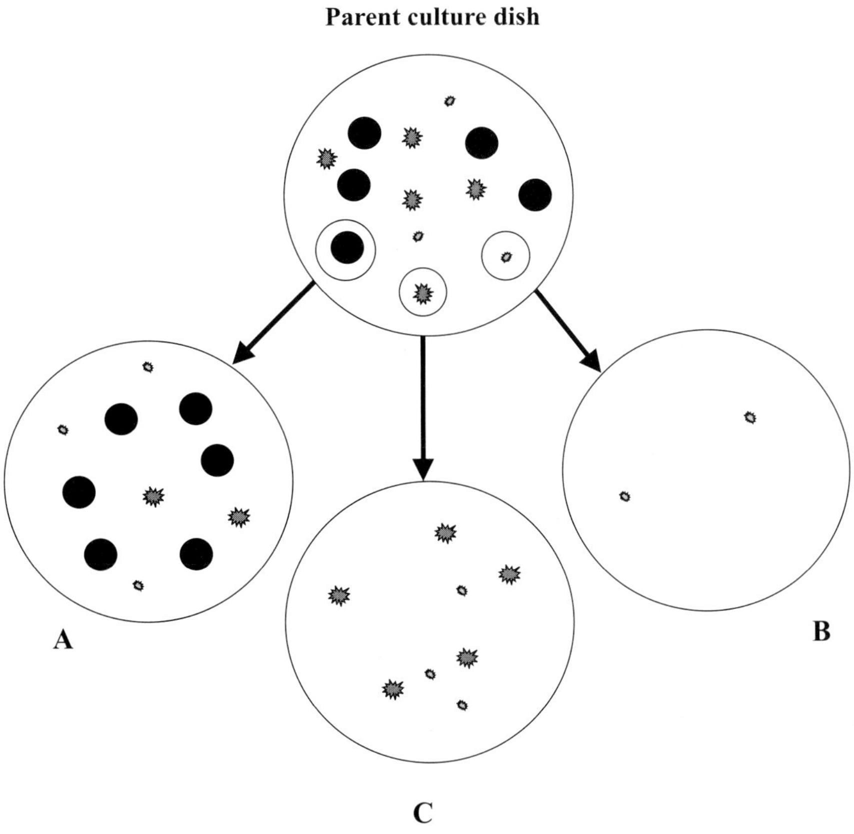

Parent culture dish

Figure 1. When keratinocytes are cultured at low density three types of colony are seen in the parent culture dish. Subcloning of the largest type of colony (A) yields many large colonies, along with several small and some abortive colonies. Subcloning a small colony (C) produces only small and abortive colonies, whereas an abortive colony (B) will produce no colonies. Large colony-forming cells behave like stem cells whereas small colony-forming cells behave like transit amplifying cells.

there are three types of colony formed when these cells are placed in culture, in high calcium medium, in the presence of a 3T3 feeder layer. The three colony types formed are called paraclones, meroclones, and holoclones, and these classifications are based on the type of progeny produced by the colonies when they are passaged into fresh dishes (*Figure 1*).

Paraclones form small, irregular colonies, composed almost entirely of differentiated cells. Colonies produced by holoclones are large and round, consisting mostly of small cells, surrounding a central region of stratified differentiated cells. Meroclones produce colonies which do not have a smooth outline and are smaller than those produced by holoclones. It is believed that the stem cells produce holoclones and transit amplifying cells, paraclones. The origin of meroclones in this definition is not clear, although it has been speculated that they

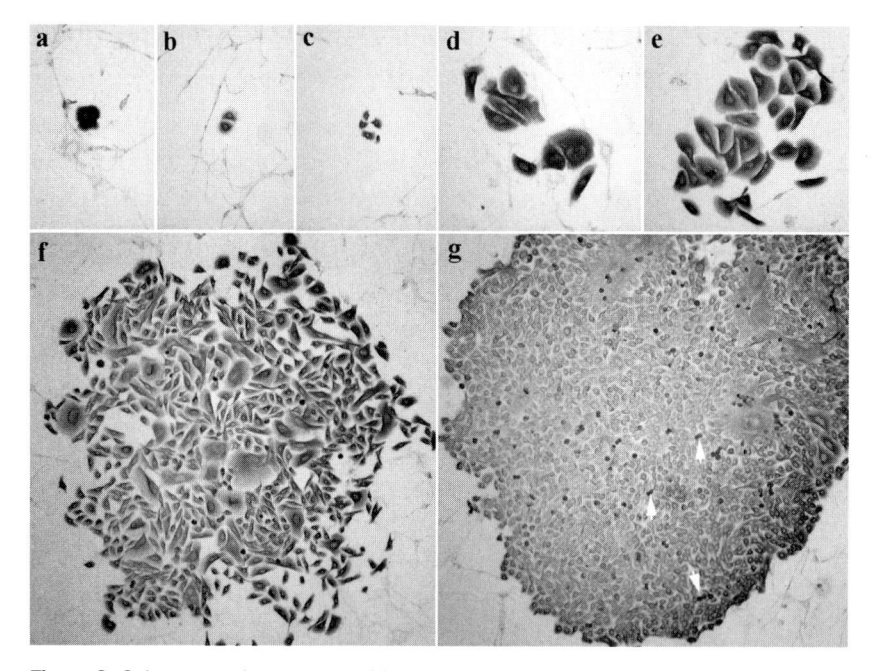

Figure 2. Colony types in prostate epithelial cell culture. Cells that attach to collagen coated dishes show wide variation in their ability to divide and form colonies. Freshly isolated prostate epithelial cells were plated as a single cell suspension in the presence of a 3T3 feeder layer and allowed to grow for ten days. The cultures were fixed and stained with LP34, an anti-cytokeratin monoclonal antibody. All colonies are photographed from the same dish, at the same magnification. Examples are shown of cells that have attached and either failed to divide (a), or which have only divided from one to five times (b–e). (f) A large but loosely packed colony which includes many large differentiating cells. (g) A colony consisting almost exclusively of small tightly packed cells with a high proportion of mitotic figures (arrowheads).

are produced by stem cells that generate transit amplifying cells at a higher rate than those producing holoclones (13). Epithelial cells from other tissues, including prostate, also show this heterogeneity in clonal growth properties and this type of experimental approach could well be of use in stem cell studies in the breast, gut epithelium, and liver. *Figure 2* shows different colony types formed when primary prostate epithelial cells are plated in the presence of a 3T3 feeder layer and allowed to grow over a 14 day period. As for epidermal keratinocytes, three colony types are formed: small abortive clones, large, round, fast growing colonies made up almost exclusively of small cells, and an intermediate colony type which may grow rapidly but contains larger differentiated cells and is irregular in shape.

The following protocols give examples for establishing cultures to examine proliferative heterogeneity and to assess the colony-forming efficiency (CFE) of cell populations. Cloning is a useful tool for the assessment of proliferative capacity and differentiation state of different colonies. *Protocol 2* details a ring cloning method which can be used both for keratinocytes and for cells from other epithelial tissues. If the cell type can be cloned from single cells it is possible to avoid ring cloning altogether and grow single colonies in 32 mm dishes (12).

Protocol 1
Determination of proliferative heterogeneity and colony-forming efficiency (CFE)

Equipment and reagents

- 6 cm tissue culture plates (Nunc or Gibco) seeded with lethally irradiated or mitomycin C-treated 3T3 or feeder cells[a]
- Epithelial growth medium normally used for the cells under investigation
- Freshly isolated primary cells or subconfluent cell cultures
- 0.05% (w/v) trypsin, 0.02% (w/v) EDTA solution (Gibco)
- 1% rhodanile blue solution: dissolve 2 g each of rhodamine powder and Nile blue (Gurr stains, BDH) into 50 ml of distilled water, filter each through Whatman filter paper, and combine (14)

Method

1. Remove the culture medium from subconfluent dishes of primary cultures of epithelial cells. Rinse twice with sterile PBS, then incubate with trypsin/EDTA for between 5–15 min at 37°C. Check cells at 5 min intervals until all cells are detached.

2. Transfer trypsinized cultured, or freshly isolated, cells into a 15 ml centrifuge tube.

3. Inhibit trypsin with soybean trypsin inhibitor (1 mg/ml) (Sigma) and centrifuge cells at 170 g for 5 min.

4. Resuspend the pellet in 5 ml of epithelial cell culture medium and count the cell number using a haemocytometer. A smaller volume may be necessary if the cell number is low.

5. Seed between 10^3 and 10^4 cells per dish, with at least three dishes per density, and allow to grow for 14 days at 37°C, changing the medium three times per week.[b]

6. After 14 days wash the dishes with PBS and fix the cultures with 3.7% formaldehyde for 10 min at room temperature.

7. Wash the dishes with distilled water and stain for 30 min with 3 ml rhodanile blue.

8. Wash dishes in running water to remove excess stain and air dry.

9. Under a microscope count the number of colonies containing 32 or more cells. The CFE is expressed as the percentage of cells plated that form colonies.[c]

Protocol 1 continued

10. Different colony types can also be scored if consistent morphological differences can be seen (see *Figure 2*).

[a] Protocols for the use of 3T3 feeder layers and other keratinocyte culture techniques are detailed in ref. 1. Use of mitomycin C to inhibit growth of 3T3 cells for use as feeder layers has the advantage of not needing specialist irradiation equipment. However, it does need confluent dishes of 3T3 cells to be ready when required. If a source of irradiation is available this is preferable since cells can be bulk cultured, treated in advance, and frozen in liquid nitrogen until needed.

[b] The cell density suitable for the establishment of the CFE will vary between cultures and it is advisable to set up a range of densities.

[c] With epithelial cells this can be adjusted to a percentage of the basal or undifferentiated cells plated by immunostaining a sample of the plated cells (see *Protocol 7*). Suitable antibodies to assess basal or differentiated phenotypes are discussed later (see *Table 2*).

Protocol 2
Cloning of epithelial colonies

Equipment and reagents

- Cloning rings 8 mm diameter (Sigma) or 'home-made' using straight upper region of 1 ml Gilson tips cut with hot scalpel: rings should be placed in a glass Petri dish on a thin layer of vacuum grease and sterilized by autoclaving prior to use
- Sterile forceps
- Trypsin solution, 0.05% in 0.02% versene (Gibco BRL)
- Microscope with camera attachment
- 1 mm grid photocopied onto overhead projector acetate, cut into 53 mm circles
- 7–14 day-old epithelial cell cultures seeded at a density allowing colonies to grow to several mm^2 in area without coalescing[a] (see *Protocol 1*)
- 1% rhodanile blue solution (see *Protocol 1*)

Method

1. Count the number of cells in the selected colony. This can be estimated in one of several ways. The area of a round colony can be measured by placing a circle of transparency film bearing a 1 mm grid on the base of the dish and measuring the diameter of the colony while observing the dish through a microscope. An estimate of the cell density can be made by counting the number of cells per linear mm. At least three estimates should be made in different areas and the mean number of cells/mm and the mean number of cells/mm^2 calculated. Multiplying the density (cells/mm^2) by the colony area (mm^2) will give an estimate of the number of cells in the colony. A more uneven colony can be measured by counting the number of 1 mm squares occupied by cells. Alternatively, if a microscope mounted camera is available, the colonies can be photographed, and the number of cells counted from the photographs. Mark the selected colonies on the base of the dish with an identifying letter and draw a ring around each one with a marker pen.

Protocol 2 continued

2. Remove the medium from the dish and wash twice with PBS to remove any trace of serum. Carefully take a cloning ring with a thin coating of vacuum grease on the base and place it over the selected colony. Press down gently with the sterile forceps to ensure a good seal between the tissue culture dish and the ring.

3. Using a Gilson pipette slowly add 75 µl of trypsin solution to the centre of the ring. Replace the dish lid and return the dish to an incubator at 37 °C.

4. After 15 min check the colony under the microscope to ensure that the cells have rounded up and are beginning to detach.

5. Pipette up and down gently to detach all cells and transfer to a 15 ml centrifuge tube.

6. Add 75 µl of serum-containing medium to the ring to collect any remaining cells. Add this to the centrifuge tube to neutralize the trypsin.

7. Dilute the cells in culture medium to a suitable seeding density and plate cells, in 1 ml, onto previously prepared dishes of feeder cells.[a]

8. After 14 days the dishes should be fixed in 3.7% formaldehyde and stained with 1% rhodanile blue (BDH). The colonies can be counted to assess the colony-forming efficiency of the original colony and scored for colony type.

[a] The seeding density will depend on the colony-forming efficiency of the culture and growth rate. Ideally several different densities should be used and the optimum selected at the time of cloning.

4 Methods for the separation of different cell populations

The colony-forming cells of freshly isolated and cultured epidermal keratinocytes were shown to be more adhesive than non-colony-forming cells to certain extracellular matrix molecules (13, 15). This increased adhesiveness was shown to be mediated by higher levels of the cell surface adhesion molecules, the β1 integrins. Integrins are involved in the attachment of many cell types, including epithelial cells, to basement membranes and an early stage of terminal differentiation is the down-regulation of the function of these molecules (16). The functional down-regulation is followed by the loss of integrin expression as cells differentiate and move away from the basement membrane, in the case of the skin into the stratified upper layers.

Integrins are expressed as heterodimeric molecules with an α and a β subunit joined together in a transmembrane matrix receptor that links the actin cytoskeleton inside the cell with the extracellular matrix outside. The composition of the dimer confers functional specificity on the protein, in that each pair has affinity for a particular extracellular matrix molecule, or in the case of non-epithelial cells, such as haemopoietic cells, for another cell surface protein.

Epithelial basement membranes are composed of a complex structure con-

Table 1. Integrins expressed by keratinocytes and their ligands

Integrin	Ligand	Reference
α2β1	Collagen, laminin 1	19
α3β1	Laminin 5, collagen	20
α5β1[a]	Fibronectin	21
α6β1[b]	Laminin 5	22
α6β4	Laminin 1 and 5	23
α9β1[c]	Tenascin	24
αvβ5[d]	Vitronectin	25
αvβ6[e]	Fibronectin	26

[a] α5β1 is expressed at detectable levels in culture but is either undetectable or at very low levels in human epidermis.

[b] α6β1 is not found in normal skin but is expressed in culture and in cells from β4 deficient patients (22).

[c] α9β1 is only expressed in tissue and not in culture.

[d] αvβ5 is expressed at low levels in normal tissue but is highly up-regulated in wound healing and at sites of inflammation.

[e] αvβ6 is not expressed in normal skin but is up-regulated during wound healing.

taining collagens, laminins, and fibronectin and it is to these proteins that the basal face of an epithelial cell attaches via integrins in focal adhesions and hemidesmosomes. Although there are now twelve known α subunits and six β subunits, keratinocytes have been shown to express only a subset, which is listed in *Table 1*. A similar repertoire of integrin expression has also been seen in human prostate (17) and breast epithelium (18). Unlike prostate, however, where integrins are mostly restricted to the basal cell layer, breast myoepithelial and lumenal cells both have cell surface integrin expression. Additionally, stromal cells also express β1 integrins and this should be taken into account when carrying out cell isolations based on cell surface integrins. For example, breast smooth muscle cells express β1 integrins in the form of both α1β1—a laminin receptor, αvβ1—the vitronectin receptor, and α5β1—the fibronectin receptor (18).

Separation of proliferative from non-proliferative epithelial cells has been achieved using differential expression of adhesion molecules in two ways (13). First, by selecting those cells which attach most rapidly to extracellular matrix proteins and secondly, by the use of antibodies raised against cell surface proteins, for example using fluorescence activated cell sorting (FACS) to separate cells with high or low levels of cell surface integrin expression. The basic protocols for cell attachment experiments, together with details of matrix coated dish preparation is given in *Protocols 3-5*. These are followed by a comparison of methods used to separate cells based on their cell surface protein expression.

4.1 Isolation of cells by differential adhesion

4.1.1 Choice of extracellular matrix proteins for attachment assays

Epithelial cells sit on a basement membrane consisting of various proteins produced by both stromal cells and epithelial cells (27). The basement membrane of skin and many other tissues consists of collagens IV and VII, K laminin, nidogen, and laminin 5. Under wound healing conditions cells will initially be exposed to fibrin and fibrinogen followed by a wound matrix laid down by the fibroblasts, the major components of which are collagen I and fibronectin. Since epithelial cells are able to attach to any of the above components they may each be used to select a subpopulation of cells with the highest affinity for that protein. In the epithelium of the intestinal crypts there is a region or niche where stem cells are believed to be located (28). This region has also been suggested to be delineated by a change in the composition of the basement membrane, with different isoforms of laminin expressed in different areas of the crypts (29). It is therefore possible that different matrix proteins could be used to isolate specific cell populations.

Protocol 3
Preparation of extracellular matrix coated dishes

Equipment and reagents

- 6 cm bacteriological plastic dishes (e.g. Sterilin or Falcon)
- Bovine serum albumin (BSA) Fraction V (Sigma): 10 mg/ml solution in PBS, filter sterilized
- Extracellular matrix proteins in PBS: fibronectin (100 µg/ml), collagen I (20 µg/ml), collagen IV (100 µg/ml), Engelbreth–Holm–Swarm sarcoma laminin (25 µg/ml) (Sigma)[a]

Method

1. Coat dishes by adding 3 ml of the required matrix protein solution per dish, three dishes per condition/time point. Swirl the solution around the dishes to ensure even coverage and incubate plates in humid conditions overnight at 4°C, or at 37°C for 1 h.
2. Heat denature the BSA solution by heating 0.5 ml aliquots to 80°C in a dry block for 3 min. Cool on ice and dilute the BSA to 0.5 mg/ml in sterile PBS.
3. Remove coating solution from the dishes and wash three times with PBS.
4. Block non-specific adhesion by incubating the dishes with 0.5 mg/ml BSA solution, at 37°C, for 1 h.
5. Wash the dishes twice with PBS, then add 2 ml of serum-free medium (i.e. DMEM) and warm to 37°C.

[a] 15–100 µg/ml is sufficient for maximal cell attachment (13).

4.1.2 Production of epithelial extracellular matrix

While collagens and fibronectin are readily available commercially (e.g. Sigma), laminin 5, the most abundant basement membrane component in epidermis, is not. A good alternative can be readily prepared in the laboratory. Epithelial cells in culture, particularly when plated with feeder cells, coat tissue culture dishes with a matrix which is mainly laminin 5 (30). A description of this procedure is given in *Protocol 4*.

Protocol 4

Preparation of epithelial cell extracellular matrix coated dishes

Equipment and reagents

- Epithelial cell cultures, keratinocytes, or cell type under investigation
- 10 mM EDTA in 25 mM Tris–HCl plus 1% Triton X-100

Method

1. Grow the epithelial cell cultures under normal conditions until confluent (approx. 10–14 days).
2. Wash dishes three times with sterile PBS.
3. Incubate the dishes at 37 °C for 30 min with the EDTA/Tris/Triton X-100 solution to remove the cells. Check under a microscope that the cells have been lysed.
4. Wash the dishes at least three times with sterile PBS.
5. Dishes can be used immediately for adhesion assays or stored wrapped in Parafilm at −80 °C until required.
6. Before use the dishes should be thawed at room temperature for 30 min and rinsed twice with sterile PBS.

4.1.3 Use of extracellular matrix coated dishes to select subpopulations of epithelial cells

Having prepared coated dishes it is now possible to use these to separate cells with the ability to stick to ECM proteins. Rapidly adherent cells, those with the highest levels of cell surface adhesion molecules, can be selected by allowing cells to settle for varying periods of time and removing by washing the unattached cells. The procedure for this is given in *Protocol 5*.

Protocol 5

Adhesion of rapidly attaching cells to extracellular matrix (ECM) proteins

Equipment and reagents

- 6 cm bacteriological plastic Petri dishes coated with ECM (*Protocol 3* or *4*) and blocked with heat denatured BSA (*Protocol 3*); control dishes blocked directly without prior ECM coating
- Irradiated or mitomycin C-treated 3T3 feeder cells (see *Protocol 1*)
- Epithelial cell growth medium
- DMEM (e.g. Gibco BRL)
- Trypsin solution, 0.05% in 0.02% versene (e.g. Gibco BRL)
- Soybean trypsin inhibitor (Sigma)

Protocol 5 continued

A. Time course of cell adhesion

1. Harvest epithelial cells, either from fresh tissue or subconfluent primary cultures.
2. Transfer cell suspension into a 15 ml centrifuge tube. Neutralize trypsin with soybean trypsin inhibitor (1 mg/ml) and pellet cells by centrifugation for 5 min at 170 g.
3. Add 5 ml of serum-free medium and count cells using a haemocytometer.
4. Pellet cells as described in step 2 and resuspend in serum-free medium at a density between 500–5000 cells/ml. The serum-free medium should be pre-warmed to 37 °C before use.
5. Add 2 ml of cell suspension to matrix coated dishes and return dishes to the incubator for the required time (between 10^3 and 10^4 cells/dish).
6. After suitable times (e.g. 5, 20, and 60 min) remove the dishes from the incubator.
7. Remove unattached cells by pipetting up and down gently three times with a 5 ml pipette. Collect the medium from each dish and transfer into centrifuge tubes.
8. Wash the dishes by pipetting gently with 3 ml of PBS and pool with the first washes from step 7.
9. After thorough washing there should be no moving cells seen under the microscope. Attached cells should be visible as small rounded cells.
10. Add 3T3 feeder cells to each dish, including a set of dishes from which unattached cells have not been washed, as a control. If the growth medium contains serum, extra serum should be added to these dishes at this time.
11. Change the medium the following day and incubate the dishes for 14 days, changing the medium three times per week.
12. The washes from steps 7 and 8 can be centrifuged and the cells plated onto fresh feeder plates to assess the CFE of the non-adherent cells at each time point.

B. Assessment of colony-forming ability of rapidly attaching cells

The above method allows the relative rate of attachment of colony-forming cells to be assessed and gives the CFE in relation to the number of cells plated. The CFE of attaching cells can be calculated if part A is followed up to step 9^a and then the following steps taken.

1. Add 2 ml of trypsin solution to each washed dish and incubate at 37 °C for 10 min.
2. Remove cells from dish with gentle pipetting and transfer cell suspension to a 15 ml centrifuge tube.
3. Wash dish with medium containing trypsin inhibitor at 1 mg/ml trypsin, add this to the cells in the centrifuge tube, and pellet cells by centrifugation for 5 min at 170 g.
4. Suspend cells in growth medium and count cell number.
5. Plate cells onto 6 cm dishes with 3T3 feeders and grow for 14 days before fixing and counting.

^{a} To allow for sufficient cell numbers for counting, more cells will need to be added at step 5, between 10^5 and 10^6 cells on a 9 cm dish.

4.2 Separation of cultured and primary cells by flow and immunomagnetic sorting

The previous section describes ways in which cell attachment can be used to select subpopulations. Other methods to separate groups of cells involve the use of differential cell surface expression of adhesion molecule receptors, such as the integrins. This can be done in two basic ways; qualitatively, whereby all cells with expression of a particular marker are selected, or quantitatively, where cells are separated on the basis of relative amounts of cell surface protein expression. The latter technique involves the use of fluorescence activated cell sorting (FACS) which measures the relative intensity of a fluorescent signal on the surface of each cell and can collect a group of cells having the required characteristic. This group may either be all stained positively or negatively for a particular marker or can select the cells with lowest or highest expression levels. FACS sorting is a complex procedure requiring expensive specialist equipment and training, and other techniques such as immunomagnetic bead sorting may be more appropriate.

4.2.1 Comparison of fluorescence activated cell sorting and immunomagnetic bead sorting

An alternative to FACS sorting is separation using magnetic beads coated with various secondary antibodies. After incubating cells with a primary antibody, the beads are simply incubated with the cells for a short period of time and, together with attached cells, isolated using a magnet. The cells can either be directly plated in culture or the beads can be removed by a short incubation with an enzyme that breaks the bond between the bead and the antibody. Alternatively the selection can be 'negative', whereby the cells required are those remaining after magnetic sorting.

There are several advantages to the use of beads. The technique is relatively cheap and does not require specialist training or equipment. Bead methods can be faster than FACS, as the cells attached to the beads are separated within seconds by the magnet. It is possible to get greater cell recovery with beads, as the FACS machine will tend to reject cells at too high a density to give clear single cell measurements. This may allow sufficient cells to be isolated from smaller tissue samples. FACS does, however, have the advantage of producing a very pure cell population. Positive cells can occasionally be paired or clumped with negative cells and FACS machines are normally set to discard anything other than single cells, including such clumps, while magnetic beads may trap them. Another disadvantage of the use of magnetic beads is the lack of quantitative selection and is, therefore, only of use for isolating or excluding all cells expressing a particular marker. This may involve the separation of differentiated cells from basal cells using the loss of integrin expression, or of basal cells from lumenal cells, in a tissue such as the prostate, where the basal cells express CD44 and the luminal cells CD57 (31).

4.2.2 Comparison of immunomagnetic bead systems

There are two main suppliers of bead technologies. One obvious deciding factor in selecting the appropriate system to use will be the local experience. Although there are differences, both systems can be optimized for both positive and negative sorting. The MACS system (Miltenyi Biotec Ltd.) uses very small beads, only 50 nm in diameter, which do not interfere with cell attachment and do not, therefore, need to be removed before plating the cells. The system does, however, involve passing the cells through a separation column and this makes individual cell isolations more expensive than the Dynabead method (Dynal UK). Although cells will attach and proliferate without the removal of selecting Dynabeads, it is advisable to do so for adhesion assay type experiments. This is a problem with the basic Dynabeads, but a new system, called CELLection, has overcome this by the inclusion of a short length of DNA as a linker between the bead and the antibody. This allows the beads to be detached from the selected cells using a DNase-based detachment solution. Another disadvantage of Dynabeads relating to their size is that the force of the magnet can cause some antigens to detach from the cell surface, preventing the capture of the cells, although this is rare.

The method given in *Protocol 6* is an outline of the Dynabeads technique and is designed to give an idea of the procedures involved. More detailed instructions are given with each kit purchased, optimized for the antibody isotype used. The method for MACS-based separation is little different and should give a similar degree of success.

Protocol 6
Positive cell sorting with immunolabelled magnetic beads

Equipment and reagents

- CELLection Pan Mouse IgG Kit[a] containing IgG Dynabeads at 4×10^8 beads/ml, releasing buffer, and freeze-dried DNase
- Magnetic particle concentrator (Dynal) MCP-1
- Platform rocker shaker at 4 °C
- 15 ml conical plastic centrifuge tubes (i.e. Falcon)
- Mouse IgG primary antibody
- Sterile PBS with 0.1% BSA
- RPMI 1640 containing 1% FCS

A. Preparation of releasing buffer

1. Add 320 µl of buffer labelled component 2 into tube labelled component 1 containing the DNase.
2. Divide into small aliquots and store at -20 °C.[b]

B. Dynabeads antibody coating procedure

1. Resuspend Dynabeads thoroughly and transfer 1 ml into a 15 ml centrifuge tube.

Protocol 6 continued

2. Suspend beads in 7 ml PBS, place tube into MCP-1 magnet and, after 1 min, remove buffer by gentle pipetting.
3. Remove from magnet and add 7 ml PBS to resuspend the beads. Replace into magnet and remove buffer.
4. Repeat step 3 and resuspend in a final volume of 1 ml.
5. Add 40 µg of monoclonal antibody of interest, giving a final concentration of 1 µg per 10^7 Dynabeads.
6. Incubate the mixture at room temperature for 30 min rotating 'end-over-end' or shaking gently.
7. Collect the beads using the magnet as before and wash three times using PBS with BSA.
8. Resuspend the beads in 1 ml of PBS/BSA giving a concentration of 4×10^8 cells/ml. The beads are ready for direct selection of cells from a cell suspension.

C. Selection of target cells

1. Suspend cell suspension of freshly isolated primary, or harvested cultured cells, at 10^7/ml, or 10^6 in 100 µl of PBS/BSA, cool to 4°C in a cold room or on ice.
2. Add beads to the cell suspension at a concentration of at least five Dynabeads per target cell. For a sample of 10^7 cells add 5×10^7 or 125 µl of antibody pre-coated Dynabeads.
3. Mix cells and beads with gentle whirl-mixing, and incubate at 4°C on a rocking platform for between 15–30 min.
4. Place the tube in the magnet and allow the suspension to clear for at least 1 min.
5. Remove the supernatant gently while the tube is in the magnet, taking care not to disturb the cells attached to the tube wall.
6. Remove the tube from the magnet and resuspend the cells in 5 ml RPMI with 1% FCS.
7. Place the tube in the magnet and clear for 1 min before removing the supernatant.
8. Repeat steps 6 and 7 twice more.
9. After the final wash resuspend the cells in 200 µl of RPMI with 1% FCS.
10. Add 4 µl of DNase solution.
11. Incubate at room temperature for 15 min with gentle rocking, ensuring that the cells remain at the bottom of the tube.
12. Pass the cell suspension through a 1 ml Gilson pipette tip several times to release cells from the beads.
13. Clear the solution in the magnet for 1 min and pipette the supernatant into a new tube containing 200 µl of RPMI with 1% FCS.
14. Remove the tube from the magnet and add a further 200 µl of RPMI with 1% FCS.

> **Protocol 6** continued
>
> 15. Repeat step 12 and add the residual cells to the tube from step 13.
> 16. The released cells are in 600 µl and should be counted and diluted to the required density for cell culture.
>
> [a] Kits are also available for use with biotinylated antibodies and lectins.
> [b] 4 µl of DNase solution will be required per sample.

5 Long-term maintenance of stem cells in culture

In an ideal model system, there should be a minimum of changes in the growth properties of the cells *in vitro* and the cells should maintain the ability to differentiate into the same cell types as the tissue *in vivo*.

5.1 Keratinocyte stem cells in culture

The most successful system used for epidermal stem cells has been the co-culture method of Rheinwald and Green, which uses a 3T3 feeder layer (14). There are several pieces of evidence which suggest that the stem cells are conserved in cultures grown in this way, including long-term maintenance of the capacity for both proliferation and differentiation. Jones and Watt (13) showed, using cells that had been in culture for as many as 12 passages, that cells could still be found that behaved as stem cells. Furthermore, at confluence, the epidermal cell sheets developed a pattern of alternating integrin bright and dull patches, indicating that, not only were stem cells present, but also that they are capable of spatial organization. Jones *et al.* (15) also showed that these cells, when placed into nude mice, would form a fully stratified, differentiated epithelium. This capacity not only to continue to divide, but also to differentiate normally was shown in grafting studies, where cultured epithelial cell sheets, applied to patients, persisted for many years (5, 6). The Rheinwald and Green culture method was also used by Li *et al.* (32), who isolated putative stem cells from a primary keratinocyte suspension by FACS sorting for cells with high α6 integrin expression and low expression of a proliferation marker, 10G7. They showed that these cells could be serially passaged to give continuous growth for up to 95 days and had an estimated proliferative output of 5.8×10^8 cells per candidate stem cell. While there are serum-free media that can make epithelial cell cultivation simpler, it appears that feeder cells, presumably by adding some form of stromal environmental factors, are essential for stem cell maintenance in long-term culture.

5.2 Maintenance of non-epidermal epithelial cells in long-term culture

While many epithelial cell types can be grown in culture, there is little in the literature relating to the maintenance of a stem cell population in these cultures. However, the methods described above involving the use of feeder cells may work for many other epithelial cell types and would be a good starting point for stem cell studies in other tissues. As described earlier, prostate epithelial cells will grow from a single cell suspension in the presence of a 3T3 feeder layer and this method has also been shown to be beneficial in the study of clonal growth of human breast epithelial cells (33) and colorectal carcinoma cell lines (34).

6 Stem cell characterization by immunocytochemistry

Following the isolation of cell populations, using differential adhesion or magnetic beads, the cell phenotypes can be characterized immunocytochemically. This can be done in one of several ways, including FACS analysis if available, or the cells can be fixed onto microscope slides, by air drying or use of a cytospin, stained, and examined by fluorescence microscopy. Alternatively, in the absence of a fluorescence microscope, staining can be carried out using a staining method whereby the location of the antibody is visualized by a colour reaction, with a substrate such as 3,3'-diaminobenzidine. One reliable way to use the latter method is with a kit such as the VECTASTAIN Elite ABC kit, from Vector Laboratories. There follows a description of proteins that are useful as markers to distinguish subpopulations of some epithelial cell types, along with methods for the preparation of samples of cells for staining.

6.1 Antibody markers of differentiated cell phenotypes

Cells of all tissues express a wide repertoire of cell surface and cytoplasmic proteins and many of these are cell type-specific, while others show changes of expression during the differentiation of the cells in an epithelium, or *in vitro*. The cell surface proteins include the integrins (see Section 4), together with a wide number of other members of the CD (cluster of differentiation) antigens such as CD10, CD44, CD31, CD57. One of the major groups of cytoplasmic proteins expressed by epithelial cells is the cytokeratin family and this has proved a useful set of markers of epithelial tissue type.

6.1.1 Keratin markers of epithelial cell type

Cytokeratins form intermediate filaments and are classified as members of either the basic or acidic subfamilies, with one basic cytokeratin (numbered 1-8) always expressed together with an acidic partner (numbered 9-20). For example keratin 14 is always found co-expressed with cytokeratin 5 (35). In various basal cell layers K5 is also found pairing K15, 17, and 19, while K8 pairs with K7, 18, and 19. Other pairs include K1 with 10, 4 with 13, and 6 with 16. K19 is unusual in that it can pair with either K5 or K8, depending on its location.

The cytokeratins are all relatively abundant and stable, which makes them suitable for immunological detection and, although the members of the keratin family share a high degree of homology, there are sufficient differences for specific antibodies to have been raised against most of them. Some antibodies also recognize a more widely expressed epitope and will see several different cytokeratins. These antibodies include LP34, which will positively stain almost all cells of epithelial origin and can be considered a pan-cytokeratin antibody. *Table 2* lists the basic distribution of cytokeratin expression and antibodies that are known to have a high degree of specificity for single cytokeratins and low

Table 2. Cytokeratin expression in epithelial tissues

Cytokeratin	Typical tissue expression pattern	Antibody clone name	Supplier
K1	Suprabasal cells of cornified epithelia	34βB4	Novocastra
K2	Suprabasal granular layers in epidermis	None commercially available	
K3	Corneal epithelium	AE5	Biogenesis
K4	Suprabasal cells of non-cornifying epithelia	6B10	Sigma/Novocastra
K5	All stratifying epithelia	None commercially available	
K6	Hyperplastic epidermis and some mucosal epithelia	LHK6B	Novocastra
K7	Some luminal cells in prostate	RCK105	Dako/ICN
K8	Luminal cells of prostate and breast	35βH11	Dako
K9	Suprabasal palmoplantar epidermal layers	None commercially available	
K10	Suprabasal layers of epidermis	LHP1	Novocastra
K11	Suprabasal layers of epidermis	K8.60[a]	Sigma
K13	Suprabasal cells of non-cornifying epithelia	1C7	ICN
K14	All stratifying epithelia	LL002	Novocastra/Serotec
K15	Epidermis,[b] staining patchy	C8/144B	Dako
		K8.12	Sigma
K16	Hyperplastic epidermis and some mucosal epithelia	LL025	Novocastra
K17	Some basal cells in prostate	E3	Sigma/Novocastra
K18	Luminal cells of prostate and breast	RGE53	ICN
K19	Some basal prostate cells and luminal cells in breast and prostate	LP2K	Amersham
K20	Umbrella cells of urothelium and Merkel cells	K$_s$20.8	Novocastra
Pan–cytokeratins	5/6/18 on frozen tissue, not 18 on paraffin embedded	LP34	Novocastra

[a] This antibody also recognizes K10.
[b] C8/144B was raised against CD8, a T cell protein but reportedly has cross-reactivity with cytokeratin 15 in a subset of human keratinocytes, particularly in the bulge region of hair follicles, making it a possible stem cell marker (36). K8.12 reacts with cytokeratins 13, 15, and 16 and so as a marker for K15 is useful in tissues lacking K13 and K16, such as the prostate, where it stains a subset of the basal epithelial layer.

cross-reactivity with other proteins. These are now available from a variety of commercial sources. Two cytokeratins have been shown to stain distinct subpopulations of basal keratinocytes and may prove to be markers for epidermal stem cells. Cytokeratin 19 was shown to stain regions of the hair follicle, individual integrin-bright interfollicular cells, and some cells in culture (36). As the hair follicle is reported to contain stem cells this may be useful. More recently Lyle et al. (37) have shown that C8/144B, an antibody originally raised against CD8, a T cell marker, recognizes cytokeratin 15 in a restricted population of hair follicle cells with stem cell characteristics such as label retention and high $\beta1$ integrin expression. The lack of cell type specificity of this antibody makes it potentially problematic for use in other tissues and other cytokeratin 15-specific antibodies have shown a wider distribution of staining in the skin (38).

6.1.2 Cell surface markers of epithelial cell type

There are many different proteins found on the cell surface, ranging from highly abundant adhesion molecules to growth factor receptors and proteins of as yet undetermined function. Cell surface markers are useful tools for the cell biologist since they can be detected by antibodies without prior fixation or permeabilization of the cells. This allows the use of flow cytometry or immunomagnetic beads for cell separation to isolate viable cell populations from tissue. The choice of a suitable cell surface marker for the characterization of experimentally isolated cells depends on several criteria. Some markers will detect all

Table 3. Non-cytokeratin markers of epithelial cells

Marker	Typical expression pattern	Antibody	Source
Androgen receptor	Luminal marker in prostate	F39.4.1	Biogenex
CD29—$\beta1$ integrin[a]	Basal layer of most epithelia	P5D2	Chemicon
CD49b—$\alpha2$ integrin	Basal layer of most epithelia	P1E6	Gibco BRL
CD49c—$\alpha3$ integrin	Basal layer of most epithelia	VM2	Novocastra
CD49e—$\alpha5$ integrin	Weak in tissue, stronger in cultured cells	AB1928	Chemicon
CD49f—$\alpha6$ integrin	Basal layer of most epithelia	GOH3	Serotec
CD104—$\beta4$ integrin	Basal layer of most epithelia	3E1	Gibco BRL
CD10—CALLA	Myoepithelial breast epithelia	MAB423	Chemicon
CD44—Hyaluronan receptor	Basal layer stronger in most epithelia	F10-44-2	Chemicon
CD57/HNK-1	Luminal cells in prostate	VC1.1	Sigma
Epithelial-specific antigen	Luminal breast epithelia	VU-1D9	Novocastra
Involucrin	Suprabasal cells in epidermis	SY5	Novocastra
MUC-1 glycoprotein	Luminal breast epithelia	Ma695	Novocastra
PAP	Luminal marker in prostate	PASE/4LJ	Dako
Prostate-specific antigen	Luminal marker in prostate	ER-PR8	Dako
Serotonin (5-HT)	Neuroendocrine cells in epithelia	NCL-SEROTp	Novocastra

[a] Integrin staining of breast epithelia shows strong expression in the myoepithelial cells while there is also weaker, basolateral staining of luminal cells (18).

cells of one type, while others may recognize only a differentiated or undifferentiated subset. Since the cells will normally have been isolated by trypsinization, the target marker epitope must not be sensitive to trypsin cleavage, a problem with particular antibodies recognizing some transmembrane proteins, such as CD44 and syndecan. *Table 3* lists commonly used markers of epithelial cell types for which there are commercially available antibodies. Most markers of this type are differentially expressed between cell layers in epithelial tissues such as the skin, breast, and prostate, while some of them are more specific for tissue type, such as MUC-1 in the breast luminal epithelial cells.

6.2 Staining cell suspensions using cytospin preparations

The following protocol is a rapid method to fix cells onto a glass microscope slide for immunostaining. The second part details a method to fix cells onto coverslips for staining without the use of a cytospin. While the second method takes a longer time it is as effective as the first.

Protocol 7

Preparation of cells for use in immunocytochemistry: cytospin cell preparation system

Equipment and reagents

- Shandon Cytospin^R 3
- Glass microscope slides
- Subconfluent cell cultures or primary cells
- 13 mm glass coverslips (Chance)
- 9 cm Petri dishes (e.g. Nunc, Sterilin)
- PBS
- Acetone/methanol (1:1) chilled to −20 °C
- 3.7% formaldehyde in PBS
- DMEM containing 10% FCS

A. Cytospin method

1. If using cell cultures, trypsinize cells and resuspend in 10 ml of medium containing 10% fetal calf serum to inhibit trypsin.
2. Transfer cells to 15 ml Falcon tube.
3. Count cells using a haemocytometer.
4. Pellet cells by centrifugation for 5 min at approx. 170 g.
5. Resuspend cells at 5×10^4/ml in PBS.
6. Assemble the cytospin sample chamber according to the manufacturer's instructions. Place the glass slide in the holder followed by the filter paper, smooth side down, and finally the plastic funnel. Clip all the parts into place and place in the centrifuge, balanced in pairs.
7. Add 300 μl of the cell suspension to each sample chamber.
8. Run centrifuge for 4 min at 400 r.p.m.

Protocol 7 continued

9. Remove the slide, discard the filter paper, and allow slide to air dry for 5 min.
10. Fix slides in acetone/methanol[a] for 10 min at −20 °C.
11. Air dry the slides for 10 min and store at −20 °C if not used immediately.

B. Air drying alternative method

1. Follow part A, steps 1-5, when the cells should be resuspended at 3×10^5/ml in DMEM medium containing 10% FCS.
2. Place required number of 13 mm coverslips into a 9 cm plastic Petri dish.
3. Using a Gilson pipette add 50 µl of cell suspension to each coverslip and spread evenly over the glass surface by pipetting gently up and down.
4. Place the Petri dish uncovered at 37 °C in a dry incubator or oven.
5. Check regularly until the medium has evaporated leaving only the very centre of the coverslip moist.
6. Remove the dish from the incubator and slowly add 10 ml of 3.7% formaldehyde to cover the coverslips.
7. Allow to fix at room temperature for 10 min.
8. Rinse the coverslips three times with PBS and leave the cells immersed in PBS until required for staining.
9. If the antibody requires the permeabilization of the cells, such as for cytokeratins, this can be done prior to staining using ice-cold methanol for 5 min.

[a] The use of methanol/acetone as a fixative is ideal for anti-cytokeratin staining. This treatment may affect epitopes of some cell surface markers, such as the integrins, which may be better preserved by fixing the slides with a 3.7% solution of formaldehyde for 10 min at room temperature. The slides will not require permeabilization.

References

1. Shaw, A. J. (ed.) (1996). *Epithelial cell culture: a practical approach*. IRL Press, Oxford.
2. Thomson, J. A., Itskovitz-Eldor, J., Shapiro, S. S., Waknitz, M. A., Swiergiel, J. J., Marshall, V. S., *et al.* (1998). *Science*, **282**, 1145.
3. Vogel, G. (1999). *Science*, **283**, 1432.
4. Bjornson, C. R. R., Rietze, R. L., Reynolds, B. A., Magli, M. C., and Vescovi, A. L. (1999). *Science*, **283**, 534.
5. Gallico, G. G., III, O'Connor, N. E., Compton, C. C., Kehinde, O., and Green, H. (1984). *N. Engl. J. Med.*, **311**, 448.
6. Compton, C. C. (1992). *Eur. J. Pediatr. Surg.*, **2**, 216.
7. Gerrard, A. J., Hudson, D. L., Brownlee, G. G., and Watt, F. M. (1993). *Nature Genet.*, **3**, 180.
8. Roberts, D. and Marks, R. (1980). *J. Invest. Dermatol.*, **74**, 13.
9. Wright, N. A. and Alison, M. (ed.) (1984). *The biology of epithelial cell populations*. Clarendon Press, Oxford.
10. Loeffler, M. and Potten, C. S. (1997). In *Stem cells* (ed. C. S. Potten), p. 1. Academic Press.

11. Potten, C. S. and Morris, R. J. (1988). *J. Cell Sci.*, Suppl., **10**, 45.
12. Barrandon, Y. and Green, H. (1987). *Proc. Natl. Acad. Sci. USA*, **84**, 2302.
13. Jones, P. H. and Watt, F. M. (1993). *Cell*, **73**, 713.
14. Rheinwald, J. G. and Green, H. (1975). *Cell*, **6**, 331.
15. Jones, P. H., Harper, S., and Watt, F. M. (1995). *Cell*, **80**, 83.
16. Hotchin, N. A., Kovach, N. L., and Watt, F. M. (1993). *J. Cell Sci.*, **106**, 1131.
17. Nagle, R. B., Hao, J., Knox, J. D., Dalkin, B. L., Clark, V., and Cress, A. E. (1995). *Am. J. Pathol.*, **146**, 1498.
18. Koukoulis, G. K., Virtanen, I., Korhonen, M., Laitinen, L., Quaranta, V., and Gould, V. E. (1991). *Am. J. Pathol.*, **139**, 787.
19. Adams, J. C. and Watt, F. M. (1991). *J. Cell Biol.*, **115**, 829.
20. Carter, W. G., Kaur, P., Gil, S. G., Gahr, P. J., and Wayner, E. A. (1990). *J. Cell Biol.*, **111**, 3141.
21. Adams, J. C. and Watt, F. M. (1990). *Cell*, **63**, 425.
22. Niessen, C. M., van der Raaij-Helmer, L. M. H., Hulsman, E. H. M., van der Neut, R., Jonkman, M. F., and Sonnenberg, A. (1996). *J. Cell Sci.*, **109**, 1695.
23. Niessen, C. M., Hogervorst, F., Jaspars, L. H., de Melker, A. A., Delwel, G. O., Hulsman, E. H. M., *et al.* (1994). *Exp. Cell Res.*, **211**, 360.
24. Wang, A., Patrone, L., McDonald, J. A., and Sheppard, D. (1995). *Dev. Dyn.*, **204**, 421.
25. Cheresh, D. A., Smith, J. W., Cooper, H. M., and Quaranta, V. (1989). *Cell*, **57**, 59.
26. Zambruno, G., Marchisio, P. C., Marconi, A., Vaschieri, C., Melchiori, A., Giannetti, A., *et al.* (1995). *J. Cell Biol.*, **129**, 853.
27. Marinkovich, M. P., Keene, D. R., Rimberg, C. S., and Burgeson, R. E. (1993). *Dev. Dyn.*, **197**, 255.
28. Potten, C. S. and Loeffler, M. (1990). *Development*, **110**, 1001.
29. Simon-Assmann, P., Duclos, B., Orian-Rousseau, V., Arnold, C., Mathelin, C., Engvall, E., *et al.* (1994). *Dev. Dyn.*, **201**, 71.
30. Rouselle, P., Lunstrum, G. P., Keene, D. R., and Burgeson, R. E. (1991). *J. Cell Biol.*, **114**, 567.
31. Liu, A. Y., True, L. D., LaTray, L., Nelson, P. S., Ellis, W. J., Vessella, R. L., *et al.* (1997). *Proc. Natl. Acad. Sci. USA*, **94**, 10705.
32. Li, A., Simmons, P. J., and Kaur, P. (1998). *Proc. Natl. Acad. Sci. USA*, **95**, 3902.
33. Stingl, J., Eaves, C. J., Kuusk, U., and Emerman, J. T. (1998). *Differentiation*, **63**, 201.
34. Kirkland, S. C. (1988). *Cancer*, **61**, 1359.
35. Moll, R., Franke, W. W., Geiger, B., Krepler, R., and Schiller, D. L. (1982). *Cell*, **31**, 11.
36. Michel, M., Torok, N., Godbout, M.-J., Lussier, M., Gaudreau, P., Royal, A., *et al.* (1996). *J. Cell Sci.*, **109**, 1017.
37. Lyle, S., Christofidou-Solomidou, M., Liu, Y., Elder, D. E., Albelda, S., and Cotsarelis, G. (1998). *J. Cell Sci.*, **111**, 3179.
38. Waseem, A., Dogan, B., Tidman, N., Alam, Y., Purkis, P., Jackson, S., *et al.* (1999). *J. Invest. Dermatol.*, **112**, 362.

Chapter 11

Senescence, apoptosis and necrosis

Ian R. Kill and Richard G. A. Faragher[†]

Department of Biological Sciences, Brunel University, Uxbridge, Middlesex UB8 3PH, UK.

[†]Department of Pharmacy and Biomolecular Sciences, University of Brighton, Cockcroft Building, Lewes Road, Brighton BN2 4GJ, UK.

1 Introduction

Populations of cells which will divide forever in culture (continuous cell lines) differ from those which cease to expand after several weeks or months of growth (finite cell lines). The end of the proliferative lifespan of a cell culture has been loosely termed as either its 'senescence' or its 'death'. These states are distinct biochemical entities and the causes of any failure to thrive in a cell population may involve both processes. The purpose of this chapter is to describe methods for studying senescence, apoptosis, and necrosis.

1.1 Cellular senescence

The term senescence is used in two distinct ways, applying either to individual cells or to entire cultures, and this can result in confusion. Senescence of an entire culture, sometimes called 'the Phase III phenomenon' (1), is a failure of the culture to proliferate under conditions which had previously allowed sustained cell growth. This growth is generally measured as population doublings (PDs), calculated as the number of times the cell population doubles in number during the course of culture.

$$PD = \frac{\log 10 \text{ (number cells harvested)} - \log 10 \text{ (number cells seeded)}}{\log 10^2}$$

Senescence has been studied mainly in normal human fibroblasts, cultures of which go through 30–60 population doublings, by which time virtually the entire culture is composed of senescent cells. These senescent fibroblasts are responsible for the decline in growth potential of ageing cell populations. The characteristic feature of senescent cells is their failure to divide in response to a mitotic stimulus. Senescent cells do not appear suddenly at Phase III, but instead compose a gradually increasing percentage of the population throughout the lifespan of the culture.

Senescence is distinct from quiescence, a transiently growth arrested state which can be induced by reducing the concentration of serum or through contact inhibition. Quiescence can be reversed by passaging the cells or by the re-addition of serum. Confusingly, both quiescence and senescence are referred to as the G_0 phase of the cell cycle (sometimes distinguished as G_0^Q or G_0^S respectively). For a detailed review of terminology and its relevance see refs 2 and 3.

1.2 Cell death

Necrosis can be triggered by simple physical trauma or gross damage. Programmed cell death or apoptosis is initiated by an individual cell in response to a specific stimulus or as a result of an inappropriate set of signals received from the external environment. There are several subtypes of apoptosis utilizing distinct pathways and patterns of gene expression. However, they converge on a common set of phenotypic markers and behaviours, thus allowing the fraction of apoptotic cells to be defined regardless of the pathway by which cell death is triggered.

1.3 Differentiation and de-differentiation

Primary keratinocytes in culture retain the ability to differentiate under the correct culture conditions (4). Similarly, cultured primary T cells retain the appropriate CD markers and the ability to kill target cells as measured by ^{51}Chromium release assays (5). The process of differentiation can also result in growth arrest and contribute to the failure of a culture to expand. The contribution of differentiation to lack of growth within a culture must be addressed on a cell type by cell type basis using known markers of the process (such as involucrin for keratinocytes), where available. Differentiation and senescence have been shown to be separate processes in some cell types.

Cell cultures can also appear to de-differentiate or lose the phenotype of the tissue from which they were derived (6). *In vitro* culture favours the relatively undifferentiated proliferating cells, which may not express tissue-specific markers either *in vivo* or *in vitro*, and this is the usual reason why primary cultures and the cell lines derived from them tend to progressively lose the differentiated phenotype.

In contrast to de-differentiation, de-adaptation is defined as the loss of the stimulus to make tissue-specific proteins rather than the capacity to do so. Detailed study may allow the missing stimulus to be identified and thus restore the specific phenotype.

2 Simple measures of the population dynamics of primary cultures

The simplest static measure of growth, increase in cell number with time, provides no clues to the fraction of actively dividing, reproductively sterile, or

SENESCENCE, APOPTOSIS, AND NECROSIS

dying cells within a given population (7). More complex static measures such as the determination of DNA content, RNA content, cellular ATP, or protein concentration provide more information, but usually require more work.

The following techniques (used in combination on the same culture) allow the determination of the fraction of growing senescent, necrotic, and apoptotic cells and thus a complete picture of that culture's population dynamics. All depend on the culture of cells on 13 mm diameter Number 1 thickness coverslips cleaned according to the following protocol.

Protocol 1
Cleaning of coverslips for cell culture

Equipment and reagents
- 13 mm, Number 1 thickness glass coverslips (Merck)
- Fine forceps
- 100 mm diameter glass Petri dishes
- 100 mm diameter filter paper (Merck)
- Standard laboratory autoclave
- Benchtop heater
- Absorbent paper
- Flow 7X tissue culture grade detergent (ICN-Flow)
- Double distilled water
- Absolute ethanol

Method
1. Add 100 coverslips to 1 litre of a 5% (v/v) solution of Flow 7X in double distilled water. This will be cloudy.
2. Heat until the solution clears (usually 40–50 °C). Maintain the coverslips at this temperature for a minimum of 1 h.
3. Wash the coverslips seven times in fresh doubled distilled water and once in 100 ml of absolute ethanol.
4. Shake the coverslips onto a sheet of absorbent paper. This absorbs the ethanol.
5. Separate the coverslips using fine forceps and place individually on a piece of filter paper within a glass Petri dish. Approx. 20 coverslips can be placed into each dish in this fashion.
6. Seal the Petri dish using autoclave tape and autoclave.
7. Open the autoclaved dish only under sterile conditions.

2.1 Measurement of the growth fraction

Normal somatic cells do not divide in synchrony during normal growth. Primary fibroblast cultures are composed of a mixture of clones with very variable growth potentials (8, 9). This behaviour was shown not to be an inherent property of individual clones, because re-cloning of a clone with a long lifespan gave subclones with a range of division potentials. These cloning experiments demonstrated that the reproductive ability of fibroblasts was to some degree determined by a chance (stochastic) process, and that smaller clones predominated as the culture aged. Some evidence suggests that this process in human cells is due to the gradual attrition of telomeric sequences.

Work on bulk cultures complemented these single cell analyses. Cristofalo and Scharf (10) carried out an analysis of the cell kinetics using embryonic fibroblasts, pulse labelling the cells with [^{3}H]thymidine at every passage throughout the lifespan of the cultures. The fraction of cells that entered S phase was then estimated by autoradiography. The assumption was made that an unlabelled cell is senescent. It was observed that unlabelled cells were present in cultures of embryonic cells and that a few labelled cells were present in even the latest passage cultures. It was also shown that the fraction of unlabelled cells increased smoothly with serial passage. Additional experiments excluded lengthening of the cell cycle as a property of senescence (11). Thus, primary cultures are mixtures of senescent and growing cells, the proportions of which alter as the cultures age. Such kinetic analyses can provide valuable data on the rates of senescence in both normal cell cultures derived from different tissues and can be performed in three ways:

(a) Short pulse label experiments designed to detect a representative growth fraction for comparative studies.

(b) Cristofalo-Scharf style long label experiments designed to detect the senescent fraction of the culture.

(c) Direct detection of cell cycle-related proteins which may be used as proliferation markers. This last technique has the advantage that labelling time is not a factor.

Protocol 2
Labelling using 5-bromo-2'-deoxyuridine (BrdU)

Equipment and reagents

- Epifluorescence microscope with filters suitable for detection of fluorescein isothiocyanate (FITC) and 4-6-diaminophenylindole (DAPI)
- Fine forceps
- Parafilm
- 100 mm diameter glass Petri dishes
- 100 mm diameter filter paper (Merck)
- Glass slides
- Vectashield mountant containing DAPI (Vector Laboratories)
- Labelling reagent: BrdU, 3 mg/ml; 5-fluoro-2'-deoxyuridine, 0.3 mg/ml (Amersham)
- Mouse monoclonal antibody (IgG) to BrdU pre-diluted with DNase I (Amersham)
- Polyclonal rabbit antiserum to mouse IgG conjugated with FITC (Dako)
- Cell type of interest cultured to subconfluence on coverslips (prepared according to *Protocol 1*)
- Phosphate-buffered saline (prepared using Oxoid PBS tablets) (Oxoid)
- PBS-FCS: 1% (v/v) FCS diluted in PBS
- 95%-5% (v/v) ethanol-acetic acid for fixation

Protocol 2 continued

A. Labelling of cells

1. Add labelling reagent at a dilution of 1:1000 to the culture medium. (Do not use medium which contain thymidine, such as Ham's F12, since this results in poor uptake of BrdU.)
2. Allow to pulse label for a suitable period. The labelling time of choice is dependent on the style of assay to be undertaken. Experiments designed to detect a representative growth fraction for comparative studies require a labelling time less than the S phase duration (we use 1 h labels). Experiments designed to detect the senescent fraction of the culture require a labelling time of 72 h or more (less than 5% label-incorporating nuclei can be taken to indicate culture senescence). However, the absolute growth fraction will be inaccurate since some cells will have divided during the labelling period. Intermediate labelling times (i.e. 24 h) are quite difficult to interpret and should be avoided.
3. Aspirate the medium. Quickly wash the cells in three changes of PBS.
4. Aspirate the last change of PBS. Flood the coverslip with ethanol–acetic acid. Fix at $-20\,°C$ for 20 min.
5. Wash a further three times in PBS. Leave the third wash in contact with the coverslip.

B. Detection of BrdU incorporation

1. Prepare a 'humidified chamber'. This consists of a glass Petri dish containing a water saturated filter paper on which is placed a square of Parafilm. This prevents desiccation during the assay.
2. Place 50 µl of anti-BrdU on the Parafilm, then using fine forceps gently place the coverslip face down on the drop of antibody (this ensures even spreading).
3. Incubate for a minimum of 90 min at room temperature (never at 4 °C since DNase action is required to allow the antibody access to the DNA).
4. Wash three times in PBS (by dipping the coverslip at least ten times into each of a series of three Universal tubes).
5. Repeat step 2 substituting a 1:20 dilution of anti-mouse IgG in PBS–FCS.
6. Incubate overnight at 4 °C

C. Quantification of label-incorporating cells

1. Wash the coverslip three times in PBS as above, in addition include a final wash in double distilled water. This removes salts from the coverslip.
2. Place a drop of Vectashield mount onto a clean glass slide and place the coverslip face down onto the mountant. The slide may be sealed with clear nail varnish if required.
3. View under epifluorescence using the DAPI filter to focus on the cell nuclei to prevent quenching (the DAPI signal may be low in intensity due to nuclease activity). Count the total number of nuclei within the field of view.

> **Protocol 2** continued
>
> 4. View the same field under FITC filters. A distinct punctate staining reveals incorporation of BrdU (*Figure 1*). Count the number of positive cells within the field of view.
> 5. Repeat steps 3 and 4 until the count has reached 400 positive or 1000 total nuclei, whichever comes first. The percentage of labelled to unlabelled cells is the 95% confidence limit of the growth fraction (corrected for labelling time).

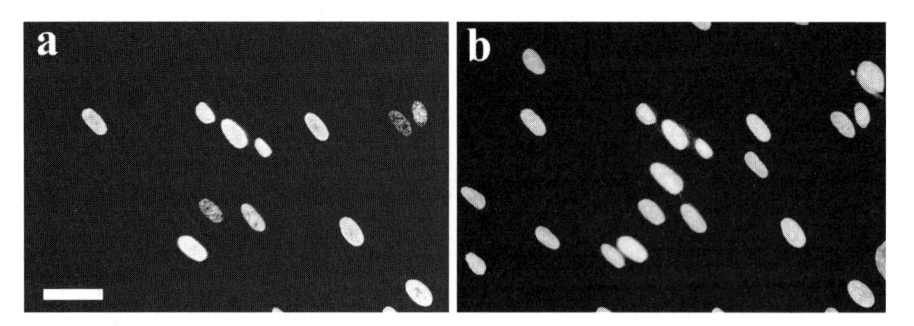

Figure 1. Shows staining for incorporation of 5-bromo-2′-deoxyuridine (BrdU) by 2DD.5 normal human dermal fibroblasts (Section 2.1). The cells were incubated in the presence of BrdU for 1 h and then fixed and stained. (a) BrdU incorporation; (b) the same field of view revealing DAPI staining. Bar = 20 μm.

Protocol 3
Detection of total growth fraction using antisera to pKi-67

Equipment and reagents

- Epifluorescence microscope with filters suitable for detection of fluorescein isothiocyanate (FITC) and 4-6-diaminophenylindole (DAPI)
- Fine forceps
- Parafilm
- 100 mm diameter glass Petri dishes
- 100 mm diameter filter paper (Merck)
- Glass slides
- Vectashield mountant containing DAPI (Vector Laboratories)
- Mouse monoclonal antibody directed against human Ki-67 (Dako)[a]
- Polyclonal rabbit antiserum to mouse IgG conjugated with FITC (Dako)[b]
- Cell type of interest cultured to subconfluence on coverslips (prepared according to *Protocol 1*)
- PBS (see *Protocol 2*)
- PBS–FCS (see *Protocol 2*)
- Methanol/acetone (1:1) for fixation

Method

1. Aspirate the medium. Quickly wash the cells in three changes of PBS.
2. Aspirate the last change of PBS. Flood the coverslip with methanol/acetone. Fix at 4°C for 5 min.

Protocol 3 continued

3. Wash a further three times in PBS. Leave the third wash in contact with the coverslip.
4. Perform immunocytochemical staining as described in *Protocol 2*, part B, substituting 15 μl of anti-Ki-67 for anti-BrdU.
5. Visualize the Ki-67 positive fraction of the culture as described in *Protocol 2*, part C. Ki-67 positive cells show a highly distinctive punctate nuclear staining under FITC excitation (*Figure 2*). When a culture is moving from G_0^Q, Ki-67 staining first becomes detectable in late G_1. However it remains present throughout the cell cycle for cultures maintained in log phase growth.

[a] Alternatively, a rabbit polyclonal antibody (Dako) may be used for dual labelling procedures in combination with, for example, a mouse monoclonal antibody.

[b] Or appropriate secondary antiserum if the rabbit polyclonal anti-Ki-67 is used.

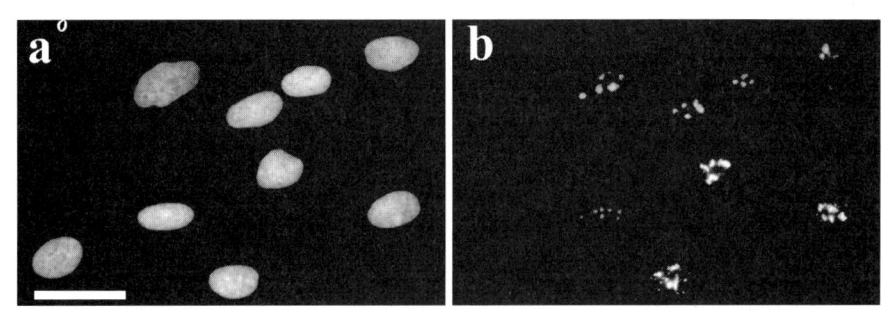

Figure 2. Detection of the total growth fraction using Ki-67. 2DD.5 cells were fixed and stained for the presence of Ki-67, an antigen detectable in proliferating cells only (Section 2.1). (a) DAPI staining used to reveal all nuclei within the field of view. (b) Typical Ki-67 staining restricted to nucleoli of proliferating cells. Bar = 20 μm.

2.2 Determination of the necrotic fraction of the population

Several techniques are available for this purpose, of which the simplest remains trypan blue staining. This dye exclusion assay can be used either as part of a routine passage protocol or as a simple washing step on coverslips of unfixed material. However, trypan blue exclusion provides only a rough estimate of viability and needs independent validation for the particular application before being used routinely.

A visually pleasing and highly discriminatory alternative is afforded by calcein AM–ethidium bromide staining, which we have employed to estimate the fraction of living and necrotic cells both on coverslips and in collagen matrices. Calcein AM is membrane permeant and is cleaved by active esterases to yield a green cytoplasmic fluorescence whilst ethidium is membrane impermeant and labels only the DNA of cells with a compromised outer membrane. Unfixed material is required in both cases.

Protocol 4
Determination of viable and necrotic fractions using calcein AM–ethidium bromide staining

Equipment and reagents
- Epifluorescence microscope with filters suitable for detection of fluorescein isothiocyanate (FITC) and tetramethyl rhodamine isothiocyanate (TRITC)
- Adherent live cells growing on coverslips prepared according to *Protocol 1*
- Calcein AM fluorescent esterase substrate (Molecular Probes)
- Ethidium homodimer 1 (Molecular Probes)
- Phosphate-buffered saline (prepared using Oxoid PBS tablets) (Oxoid)

Method
1. Wash the cells three times in PBS.
2. Add 0.05 mg/ml each of calcein AM and ethidium homodimer in PBS.
3. Incubate for a minimum of 10 min at 37 °C.
4. View under a fluorescence microscope. Live cells show strong cytoplasmic fluorescence under FITC excitation. Dead cells show nuclear fluorescence under TRITC filters.
5. Score 200 cells first under FITC and then TRITC filters. Express viability as a percentage.

2.3 Determination of the senescent fraction of the population

Until recently there was no simple way to directly demonstrate that a single cell was senescent other than a combination of lack of label incorporation and metabolic viability. However a variation of the standard catalytic histochemical staining technique for β-galactosidase activity has been shown to discriminate between senescent cells and their growing counterparts (12, 13). The mechanistic basis of this 'senescence-associated β-galactosidase' (SA-β) assay is unknown and may be related to the overexpression of the enzyme at senescence or to a simple increase in cell size correlating with entry into the permanently non-dividing state. It may be that the majority of lysosomal enzymes are amenable to histochemical visualization by related techniques.

A series of cell types on which the assay has been validated has been published (12). Variants of the technique have also been employed. Although we do not routinely undertake dual labelling studies, SA-β staining is compatible with [^{3}H]thymidine pre-labelling (10 mCi/ml) and we have had success combining SA-β with BrdU labelling (72 hours) and antisera to Ki-67 (rabbit polyclonal). SA-β staining also supports cell permeabilization with 0.2% Triton X-100.

Protocol 5

Demonstration of senescent cells by senescence-associated β-galactosidase (SA-β) activity

Equipment and reagents

- Light microscope
- Glass slides
- Other equipment for preparing glass coverslips as given in previous protocols
- 0.1 M citric acid solution: 2.1 g citric acid monohydrate in 100 ml distilled water
- Sodium phosphate solution: 2.84 g/100 ml sodium dibasic phosphate or 3.56 g/100 ml sodium dibasic phosphate dihydrate
- Citric acid/sodium phosphate buffer (100 ml): 36.85 ml of 0.1 M citric acid solution and 63.15 ml of 0.2 M sodium phosphate (dibasic) solution; verify that the pH is 6.0
- X-gal powder (Melford Laboratories Ltd.); stored in dark at −20°C
- 5 M sodium chloride
- Dimethylformamide (DMF)
- Potassium ferrocyanide
- Potassium ferricyanide
- 1 M magnesium chloride
- Apathy's syrup, histamount, or similar aqueous mounting medium (Merck)
- Fixative: 2% paraformaldehyde/0.2% glutaraldehyde in PBS; SA-β staining is also compatible with 3% formaldehyde

A. Preparation of staining solution

1. For 20 ml staining solution combine 4 ml citric acid/sodium phosphate buffer with 20 mg of X-gal (in a few drops of DMF only since this inhibits the β-galactosidase). Then add:
   - 1 ml of 100 mM potassium ferrocyanide
   - 1 ml of 100 mM potassium ferricyanide
   - 0.6 ml of 5 M sodium chloride
   - 40 μl of 1 M magnesium chloride
   - 13.4 ml double distilled water

B. Detection of SA-β activity in adherent cells

1. Wash material twice in PBS.
2. Fix for 3–5 min at room temperature. Prolonged fixation will inactivate the enzyme.
3. Repeat washes in PBS.
4. Add staining solution (1–2 ml per 35 mm diameter dish).[a]
5. Incubate at 37°C.
6. Senescent cells display a dense blue colour. This is detectable in some cells within 2 h, staining is usually maximal after 12–16 h (Figure 3).
7. Material can be mounted in aqueous mountains (e.g. Apathy's syrup) and can be counterstained with fast green or similar colour compatible stain.

[a] A useful positive control for the assay is to shift the pH of the staining solution to pH 5.0. This detects all mammalian cells with working endogenous β-galactosidases.

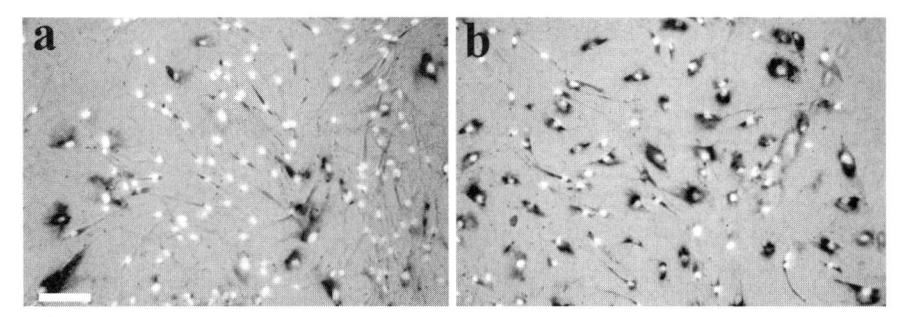

Figure 3. Detection of senescent cells using the senescence-associated β-galactosidase assay. 2DD human dermal fibroblasts at passage 6 (a) and 2DD cells at passage 26 (b) were stained to reveal the presence of senescence-associated β-galactosidase activity. Positive cells stain a deep blue colour in the cytoplasm (seen as a dense stain in the figure). The figure also shows DAPI staining used to reveal all nuclei within the field of view (light blue nuclear staining—seen as bright regions on the figure). Note that even the early passage cultures contain senescent cells. Bar = 50 µm.

2.4 Determination of the apoptotic fraction of the population using TUNEL

Activation of endogenous nuclease activity is considered to be a key biochemical event in apoptosis, leading to the cleavage of DNA into nucleosome-sized fragments. This process gives rise to free 3′ hydroxyl groups which can be extended using terminal transferase or DNA polymerase. If labelled nucleotides are included in the end-labelling mixture, apoptotic nuclei can be directly visualized. This TdT-mediated dUTP nick end-labelling assay (TUNEL) is fast and effective but should not be employed on any coverslips on which the cells have been pre-pulsed with BrdU since this undergoes photolysis in daylight. This can result in a large number of free 3′ hydroxyl groups within the DNA of any labelled cells and thus a falsely high TUNEL reading.

Protocol 6

Demonstration of apoptotic cells by TdT-mediated dUTP nick end-labelling (TUNEL)

Equipment and reagents

- Epifluorescence microscope with filters suitable for detection of fluorescein isothiocyanate (FITC) and 4-6-diaminophenylindole (DAPI)
- Vectashield mountant containing DAPI (Vector Laboratories)
- Glass slides
- Other equipment for handling glass coverslips as given in previous protocols
- Phosphate-buffered saline (prepared using Oxoid PBS tablets) (Oxoid)
- 3% (w/v) paraformaldehyde in PBS for fixation[a]
- 0.1% (v/v) Triton X-100™ in PBS
- Pre-diluted terminal transferase solution and nucleotide labelling solution containing fluorescein-conjugated dUTP (sold together by Boehringer Mannheim)

Protocol 6 continued

Method

1. Aspirate the medium and wash the cells as described in *Protocol 2*.

2. Aspirate the last change of PBS. Flood the coverslip with the paraformaldehyde solution. Incubate at room temperature for 10 min to fix. Wash with PBS as described in *Protocol 2*.

3. Add 0.1% Triton X-100 and incubate for 5 min at room temperature, this permeabilizes the cells and allows the terminal transferase access to the nuclear DNA. Wash a further three times with PBS.

4. Add 50 μl of terminal transferase solution to 450 μl of labelling solution to reconstitute a working TUNEL solution.[b]

5. Spot 50 μl of TUNEL solution onto Parafilm in a humidified chamber and incubate the coverslip with the solution as described in *Protocol 2*, part B.

6. Incubate the humidified chamber for 90 min at room temperature then wash, mount, and quantify as described in *Protocol 2*, part C. Apoptotic nuclei show strong but diffuse FITC fluorescence.

[a] Add 9 g paraformaldehyde to 300 ml PBS warmed to 60°C on a heated stirrer in a fume hood. Stir for 10 min and then add three drops of 1 M NaOH and stir until clear. Cool, adjust pH to 7.4, and filter through Whatman No. 1 paper.

[b] Controls should be included at this stage. We routinely include a negative control in which terminal transferase has been omitted and a positive control in which the DNA has been nicked by exposure to DNase for 30 min.

3 Other techniques for analysing population dynamics

Growth and apoptosis require ATP. Apoptosis, but not growth arrest, is accompanied by an elevation in ADP levels. Therefore measuring the relative levels of ATP and ADP in cultures, and calculating the ADP/ATP ratio, can be informative. Necrosis is characterized by a rapid decline in ATP levels, so it may be possible to discriminate between the forms of cell death. A novel assay system has been developed in kit form which is based upon estimation of ATP/ADP levels by luminometry (Lumitech). The assay is performed in 96-well plates, allowing high throughput.

The simplest method for detecting dead or dying cells within a population is by microscopic observation. Phase-contrast will reveal floating or dying cells as rounded, dense bodies (*Figure 4a*). Fluorescence microscopy allows for staining of cultures with dyes such as DAPI or Hoechst 33258. Cells undergoing apoptosis have fragmented or irregularly stained chromatin whereas unaffected cells display relatively homogeneous staining (*Figure 4b*).

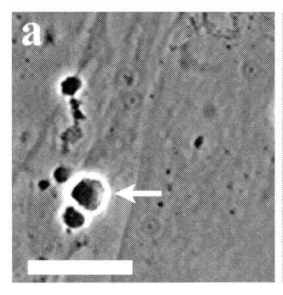

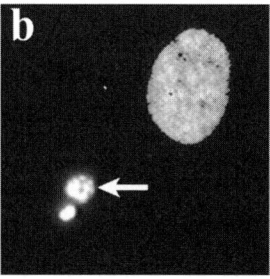

Figure 4. Visualization of apoptotic cells. (a) Apoptotic cells viewed using phase-contrast optics appear as phase-dense 'apoptotic bodies' (*arrow*). (b) The same field stained with DAPI and viewed using fluorescence optics. The chromatin in the apoptotic cells is dense and irregular (*arrow*) whilst the chromatin of an unaffected cell appears relatively homogeneous. Bar = 10 μm.

3.1 Detection of apoptosis using DNA laddering

Some apoptotic pathways cleave genomic DNA into mono- and oligonucleosomal DNA fragments. Visualization of this process following agarose gel electrophoresis of DNA reveals multiples of the 180–200 bp nucleosomal fragment in a characteristic 'ladder' pattern. The occurrence of DNA fragmentation should not be considered as definitive for cells undergoing apoptosis, since some cells undergoing necrosis also display 'ladders'. Conversely, several cell types do not show ladders despite undergoing apoptosis. Thus, detection of apoptosis by DNA laddering should be qualified by other tests.

Protocol 7
Analysis of DNA fragmentation

Reagents

- Buffer 1: 10 mM EDTA, 0.5% (w/v) sodium lauryl sarkosinate, 50 mM Tris–HCl pH 8.0
- Proteinase K: 20 mg/ml stocks stored at −20 °C; prepare a fresh working solution (5 μl of stock plus 195 μl of buffer 1) for each assay and keep on ice before use
- RNase A: 10 mg/ml stock in water, boiled for 10 min to destroy DNases; dilute 10 μl of stock with 190 μl of buffer 1 for use
- Ethidium bromide: 10 mg/ml stock in TE
- TPE buffer: 342 g of Tris base added to 1 litre of dH$_2$O, add 46.5 ml phosphoric acid, 240 ml of a 0.25 M solution of EDTA pH 8.0, mix to dissolve, then make up to 3 litres
- Loading buffer: 10 mM EDTA pH 8.0, 1% (w/v) low melting point agarose, 40% (w/v) sucrose, heat to dissolve, aliquot into 0.5 ml

Method

1. Transfer 5×10^5 to 10^6 cells to a 1.5 ml microcentrifuge tube and pellet by centrifugation at 4000 r.p.m. for 5 min at 4 °C.

Protocol 7 continued

2. Remove supernatant and recentrifuge for 15 sec. Remove the last of the supernatant.[a] Keep cells on ice.
3. Resuspend pellet by vortexing in 20 µl of buffer 1.[b]
4. Incubate at 50°C for 1 h in a dry block.
5. Spin down condensate from lid and add 10 µl of RNase solution. Vortex and incubate for a further hour at 50°C.
6. Meanwhile, make up a 2% (w/v) agarose solution in TPE buffer. Heat in a microwave to melt the agarose.[c] Cool and add 1 µl ethidium bromide for 100 ml gel solution.
7. Whilst agarose is cooling, place a well comb in a prepared gel tray about 1 cm from one end. Pour in the agarose solution ensuring no bubbles form in the gel. Allow to set on a level surface for at least 20 min.
8. Centrifuge cell suspension for 15 sec to collect condensate from the lid of the tube. Add 10 µl of loading buffer which has previously been incubated at 70°C to melt the agarose.
9. Mix by vortexing, centrifuge briefly, and reheat to 70°C.
10. Load samples[d] into dry wells[e] of the agarose gel using pipette tips just narrower than the well.
11. Carefully flood the gel tank with TPE buffer and run at 50 V for 1.5 h.
12. Visualize DNA using a UV light source (*Figure 5*).

[a] Cell pellet can be stored frozen at −20°C for up to two weeks.

[b] EDTA chelates magnesium and calcium ions thus reducing further nuclease activity. Sodium lauryl sarkosinate denatures histones thereby releasing DNA. Proteinase K degrades proteins including DNases and histones.

[c] Low power (70%) for 3–4 min. Do not boil and avoid bubbles.

[d] Include a 100 bp ladder for reference in one well.

[e] If wells are flooded prior to loading, high molecular weight DNA can float out and cause spurious banding patterns.

3.2 Inhibition of apoptosis using peptide inhibitors of caspases

A critical step in the pathway to programmed cell death involves the activation of a series of proteases known collectively as caspases (cysteinyl aspartate-specific proteinases) (14). The prototypic caspase, caspase 1, was identified as the protease responsible for cleaving interleukin 1β and was known as ICE (interleukin 1β-converting enzyme). The caspases cleave substrates to render them inactive (e.g. poly(ADP)-ribose polymerase cleaved by caspases 1, 4, 6, and 7) or to activate precursor forms (e.g. caspase 8 cleaves inactive forms of caspases 3, 4, 7, and 9). In this way caspases initiate a cascading system of proteases leading to cell death. Thus, inhibition of caspase activities, particularly at the apex of a

Figure 5. DNA fragmentation during apoptosis (Section 3.1). DNA isolated from untreated cells remains at the top of an agarose gel (a) whereas DNA isolated from cells undergoing apoptosis is fragmented into 180–200 bp repeats forming a characteristic 'ladder' (b).

cascade, will prevent apoptosis and provide strong evidence for apoptosis. Inhibition of caspases is possible using specific peptide inhibitors. These inhibitors can be either caspase-specific or general. The protocol below describes an inhibition assay using a general inhibitor of all caspases. A range of other peptide inhibitors can then be used to assess the involvement of specific caspases in the apoptotic pathway under examination.

Protocol 8
Inhibition of caspase activity by peptide inhibitors

Reagents
- 50 mM stock of peptide inhibitor in DMSO (Z-VAD.fmk or BOC-Asp.fmk, TCS Biologicals)
- 50 mM stock of negative control peptide, CBZ-Phe-Ala.fmk (TCS Biologicals)

Method
1. Pre-incubate cells for 1 h in the presence of inhibitor or negative control peptide.[a]
2. Remove medium from culture dish and treat with suspected apoptosis-inducing agent.
3. Replace medium containing inhibitor/control peptide.
4. Maintain cultures for appropriate time to assess occurrence of apoptosis.

[a] For most cellular assays the optimum concentration of each inhibitor required to block apoptosis varies from 50–200 μM dependent upon the cell type used. A preliminary dose-response study is recommended.

3.3 Determination of the non-dividing fraction of a population by simplified haptotactic assays, 'Ponten Plates'

Haptotactic island analysis is an elegant, definitive, and non-invasive, but little used, assay for the presence of senescent cells (and phenotypic heterogeneity in general) within a bulk culture. The essence of the method is to 'trap' single live cells on colonizable islands of palladium shadowed in a pattern onto a non-adhesive substrate. A larger area of palladium metal is then shadowed around the islands. First developed by Carter (15) the method was significantly improved first by Westermark (16) and then by Ponten and co-workers (17). Its advantage is that it allows the behaviour of single cells to be studied within a bulk culture without the potential pitfalls which accompany the changes of density required in a standard cloning experiment. The method described below is the original Westermark technique which, unlike that of Ponten, does not require custom-fabricated photolithographic masks but instead uses a standard copper electron microscope grid. The disadvantage of Westermark's technique is that several cell types (including many cell lines) are able to bridge the gaps between the islands because they are rather close together. We have used the technique with human glia and fibroblasts.

Protocol 9
Simplified haptotactic analysis

Equipment and reagents

- 35 mm standard tissue culture dishes (Nunc, Gibco BRL)
- 200 mesh copper electron microscope grids (Agar Scientific)
- Cloning rings
- Palladium metal and shadowing equipment
- Molecular biology grade agarose (Promega)

Method

1. Pour a boiling 1% (w/v) solution of agarose into each Petri dish.
2. Immediately suck off the surplus solution to leave a thin film behind. Leave to dry at 37°C for 1–2 h.
3. Add a standard copper EM grid (or a grid-pattern mask produced by photoetching) to the agarose-coated bottom of the Petri dish.
4. Evaporate palladium (or gold) onto the surface at 10^{-4} torr using a conventional EM evaporator. This produces a grid pattern and a thick metal 'ring'. The palladium islands should have distinct borders.
5. Place a cloning ring over the area of grids. Seed cells at cloning density (such that an average of one cell will attach to any one island) within the ring and at normal density outside it. This allows the cells on the islands to cross-feed and function at normal density.

> **Protocol 9** continued
>
> 6. View cell attachment after 24 h. Remove old medium (together with any floating cells). Note the number of 'one-cell', 'two-cell', and 'three-cell islands'.
>
> 7. Allow the culture to proliferate for seven to ten days. Note the cell number on each island daily. Senescent cells will not divide and will remain as one-cell colonies.
>
> Note: The palladium islands can be manipulated by attaching serum proteins, poly-lysine, and other attachment factors to enhance cell adhesion.

3.4 Determination of telomerase activity and telomere length

Strong evidence now exists that the loss of telomeric DNA that accompanies DNA replication in several human cell types is the primary means by which such cells count divisions (18, 19). When telomerase, an enzyme that maintains telomere length, is expressed at sufficient levels, the onset of senescence in these cells is also blocked (20). The simplest interpretation is that, in some cells, senescence is 'telomere driven'. However recent evidence suggests that telomere-driven senescence is not found in every human cell type and does not appear to exist at all in some species. In Syrian hamster fibroblasts, senescence occurs in the presence of telomerase (21). The following assays estimate telomerase activity by the telomerase repeat amplification or TRAP assay and establish whether telomere length is stably maintained by terminal restriction fragment or TRF analysis.

A TRAP assay consists of three stages:

(a) The production of a cell extract potentially containing telomerase.

(b) Extension of a target primer by the telomerase in the extract (telomerase adds the repeat sequence TTAGGG).

(c) A polymerase chain reaction step which amplifies the extended products to detectable levels.

Protocol 9 is a modified version of that of Kim *et al.* (22).

Terminal restriction fragment (TRF) length analysis detects the length of chromosomal DNA containing the telomere and undigested subtelomeric DNA. Genomic DNA is digested with a combination of restriction enzymes that cleave frequently, but do not cut within the (TTAGGG)$_n$ arrays at the telomeres, followed by agarose gel electrophoresis, Southern blotting, and finally detection of the (TTAGGG)$_n$-containing terminal restriction fragments with an appropriate hybridization probe. The main drawback of TRF analysis is the large number (10^5 to 10^6) of cells that are required. This is not a protocol that can be applied to single cells or an *in situ* format. TRF gives the mean telomere length of a population of cells and provides no information about the distributions of telomere length within the population or within the chromosomes of an individual cell. For these measurements, more sophisticated methods such as peptide nucleic acid (PNA) probes for telomere *in situ* hybridization coupled to extremely sensitive imaging equipment must be used (23). Other TRF protocols have been developed to use in-gel hybridization (which avoids technical problems related to Southern transfer of large fragments). The protocol below is of this type and is reliable.

Protocol 10
Demonstration of telomerase activity by TRAP

Equipment and reagents
All reagents are from Sigma unless otherwise stated.

- RNase-free Gilson pipette tips (Sarstedt).
- DEPC treated microcentrifuge tubes
- DEPC treated polycarbonate centrifuge tubes (Beckman)
- Beckman benchtop ultracentrifuge with TLA 100.2 rotor at 4°C
- PCR machine
- Growing cells from which to obtain S100 extract (see below); at least 10^6 cells are required
- DEPC treated PBS
- Trypsin/EDTA solution (Life Technologies)
- Wash buffer: 10 mM Hepes–KOH pH 7.5, 1.5 mM $MgCl_2$, 1 mM KCl made to 1 mL
- 0.5 M PMSF: 87 mg/ml in methanol kept on ice
- Lysis buffer: 10 mM Tris–HCl pH 7.5, 1.5 mM $MgCl_2$, 10 mM EGTA, 1 mM PMSF 10% (v/v) glycerol, 0.5% (v/v) CHAPS, 5 mM 2-mercaptoethanol; stored at -20°C

- TRAP assay reaction buffer: 20 mM Tris–HCl pH 8.3, 1.5 mM $MgCl_2$, 63 mM KCl, 0.005% Tween, 1 mM EGTA, 50 μM each dNTP, 0.1 mg/ml BSA (acetylated and nuclease-free from Gibco BRL) in DEPC treated water—this is normally made up as a double strength stock, 10 ml at a time, and frozen in 500 μl aliquots
- T4 gene 32 protein (Amersham Pharmacia Biotech)
- TS primer: 5'-AAT CCG TCG AGC AGA GTT-3'
- CX primer sequence: 5'-CCC TTA CCC TTC CCC TTA CCC TAA-3'
- 10% polyacrylamide (19:1) 20 cm gels containing 1.5 × TBE (Bio-Rad Protean II system)
- Sybr Gold (Molecular Probes)
- STORM system (Molecular Dynamics)

A. Preparation of S100 cell extract
The key to this procedure is the production of the cell extract in a RNase-free environment. *S100* refers to the fact that the extract consists of the cellular supernatant following a 100 000 g centrifugation step.

1. Detect adherent cells by trypsin dispersion.
2. Wash once in standard cell culture medium (containing serum if appropriate).
3. Wash twice in 10 ml ice-cold DEPC treated PBS.
4. Pellet the cells by centrifugation at 350 g for 5 min. This minimizes the chances of early lysis.
5. Resuspend the cells in 1 ml of wash buffer. Transfer the cell suspension to an Eppendorf tube and spin at 15 000 g for 2 min at 4°C. Resuspend the cells in 18.5 μl of lysis buffer per 10^6 cells.

Protocol 10 continued

6. Incubate the samples on ice for 30 min. The length of this incubation is crucial.
7. Load the samples into DEPC treated polycarbonate centrifuge tubes immediately after the end of the 30 min.
8. Immediately centrifuge the samples at 100 000 g for 30 min at 4°C. This equates to 53 000 r.p.m. on a TLA 100.2 rotor.
9. Following centrifugation remove the supernatant[a] and snap-freeze on dry ice in 10 μl aliquots in DEPC treated microtubes.
10. Estimate the protein concentration of a sample of supernatant (by Bradford assay). This should be approx. 5–10 mg/ml.

B. TS primer extension (RNase-free conditions should be employed throughout!)

In order to standardize this part of the assay and reduce the effect of pipetting errors, a 'master mix' or 'pre-mix' is made corresponding to the number of reactions plus two more. A four reaction mix is summarized below.

1. Combine 100 μl of 2 × TRAP assay reaction buffer with 4 μg T4 gene 32 protein.
2. Add 400 ng TS primer.
3. Add 4 μl S100 extract.
4. Incubate under oil in a PCR machine at 30°C for 30 min.
5. Heat to 92°C to destroy the telomerase.

C. PCR amplification of extension products

Again a 'pre-mix' equal to the number of reactions plus two is made to limit dilution errors.

1. Add 100 ng of the CX primer per reaction.
2. Add 2.5 U of *Taq* polymerase (PE Biosystems) per reaction.
3. Subject the sample to 31 cycles of denaturation (92°C, 30 sec), annealing (50°C, 30 sec), and extension (72°C, 90 sec).
4. Resolve the PCR products on a 10% polyacrylamide gel containing bromophenol blue. Run the gel at about 300 V until the bromophenol blue is approx. 3 cm from the bottom (this takes about 4 h).
5. Stain the gel using 1:10 000 dilution Sybr gold and view under blue fluorescence on a STORM system (Molecular Dynamics). An extract containing telomerase will produce a DNA ladder with 6 bp periodicity (*Figure 6*).

[a] The crude pellet contains DNA which can be extracted if required for TRF analysis if cell number is limiting.

An alternative strategy which may be more applicable to some laboratories is to end-label the TS primer with [γ-^{32}P]ATP using T4 polynucleotide kinase. The resulting gel can then be dried and exposed to films or phosphor screens.

SENESCENCE, APOPTOSIS, AND NECROSIS

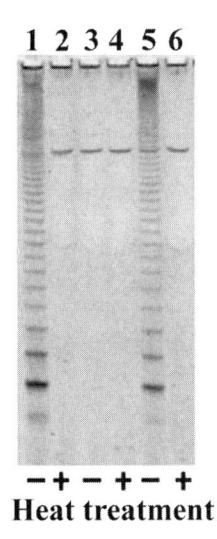

Figure 6. Demonstration of telomerase activity by TRAP (Section 3.5). A human fibroblast cell line HCA2 was infected with a retrovirus carrying an empty vector (puro) or the catalytic subunit of telomerase (hTERT). Introduction of hTERT (lanes 5 and 6), but not the empty vector (lanes 3 and 4), restores telomerase activity to HCA2 cells. Cell extracts prepared from infected fibroblasts (lanes 3–6) and a telomerase positive control cell line 293 (lanes 1 and 2) were tested in a TRAP assay. In polyacrylamide gels stained with Sybr Gold a DNA ladder signifies telomerase activity. An internal standard of 150 bp serves as a control for PCR amplification. Heat treatment of the cell extracts destroys telomerase activity. Photograph courtesy of Dr Chris Jones, University of Wales College of Medicine.

Protocol 11

Determination of terminal restriction fragment (TRF) length by in-gel hybridization

Equipment and reagents

- Standard gel electrophoresis equipment (power pack, tanks, combs)
- Gel dryer
- Hybridization oven and bottles (Hybaid)
- Genomic DNA (at least 1 µg) from test cells of interest
- HinfI and RsaI restriction enzymes (Amersham)
- TBE
- 10 × buffer M: 100 mM Tris–HCl pH 7.5, 100 mM MgCl$_2$, 10 mM DTT, 500 mM NaCl
- ^{32}P-labelled HindIII markers

- Denaturing buffer: 1.5 M NaCl, 0.5 M NaOH (to prepare add 87.66 g/litre NaCl and 20 g/litre NaOH in double distilled water)
- Neutralizing buffer: 1.5 M NaCl, 0.5 M Tris pH 8.0 (to prepare add 87.66 g/litre NaCl and 60.57 g/litre Tris base; adjust to pH 8.0 with HCl)
- Hybridization solution: 5 × SSC, 0.1 × phosphate wash (1 × contains 5 mM sodium pyrophosphate, 100 mM Na$_2$HPO$_4$)
- 5 × Denhardt's solution
- Tel 2 probe: 5'-CCCTAACCCTAACCCTAA-3'

A. Labelling of Tel 2 probe with ^{32}P

1. Mix the following:
   - 10 µl (100 µCi) [^{32}P]ATP (3000 Ci/mmol, Amersham, Cat. No. AA0068)
   - 2 µl of 10 × T4 polynucleotide kinase buffer
   - 1 µl of T4 polynucleotide kinase
   - 500 ng of unlabelled Tel 2
   - Water to 20 µl

Protocol 11 continued

2. Incubate for 1 h at 37 °C, then heat at 90 °C for 5 min.
3. Separate Tel 2 from the unincorporated label on a 2 ml volume P4 gel column (Bio-Rad). Collect 0.5 ml fractions. The first fraction is the column void volume. The second fraction contains the probe.
4. Alternatively the probe can be purified by centrifugation on Quick Spin column (Boehringer Mannheim).

B. In-gel hybridization

1. Digest 1 μg of genomic DNA in 30 μl of buffer M containing 15 U of *Hinf*I and *Rsa*I.
2. Run out 1 μl of sample on a 0.7% agarose gel to test if the DNA is fully digested.
3. If the digest has been successful then load the sample on a 15 × 15 cm 0.5% agarose gel (about 200 ml) in TBE. One lane should contain ^{32}P-labelled *Hind*III markers. Load all the samples in 6 × Ficoll loading buffer containing xylene cyanol and bromophenol blue. The samples should be run in at 100 V for 20 min followed by 20 V overnight.
4. When the bromophenol blue has run approx. three-quarters of the way down the gel remove it from the tank.
5. Denature the DNA by washing with denaturing buffer for 15 min.
6. Wash the gel in neutralizing buffer for 10 min.
7. Dry the gel at room temperature for 1 h (i.e. under vacuum but with the lid of gel drier up) followed by a further 30 min drying at 50 °C.
8. In order to remove the gel it is necessary to wet the paper with 5 × SSC. Gently lift the agarose gel off the paper. At this point the gel can be stored at −80 °C wrapped in Saran. This method does not require a pre-hybridization step.
9. Hybridize the gel at 37 °C overnight in 25 ml of hybridization solution (large Hybaid bottles are useful for this) with 500 ng of Tel 2 probe.
10. Wash the gel as below:
    (a) Three times for 7 min at 37 °C in bottles with 0.1 × SSC.
    (b) Twice for 7 min each at RT in bottles with 0.1 × SSC.
    (c) Once for 7 min in tray with 0.1 × SSC pre-heated to 37 °C.
11. Wrap the gel in Saran and place in a cassette with autoradiographic film. If high background is a problem further washes can be performed. When the gel is not in use it should be kept at −80 °C as the signal tends to become more diffuse with time.

The DNA appears as a smear, representing the telomeres from all the chromosomes from a population of cells. The smear demonstrates the heterogeneous nature of telomere length within and between cells. Cells with shorter telomeres exhibit a faint low molecular weight smear while cells with long telomeres show a more intense higher molecular weight signal (*Figure 7*). The use

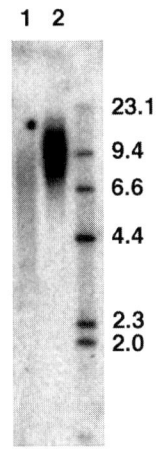

Figure 7. Estimation of telomere length by TRF (Section 3.5). Terminal restriction fragment analysis indicates that telomerase expressing cells have extended telomeres (high molecular weight smear, lane 2) compared to vector controls (low molecular weight smear, lane 1). Molecular size markers are shown in the right column. For details of the experiment see legend for *Figure 6*. Photograph courtesy of Dr Chris Jones, University of Wales College of Medicine.

of densitometry can assign a quantitative value to the mean terminal restriction fragment length (24). It should be kept in mind that the TRF is not formally equivalent to the telomere as the fragments contain subtelomeric sequences of 1–2 kb that are resistant to digestion.

Acknowledgements

We would like to thank the following colleagues who have helped us in the preparation of this chapter: Dr Chris Jones, Dr Eman Al-Baker, Prof. Gwyn Williams, Dr Jan Ponten, and Dr Kevin Slater.

References

1. Hayflick, L. (1979). *J. Invest. Dermatol.*, **73**, 8.
2. Shall, S. (1987). In *Perspectives on mammalian cell death* (ed. C. S. Potten), p. 184. Oxford University Press.
3. Grey, M. D. and Norwood, T. H. (1995). *Rev. Clin. Gerontol.*, **53**, 69.
4. Rheinwald, J. G. (1989). In *Cell growth and division: a practical approach* (ed. D. Rickwood and B. D. Hames), p. 81. IRL Press, Oxford.
5. Effros, R. B. and Walford, R. L. (1987). *Exp. Gerontol.*, **22**, 307.
6. MacDonald, C. (1994). In *Mammalian biotechnology: a practical approach* (ed. D. Rickwood and B. D. Hames), p. 149. IRL Press, Oxford.
7. Baserga, R. (1995). In *Cell growth and apoptosis: a practical approach* (ed. G. P. Studzinski), p. 1. IRL Press, Oxford.
8. Smith, J. R. and Hayflick, L. (1974). *J. Cell Biol.*, **62**, 48.
9. Smith, J. R. and Whitney, R. G. (1980). *Science*, **207**, 82.
10. Cristofalo, V. J. and Scharf, B. B. (1973). *Exp. Cell Res.*, **76**, 419.
11. Grove, G. L. and Cristofalo, V. J. (1977). *J. Cell. Physiol.*, **90**, 415.
12. Dimri, G. P., Lee, X., Basile, G., Acosta, M., Scott, G., Roskelley, C., et al. (1995). *Proc. Natl. Acad. Sci. USA*, **92**, 9362.

13. Thomas, E., Al-Baker, E., Dropcova, S., Denyer, S., Ostad, N., Lloyd, A., et al. (1997). *Exp. Cell Res.*, **236**, 355.
14. Porter, A. G., Ng, P., and Jaenicke, R. U. (1997). *BioEssays*, **19**, 501.
15. Carter, S. B. (1967). *Exp. Cell Res.*, **48**, 189.
16. Westermark, B. (1978). *Exp. Cell Res.*, **111**, 295.
17. Ponten, J., Shall, S., and Stein, W. D. (1983). *J. Cell. Physiol.*, **117**, 342.
18. Kipling, D. and Faragher, R. G. A. (1999). *Nature*, **398**, 192.
19. Faragher, R. G. A. and Kipling, D. (1998). *Bioessays*, **20**, 985.
20. Bodnar, A. G., Ouellette, M., Frolkis, M., Holt, S. E., Chiu, C. P., Morin, G. B., et al. (1998). *Science*, **279**, 349.
21. Russo, I., Silver, A. R., Cuthbert, A. P., Griffin, D. K., Trott, D. A., and Newbold, R. F. (1998). *Oncogene*, **17**, 3417.
22. Kim, N. W. and Wu, F. (1997). *Nucleic Acids Res.*, **25**, 2595.
23. Hande, P., Slijepcevic, P., Silver, A., Bouffler, S., van Buul, P., Bryant, P., et al. (1999). *Genomics*, **56**, 221.
24. Kruk, P. A., Rampino, N. J., and Bohr, V. A. (1995). *Proc. Natl. Acad. Sci. USA*, **92**, 258.

List of suppliers

Agar Scientific, 66A Cambridge Road, Stansted, Essex CM24 8DA, UK.

Alpha Labs, 40 Parham Drive, Eastleigh, Hampshire, UK.

Amicon Products, see Millipore

Anderman and Co. Ltd., 145 London Road, Kingston-upon-Thames, Surrey KT2 6NH, UK.
Tel: 0181 541 0035
Fax: 0181 541 0623

Appligene Oncor Lifescreen, Unit 15, The Metro Centre, Dwight Road, off Tolpits Lane, Watford, Hertfordshire WD1 8SS, UK.
Tel: 44 (0)1923 241515
Fax: 44 (0)1923 242215

Applikon BV, De Brauwwerg 13, PO Box 149, NL-3100 Schiedam AC, The Netherlands.

Ashby Scientific Ltd., Unit 11, Atlas Court, Coalville, Leicestershire LE67 3FL, UK.
Tel: 44 1530 832590
Fax: 44 1530 832591

ATCC
American Type Culture Collection, 10801 University Blvd., Manassas, VA 20110 2209 USA
Tel: 001 703 365 2700
Web site: www.atcc.org

BASF Corp., 8 Campus Drive, Parsipanny, NJ 07054, USA.

B. Braun Biotech International, PO Box 11 20, D-34209 Melsungen, Germany.
Tel: 49 5661/713842   Fax: 49 5661/713702

BDH/Merck, Hunter Boulevard, Magna Park, Lutterworth, Leicestershire LE17 4XN, UK.
Tel: 0800 223344
Fax: 01455 558586
Web site: www.merck-ltd.co.uk

Beckman Coulter Inc.
Beckman Coulter Inc., 4300 N Harbor Boulevard, PO Box 3100, Fullerton, CA 92834-3100, USA.
Tel: 001 714 871 4848
Fax: 001 714 773 8283
Web site: www.beckman.com

Beckman Coulter (UK) Ltd., Oakley Court, Kingsmead Business Park, London Road, High Wycombe, Buckinghamshire HP11 1JU, UK.
Tel: 01494 441181   Fax: 01494 447558
Web site: www.beckman.com

Becton Dickinson and Co.
Becton Dickinson and Co., 21 Between Towns Road, Cowley, Oxford OX4 3LY, UK.
Tel: 01865 748844
Fax: 01865 781627
Web site: www.bd.com

LIST OF SUPPLIERS

Becton Dickinson and Co., 1 Becton Drive, Franklin Lakes, NJ 07417-1883, USA.
Tel: 001 201 847 6800
Web site: www.bd.com

Bellco Glass Inc., 340 Edrudo Road, Vineland, NJ 08360-3492, USA.
Tel: 001 800 257 7043
Fax: 001 609 691 3247
Web site: http://www.ispex.ca/labware/BellcoGlasswareInc.html

Bibby Sterilin Ltd., Tilling Drive, Stone, Staffordshire ST15 0SA, UK.
Tel: 44 1785 812121   Fax: 44 1785 813748
Web site: www.bibby.sterilin.com

Bio 101 Inc.
Bio 101 Inc., c/o Anachem Ltd., Anachem House, 20 Charles Street, Luton, Bedfordshire LU2 0EB, UK.
Tel: 01582 456666
Fax: 01582 391768
Web site: www.anachem.co.uk

Bio 101 Inc., PO Box 2284, La Jolla, CA 92038-2284, USA.
Tel: 001 760 598 7299   Fax: 001 760 598 0116
Web site: www.bio101.com

Bioengeneering AG, Sagenrainstrasse 7, CH-8636 Wald, Switzerland.
Tel: 41 55 938111
Fax: 41 55 954964

Biofluids Inc., 1146 Taft Street, Rockville, MD 20850, USA.
Tel: 301 424 4140, 800 972 5200
Fax: 301 424 3619

Biogenesis
Biogenesis Ltd., New Fields, Stinsford Road, Poole BH17 0NF, UK.
Tel: 01202 660006
Fax: 01202 660020
Web site: www.biogenesis.co.uk/home/

Biogenesis Inc., PO Box 1016, Kingston, NH 03848, USA.
Tel: 603 642 8302
Fax: 603 642 8322

BioGenex
BioGenex Inc., Distributor, A. Menarini Diagnostics UK, 405 Wharfedale Road, Winnersh, Wokingham, Berkshire RG41 5RA, UK.
Tel: 0118 944 4100
Fax: 0118 944 411

BioGenex Laboratories, Inc., 4600 Norris Canyon Road, San Ramon, CA 94583, USA.
Tel: 925 275 0550
Fax: 925 275 0580
Web site: www.biogenex.co

Bio-Rad Laboratories Ltd.
Bio-Rad Laboratories Ltd., Bio-Rad House, Maylands Avenue, Hemel Hempstead, Hertfordshire HP2 7TD, UK.
Tel: 0181 328 2000
Fax: 0181 328 2550
Web site: www.bio-rad.com

Bio-Rad Laboratories Ltd., Division Headquarters, 1000 Alfred Noble Drive, Hercules, CA 94547, USA.
Tel: 001 510 724 7000
Fax: 001 510 741 5817
Web site: www.bio-rad.com

BioWhittaker, Inc., 8830 Biggs Ford Road, PO Box 127, Walkersville, MD 21793-0127, USA.
Tel: 301 898 7025, 800 654 4452
Fax: 301 845 8338
Web site: www.biowhittaker.com

Cambio Ltd., 34 Millington Road, Cambridge, UK.
Tel: 44 (0)1223 366500
Fax: 44 (0)1223 350069

LIST OF SUPPLIERS

Cellon, 204 route d'Arlon, L-8010 Strassen, Grand Duchy of Luxembourg.
Tel: 352 312313   Fax: 352 311052

Cellex Biosciences Inc., 8500 Evergreen Road, Minneapolis, MN 55433, USA.
Tel: 1 612 7860302   Fax: 1 612 7860915

Chemicon International
Chemicon International Ltd., 2 Admiral House, Cardinal Way, Harrow HA3 5UT, UK.
Tel: 44 181 863 0415
Fax: 44 181 863 0416
Web site: www.chemicon.com

Chemicon International Inc., 28835 Single Oak Drive, Temecula, CA 92590, USA.
Tel: 909 676 8080
Fax: 909 676 9209
Web site: www.chemicon.com

Clonetics, see BioWhittaker

Coletica S.A., 32 Rue Street, Jean de Dieu, 69007 Lyon, France.
Tel: 04 78 72 11 95
Fax: 04 78 58 09 71

Corning Medical, Medfield, MA 02052, USA.

Corning Science Products, 45 Nagog Park, Acton, MA 01720, USA.
Tel: 001 607 974 0353
Fax: 001 607 974 0354
Web site: www.corning.com

Corriel Cell Repository, 401 Haddon Avenue, Camden, NJ 08103, USA.
Tel: 001 856 757 4843
Web site: locus.umdnj.edu/ccr

Costar, One Alewife Center, Cambridge, MA 02140, USA.
Tel: 1 617 8686200
Fax: 1 617 8682076

CP Instrument Co. Ltd., PO Box 22, Bishop Stortford, Hertfordshire CM23 3DX, UK.
Tel: 01279 757711
Fax: 01279 755785
Web site: www.cpinstrument.co.uk

Cytocell Ltd., Unit 6, Somerville Court, Trinity Way, Banbury, Oxon OX17 3SN, UK.
Tel: 44 (0)1295 810910
Fax: 44 (0)1295 812333

Dako
Dako Ltd., Denmark House, Angel Drove, Ely, Cambridgeshire CB7 4ET, UK.
Tel: 01353 66 99 11
Fax: 01353 66 89 89
Web site: www.dakousa.com

Dako Corporation, 6392 Via Real, Carpinteria, CA 93013, USA.
Tel: 805 566 6655
Fax: 805 566 6688
Web site: www.dakousa.com

Dow Corning Corp., Dow Corning Center, Box 0994, Midland, MI 48686-0994, USA.

DSMZ
Deutsche Sammlung von Mikroorganismen und Zellkulturen, Mascheroder Weg 1b, D-38124 Braunschweig, Germany.
Tel: 0049 531 2616 319
Web site: www.dsmz.de

Dupont
Dupont (UK) Ltd., Industrial Products Division, Wedgwood Way, Stevenage, Hertfordshire SG1 4QN, UK.
Tel: 01438 734000   Fax: 01438 734382
Web site: www.dupont.com

Dupont Co. (Biotechnology Systems Division), PO Box 80024, Wilmington, DE 19880-002, USA.
Tel: 001 302 774 1000
Fax: 001 302 774 7321
Web site: www.dupont.com

LIST OF SUPPLIERS

Dynal
Dynal (UK) Ltd., 10 Thursby Road, Croft Business Park, Bromborough, Wirral, Mersyside L62 3PW, UK.
Tel: 0151 346 1234
Fax: 0151 346 1223

Dynal Inc., 5 Delaware Drive, Lake Success, NY 11042, USA.
Tel: 1 800 638 9416
Fax: 516 326 3298
Web site: www.dynal.no

Eastman Chemical Co., 100 North Eastman Road, PO Box 511, Kingsport, TN 37662-5075, USA.
Tel: 001 423 229 2000
Web site: www.eastman.com

ECACC
European Collection of Animal Cell Cultures, CAMR, Salisbury, Wilts SP4 0JG, UK.
Web site: www.camr.org.uk

First Link UK Ltd., Premier Partnership Estate, Leys Road, Brockmoor, Brierley Hill, West Midlands DY5 5UP, UK.
Tel: 01384 263862
Fax: 01384 480351

Fisher Scientific
Fisher Scientific UK Ltd., Bishop Meadow Road, Loughborough, Leicestershire LE11 5RG, UK.
Tel: 01509 231166   Fax: 01509 231893
Web site: www.fisher.co.uk

Fisher Scientific, Fisher Research, 2761 Walnut Avenue, Tustin, CA 92780, USA.
Tel: 001 714 669 4600
Fax: 001 714 669 1613
Web site: www.fishersci.com

Fluka
Fluka, PO Box 2060, Milwaukee, WI 53201, USA.
Tel: 001 414 273 5013
Fax: 001 414 2734979
Web site: www.sigma-aldrich.com

Fluka Chemical Co. Ltd., PO Box 260, CH-9471, Buchs, Switzerland.
Tel: 0041 81 745 2828
Fax: 0041 81 756 5449
Web site: www.sigma-aldrich.com

Heraeus Instruments GmbH, PO Box 1211, D-37502 Osterode, Germany.
Tel: 49 5522316212
Fax: 49 5522316211

Hybaid
Hybaid Ltd., Action Court, Ashford Road, Ashford, Middlesex TW15 1XB, UK.
Tel: 01784 425000   Fax: 01784 248085
Web site: www.hybaid.com

Hybaid US, 8 East Forge Parkway, Franklin, MA 02038, USA.
Tel: 001 508 541 6918
Fax: 001 508 541 3041
Web site: www.hybaid.com

HyClone Laboratories, 1725 South HyClone Road, Logan, UT 84321, USA.
Tel: 001 435 753 4584
Fax: 001 435 753 4589
Web site: www.hyclone.com

ICN Biomedicals
ICN Biomedicals Ltd., 1 Elmwood, Chineham Business Park, Basingstoke, Hampshire RG24 8WG, UK.
Tel: 0800 282 474
Fax: 0800-614735

ICN Biomedicals Inc., 3300 Hyland Avenue, Costa Mesa, CA 92626, USA.
Tel: 714 545 0100
Fax: 800 334 6999
Web site: www.icnbiomed.com

Innovative Chemistry, Inc., PO Box 90, Marshfield, MA 02050, USA.

Integra Biosciences
Integra Biosciences, Unit 9, Ascot Industrial

LIST OF SUPPLIERS

Estate, Icknield Way, Letchworth, Hertfordshire SG6 1TD, UK.
Tel: 44 1462 486548
Fax: 44 1462 486678

Integra Biosciences, Industriestrasse 44, CH-8304 Wallisellen, Switzerland.
Tel: 41 1877 4646
Fax: 41 18774600

Invitrogen
Invitrogen BV, PO Box 2312, 9704 CH Groningen, The Netherlands.
Tel: 00800 5345 5345
Fax: 00800 7890 7890
Web site: www.invitrogen.com

Invitrogen Corp., 1600 Faraday Avenue, Carlsbad, CA 92008, USA.
Tel: 001 760 603 7200
Fax: 001 760 603 7201
Web site: www.invitrogen.com

JCRB
Japanese Collection of Research Bioresources, 1-1-43 Hoen-Zaka, Chuo-ku, Osaka 540, Japan
Web site: cellbank.nihs.go.jp

Life Technologies
Life Technologies Ltd., PO Box 35, Free Fountain Drive, Incsinnan Business Park, Paisley PA4 9RF, UK.
Tel: 0800 269210   Fax: 0800 838380
Web site: www.lifetech.com

Life Technologies Inc., 9800 Medical Center Drive, Rockville, MD 20850, USA.
Tel: 001 301 610 8000
Web site: www.lifetech.com

Lumitech Ltd., Nottingham Business Park, City Link, Nottingham NG2 4LA, UK.

Melford Laboratories Ltd., Bildeston Road, Chelsworth, Ipswich, Suffolk IP7 7LE, UK.

Merck Sharp & Dohme
Merck Sharp & Dohme Research Laboratories, Neuroscience Research Centre, Terlings Park, Harlow, Essex CM20 2QR, UK.
Web site: www.msd-nrc.co.uk

MSD Sharp and Dohme GmbH, Lindenplatz 1, D-85540, Haar, Germany.
Web site: www.msd-deutschland.com

Meredos GmbH, Alte Dorfstrasse, D-37120 Bovenden, Germany.
Tel: 49 5503 1766   Fax: 49 55 031771

Millipore
Millipore (UK) Ltd., The Boulevard, Blackmoor Lane, Watford, Hertfordshire WD1 8YW, UK.
Tel: 01923 816375
Fax: 01923 818297
Web site: www.millipore.com/local/UK.htm

Millipore Corp., 80 Ashby Road, Bedford, MA 01730, USA.
Tel: 001 800 645 5476
Fax: 001 800 645 5439
Web site: www.millipore.com

Miltenyi Biotec
Miltenyi Biotec Ltd., Almac House, Church Lane, Bisley, Surrey GU24 9DR, UK.
Tel: 01483 799800
Fax: 01483 799311
Web site: www.miltenyibiotec.com

Miltenyi Biotec Inc., Suite 208, 251 Auburn Ravine Road, CA 95603, USA.
Tel: 530 888 8871
Fax: 530 888 8925
Web site: www.miltenyibiotec.com

Molecular Dynamics, World Headquarters, 928 East Arques Avenue, Sunnyvale, CA 94086-4520, USA.

Molecular Probes, Cambridge Bioscience, 24-25 Signet Court, Newmarket Road, Cambridge CB5 8LA, UK.

LIST OF SUPPLIERS

New England Biolabs, 32 Tozer Road, Beverley, MA 01915-5510, USA.
Tel: 001 978 927 5054

Nikon
Nikon Corp., Fuji Building, 2-3, 3-chome, Marunouchi, Chiyoda-ku, Tokyo 100, Japan.
Tel: 00813 3214 5311
Fax: 00813 3201 5856
Web site: www.nikon.co.jp/main/index_e.htm

Nikon Inc., 1300 Walt Whitman Road, Melville, NY 11747-3064, USA.
Tel: 001 516 547 4200
Fax: 001 516 547 0299
Web site: www.nikonusa.com

Novocastra Laboratories
Novocastra Laboratories Ltd., see Vector Laboratories Ltd.

Novocastra Laboratories Inc., see Vector Laboratories Inc.

Nunc A/s, PO Box 280, Kamstrup, DK-4000 Roskilde, Denmark.
Tel: 45 46359065   Fax: 45 46350105

Nycomed
Nycomed Amersham plc, Amersham Place, Little Chalfont, Buckinghamshire HP7 9NA, UK.
Tel: 01494 544000   Fax: 01494 542266
Web site: www.amersham.co.uk

Nycomed Amersham, 101 Carnegie Center, Princeton, NJ 08540, USA.
Tel: 001 609 514 6000
Web site: www.amersham.co.uk

Oxoid Ltd., Wade Road, Basingstoke, Hampshire RG24 8PW, UK.

Pall Gelman
Pall Gelman, Laboratory Products Division, 50 Bearfoot Road, Northborough, MA 01532 1551, USA.
Tel: 508 393 1800   Fax: 508 393 1874

Pall Gelman Sciences, Brackmills Business Park, Caswell Road, Northampton NN4 7EZ, UK.
Tel: 01604 704704   Fax: 01604 704724

Pan Systems (Pan Biotech GmbH), Gewerbepark 13, 94501 Aidenbach, Germany.
Tel: 49 8543 3636   Fax: 49 8543 1473
Web site: www.pan-biotech.de

Perkin Elmer Ltd.
Perkin Elmer Ltd., Post Office Lane, Beaconsfield, Buckinghamshire HP9 1QA, UK.
Tel: 01494 676161
Web site: www.perkin-elmer.com

Perkin Elmer Ltd./Applied Biosystems Division, Birchwood Science Park North, Warrington WA3 7PB, UK.

Pharmacia
Pharmacia Biotech (Biochrom) Ltd., Unit 22, Cambridge Science Park, Milton Road, Cambridge CB4 0FJ, UK.
Tel: 01223 423723   Fax: 01223 420164
Web site: www.biochrom.co.uk

Pharmacia and Upjohn Ltd., Davy Avenue, Knowlhill, Milton Keynes, Buckinghamshire MK5 8PH, UK.
Tel: 01908 661101   Fax: 01908 690091
Web site: www.eu.pnu.com

Pharmacia Biotech AB, S-751 82 Uppsala, Sweden.
Tel: 46 18165000   Fax: 46 18101403

Promega
Promega UK Ltd., Delta House, Chilworth Research Centre, Southampton SO16 7NS, UK.
Tel: 0800 378994   Fax: 0800 181037
Web site: www.promega.com

Promega Corp., 2800 Woods Hollow Road, Madison, WI 53711-5399, USA.
Tel: 001 608 274 4330

LIST OF SUPPLIERS

Fax: 001 608 277 2516
Web site: www.promega.com

PromoCell GmbH, Handschuhsheimer Landstr. 12, D-69120 Heidelberg, Germany.
Tel: 6221 439 049, 0800 960 333
Fax: 6221 484943
Web site: www.promocell.com/

Qiagen
Qiagen UK Ltd., Boundary Court, Gatwick Road, Crawley, West Sussex RH10 2AX, UK.
Tel: 01293 422911   Fax: 01293 422922
Web site: www.qiagen.com

Qiagen Inc., 28159 Avenue Stanford, Valencia, CA 91355, USA.
Tel: 001 800 426 8157
Fax: 001 800 718 2056
Web site: www.qiagen.com

Riken Gene Bank, 3-1-1 Koyadai, Tsukuba Science City, Ibaraki 305, Japan
Tel: 0081 298 36 3611
Web site: www.rtc.riken.go.jp

Roche Diagnostics
Roche Diagnostics Ltd., Bell Lane, Lewes, East Sussex BN7 1LG, UK.
Tel: 01273 484644   Fax: 01273 480266
Web site: www.roche.com

Roche Diagnostics Corp., 9115 Hague Road, PO Box 50457, Indianapolis, IN 46256, USA.
Tel: 001 317 845 2358
Fax: 001 317 576 2126
Web site: www.roche.com

Roche Diagnostics GmbH, Sandhoferstrasse 116, 68305 Mannheim, Germany.
Tel: 0049 621 759 4747
Fax: 0049 621 759 4002
Web site: www.roche.com
Roche Molecular Biochemicals (Boehringer Mannheim), PO Box 50414, Indianapolis, Indiana 46250-0414, USA.

Sarstedt, 68 Boston Road, Leicester LE4 1AW, UK.

Sartorius
Sartorius AG, PO Box 3243, Weender, Ondestrasse, 94-108, 3707 Goettingen, Germany.
Tel: 49 551 3080

Sartorius Ltd., Blenheim Road, Longmead Industrial Estate, Epsom, Surrey KT19 9QN, UK.
Tel: 01372 737100   Fax: 013727 20799

Sartorius Corp., 131 Heartland Boulevard, Edgewood, NY 11717, USA.
Tel: 516 254 4249, 800 368 7178
Fax: 516 254 4253
Web site: www.sartorius.com/

Schleicher and Schuell Inc., Keene, NH 03431A, USA.
Tel: 001 603 357 2398

Schott Glaswerke, Hattenbergstrasse, 10 Postfach 2480, D-6500 Mainz 1, Germany.

Scottish National Blood Transfusion Service, Plasma Fractionation Centre, 21 Ellen's Glen Road, Edinburgh EH17 7QT, UK.
Tel: 0131 536 5700   Fax: 0131 536 5781

Shandon Scientific Ltd., 93-96 Chadwick Road, Astmoor, Runcorn, Cheshire WA7 1PR, UK.
Tel: 01928 566611   Fax: 01928 565845
Web site: www.shandon.com

Sigma-Aldrich
Sigma-Aldrich Co. Ltd., The Old Brickyard, New Road, Gillingham, Dorset XP8 4XT, UK.
Tel: 01747 822211   Fax: 01747 823779
Web site: www.sigma-aldrich.com

Sigma-Aldrich Co. Ltd., Fancy Road, Poole, Dorset BH12 4QH, UK.
Tel: 01202 722114   Fax: 01202 715460
Web site: www.sigma-aldrich.com

LIST OF SUPPLIERS

Sigma Chemical Co., PO Box 14508, St Louis, MO 63178, USA.
Tel: 001 314 771 5765
Fax: 001 314 771 5757
Web site: www.sigma-aldrich.com

Stratagene
Stratagene Europe, Gebouw California, Hogehilweg 15, 1101 CB Amsterdam Zuidoost, The Netherlands.
Tel: 00800 9100 9100
Web site: www.stratagene.com

Stratagene Inc., 11011 North Torrey Pines Road, La Jolla, CA 92037, USA.
Tel: 001 858 535 5400
Web site: www.stratagene.com

Stratech Scientific Ltd., 61-63 Dudley Street, Luton, Bedfordshire LU2 ONP, UK.
Tel: 44 (0)1582 481884
Fax: 44 (0)1582 481895

TCS Biologicals Ltd., Botolph Claydon, Buckingham MK18 2LR, UK.

Techne Ltd., Hinxton Road, Duxford, Cambridge CB2 44PZ, UK.
Tel: 44 1223 832401   Fax: 44 12223 836838
Web site: www.techne.co.uk

Unisyn Technologies (Cell-Pharm), 25 South Street, Hopkinson, MA 017482-2217, USA.
Tel: 1 508 435-2000   Fax: 1 508 435 8111

United States Biochemical, PO Box 22400, Cleveland, OH 44122, USA.
Tel: 001 216 464 9277

Vector Laboratories
Vector Laboratories Ltd., 3 Accent Park, Bakewell Road, Orton Southgate, Peterborough PE2 6XS, UK.
Tel: 01733 237999   Fax: 01733 237119
Web site: www.vectorlabs.com
Vector Laboratories Inc., 30 Ingold Road, Burlingame, CA 94010, USA.
Tel: 800 2276666   Fax: 650 6970339
Web site: www.vectorlabs.com

Vector Labs, 16 Wulfric Square, Bretton, Peterborough PE3 8RF, UK.

Vogelbusch GmbH, Biopharma, Blechturmgasse 11, A-1050 Vienna, Austria.
Tel: 43 1 546610   Fax: 43 1 5452979

VYSIS UK Ltd., Customer Service Department, Rosedale House, Rosedale Road, Richmond, Surrey TW9 2SZ, UK.

Index

accessioning scheme 70, 71
Acusyst 59-60
ADP/ATP ratio 291
adult tissue source 4
agarose 61
 coating protocol 130-1
agar-overlay cultures 131-2, 134-6
air filters 22
airlift fermenter 57-8
alginate 61
allelic ladder 97
allogeneic cells 150, 157
amino acid labelling 188, 204
ampoules 72
anchorage-dependent cells 32-3
antibody detection 229-30
antibody markers 274-7
antimicrobials 112, 114-15
aortic smooth muscle cells 106, 108, 110
Apligraf™ 152, 157
apoptosis 125-6, 190, 282
 detection 190-1, 290-3
 inhibition 293-4
ATP, luminescence measurement 190, 209-10, 291
authentication of cell lines 78-102
AuthentiKit™ 78
autologous cells 150, 157
avidin probe detection 229, 230

basal membrane, co-cultivation on 138-9
batch culture 30-1

bead bed reactors 38-40, 48
Bellco Bioreactor 61-2
Bellco-Corbeil Culture System 36-7, 48
bladder smooth muscle cells 152-3
brain cell cultures 140
Brockmann bodies 140
bromodeoxyuridine 189, 284-6
bronchial epithelial cells, see NHBE
buffer systems 24-5

calcein AM-ethidium bromide staining 287-8
calcium ion concentration 3
calcium labelling 188
calcium phosphate-DNA co-precipitation 237-8
cancer chemotherapy 175-7
 pharmokinetics 180-1
caspase inhibition 293-4
cell adhesion 265-9
cell banks 70-1, 101-2
CellCube 37, 48
cell culture 1-18
 biology 2-3
 cell line choice 5-7
 cell type choice 3-4
 contamination 13, 69-70, 81-93, 112
  antimicrobial use 112, 114-15
  cross-contamination 13, 16, 94-101
 feeding 13
 growth curves 12-13, 14-15
 instability 16

media, see culture media
 population dynamics 282-301
 primary cultures 11-12
 procedures 10-16
 sampling 21-2
 scale of 10, 19; see also scaling-up cultures
 subculture 5-7, 12
 substrates 10-11, 21, 34-5, 154-7
 systems 9-10
 tissue source 4-5, 150, 157-8
 types 30-1
 vessels 10, 21-2
 yield 21, 22-3
cell differentiation 3, 282
cell factory units 37, 48
cell growth, see growth headings
cell lines 6
 accessioning 70, 71
 authentication 78-102
 banking 70-1, 101-2
 choice of 5-7
 contamination, see contamination
 continuous 5-7, 16
 finite 5
 species verification 78-81
cell orientation 159-63
cell polarity 3
cell proliferation, see growth headings
cell seeding 163-5
cell separation 265-73
 by cell sorting 270-3
 by differential adhesion 266-9
cell shape 3
cell sorting 270-3

311

INDEX

cell speed 160
cell surface markers 276-7
cell survival 191-2, 194-200
cell type choice 3-4
cellulose fibres 62
cell velocity 160
cell viability 20-1, 75-6, 183-91, 206-10
cell vitality 77-8
chemiluminescent detection 92-3
chemostat 56
chemotherapy 175-7
  pharmokinetics 180-1
chondrocytes 140
chromium release 183-4
chromosome preparation 224-6
chromosome transfer (microcell-mediated) 245-52
  hybrid selection 250-1, 252
  microcell fusion 250, 252
  microcell preparation 248-50
  microcell purification 248
cloning
  positional 254-5
  ring, of epithelial cells 264-5
clonogenicity 179, 191, 194-200
closed cultures, headspace volume 25
co-cultures of spheroids 133-9
  with endothelial cells 136-9
  with fibroblasts 135-6
  with monocytes 134-5
collagen
  fibril alignment 159-60
  gels 157
  lattices 157, 164
  sponges 156, 163-4, 165
colony-forming efficiency 76-7, 263-4
colorimetric assays 189-91
COMPARE 176
compression forces 162
contamination 13, 69-70, 81-93, 112
  antimicrobial use 112, 114-15
  cross-contamination 13, 16, 94-101
continuous cell lines 5-7, 16
continuous-flow culture 30, 31, 51, 56-7
Corbeil Culture System 36-7, 48

Costar CellCube 37, 48
'Courtenay' method 197
coverslip cleaning 283
cross-contamination 13, 16, 94-101
culture, see cell culture
culture contamination 13, 69-70, 81-93, 112
  antimicrobial use 112, 114-15
  cross-contamination 13, 16, 94-101
culture media 7-8, 11, 23-4
  additives 24
  changing 13
  component purity 117
  component toxicity testing 116
  for cytotoxicity testing 193
  serum-free 7-8, 105-21
  supplements 22, 24, 105-6
  temperature 22
culture sampling 21-2
culture systems 9-10
culture vessels 10, 21-2
cytogenetics 78-81
cytokeratins 274-6
Cytopilot 65
cytospin 277-8
cytotoxicity index 210-11
cytotoxicity testing 175-219
  commercial kits 185-6
  culture methods 177-9
  dose-response curves 211-13
  drug samples
    activation 193-4
    combinations 213
    concentration 179-80
    diluents 193
    exposure 180-2
    incubation 194
    storage 193
  end-points 183-92
  in vitro/in vivo correlation 212-13
  medium components 193
  recovery period 182-3
  reproducibility problems 214
  troubleshooting 214-15

de-adaptation 282
de-differentiation 282
Dermagraft™ 152
dermal fibroblasts 2

differential staining cytotoxicity (DISC) assay 207-8
differentiation 3, 282
digoxigenin antibody detection 229-30
DNA content determination 189, 205-6
DNA fingerprinting 95-101
DNA laddering 292-3
DNA preparation
  double-stranded for amplification 89
  genomic, from cell lines 98-9
dose-response curves 211-13
doublings, numbers of 23, 281
doubling time 23
drug sensitivity testing 175-219
  commercial kits 185-6
  culture methods 177-9
  dose-response curves 211-13
  drug samples
    activation 193-4
    combinations 213
    concentration 179-80
    diluents 193
    exposure 180-2
    incubation 194
    storage 193
  end-points 183-92
  in vitro/in vivo correlation 212-13
  medium components 193
  recovery period 182-3
  reproducibility problems 214
  troubleshooting 214-15
dye exclusion test 20, 75-6, 184-5
Dynabeads 271-3

electrophoretic transblotting 91
electroporation 240-3
embryoid bodies 143
embryonic cell lines 4
encapsulation 61-2
endothelial cell and spheroid co-cultures 136-9
end-points 183-92
entrapment cultures 62
enucleation 247-8
enzyme release assays 184
epidermal keratinocytes 2, 140
  proliferative heterogeneity 260-3

INDEX

sodium chloride sensitivity 110, 112
stem cell maintenance 273
epithelial extracellular matrix 266-9
epithelial stem cells 259-79
  cell separation 265-73
  characterization 274-8
  clinical applications 259-60
  identification 260
  maintenance 273-4
  proliferative heterogeneity 261-5
  purification 260
Epstein–Barr virus (EBV)-based vectors 236-7, 244-5
erythrocin B 20, 75
extracellular matrix 266-9

fatty acids 117-20
feeder cells 197
fetal tissue 4
fibrin 156
fibroblasts 2
  co-culturing with spheroids 135-6
  fatty acid composition 118, 119
  seeding protocols 163-4
  senescence 281
fibronectin fibre orientation 159, 160
fibronectin mats 156
  phenytoin release from 165-7
finite cell lines 5
fluidized beds 64-5
fluid shear 162
fluorescence activated cell sorting 270
fluorescence *in situ* hybridization (FISH) 221-33
  advantages of 221-2
  probes 222-3
    denaturation 227
    detection 228-31
    labelling 223
    precipitation 223-4
    Web sites 232-3
  resources 232
fluorescent light 116
fluorescent probes 185, 191
fluorochromes 228

freezing 71-4
  recovery from 74-8
FTA™ paper 99

β-galactosidase expression 243-4
  senescence-associated 288-90
gas phase 8-9
gel electrophoresis 91
gene mapping 253-4
gene therapy 158
gene transfer 235-58
  by irradiation fusion 252-6
  by microcell-mediated chromosome transfer 245-52
    hybrid selection 250-1, 252
    microcell fusion 250, 252
    microcell preparation 248-50
    microcell purification 248
  by transfection 235-45
    calcium phosphate–DNA co-precipitation 237-8
    electroporation 240-3
    lipid-mediated 238-40
genomic DNA preparation 98-9
gentamicin 114-15
Giemsa banding 94-5
glass ampoules 72
glass beads 38-40, 48
  porous 48, 62-4
glass substrates 10, 21, 34-5
glucose 24, 25
  labelling 188
glycolysis measurement 187
Golgi body staining 189
Gompertz equation 124
graft rejection 157-8
growth 23
growth curves 12-13, 14-15
growth factors 3, 167-8
growth fraction measurement 283-7
growth rate 23
growth yield 21, 22-3

hair follicles 140
haptotactic analysis 295-6
headspace volume 25
heart cell cultures 140

Heli-Cel 38, 48
hepatitis virus (HBV) detection 87, 88, 90, 93
hepatocytes 111-12, 113, 139-40
Hepes 115-16
heterogeneous reactors 41, 48
HIV detection 87-8, 90
hollow fibres 37-8, 48, 59-60
holoclones 261
HYAFF 155-6
hyaluronan 155
hybridization 92, 299-300; *see also* fluorescence *in situ* hybridization (FISH)

ID_{50}/ID_{90} 212
ImmobaSil 65-6
immobilized cultures 58-66
immunocytochemistry 274-8
immunomagnetic bead sorting 270-3
immurement cultures 59-62
inoculation density 22
Integra™ 152, 156
integrins 265-6
INT violet 197-8
iododeoxyuridine labelling 188
irradiation fusion gene transfer 252-6
isoenzyme profiles 78

karyology 78-81
Kenacid blue 203
keratin markers 274-6
keratinocytes 2, 140
  proliferative heterogeneity 260-3
  sodium chloride sensitivity 110, 112
  stem cell maintenance 273

lactate dehydrogenase 21, 187
laser skin™ 155-6
light 116
lipofection 238-40
liquid nitrogen refrigerators 71, 73
liquid-overlay cultures 131-2, 134-6

313

INDEX

liver, artificial 139-40
luminescence measurement of ATP 190, 209-10, 291
lysosomal staining 189

MACS system 271
magnetic bead sorting 270-3
Matrigel 156
matrix coating 11
mechanical loading 161-3
media 7-8, 11, 23-4
  additives 24
  changing 13
  component purity 117
  component toxicity testing 116
  for cytotoxicity testing 193
  serum-free 7-8, 105-21
  supplements 22, 24, 105-6
  temperature 22
melanoma 137
melatonin secreting cell cultures 140
membrane culture systems 60-1
membrane integrity 183-5
meroclones 261-2
metal substrate 35
methylene blue 202
microbial contamination 13, 69-70, 81-93, 112
  antimicrobial use 112, 114-15
microcarrier culture 41-8
microcell-mediated chromosome transfer 245-52
  hybrid selection 250-1, 252
  microcell fusion 250, 252
  microcell preparation 248-50
  microcell purification 248
micronucleation 246-7
microscopy 81-2, 87, 291-2
microtitration plate preparation 201
miniPERM Bioreactor 60-1
monocyte and spheroid co-cultures 134-5
monolayer cloning 179, 191, 195-6
monolayer culture 32-48
  advantages 33-4
  disadvantages 34

for drug sensitivity testing 178-9
MTT 190, 207
mucosa models 140-3
multicellular tumour spheroids, see spheroids
multitray units 37, 48
mycoplasma contamination 13, 82-6

necrosis 282, 287-8
neoplastic cells 4, 5
nerve implants 153-4
Neuroplast 156
neutral red 189, 206
NHBE (normal human bronchial/tracheal epithelial cells)
  gentamicin response 115
  nutrient response curves 106, 109, 111
  phenol red response 114
nucleotide labelling 187, 205
nutrient response curves 106-12
nutrient utilization 20-1

oligonucleotide labelling 92
Opticell culture system 38, 48, 62
organ culture 9-10, 177
oxidation-reduction potential 29
oxygen 26-9, 187
  in clonogenic assays 198
  transfer rate 26

pancreas, artificial 140
paraclones 261
PCR amplification 90, 100-1
perfusion culture 10, 28, 30, 31
peripheral nerve implants 153-4
PGA 155
pH 22, 24-6
  changes with contamination 13
  probes 25-6, 187
pharmokinetics 180
Phase III phenomenon 281

phenol red 113-14
phenytoin release 165-7
phosphate labelling 188
pineal gland 140
pituitary gland modelling 140
PLA 155
plasmid rescue 244-5
plastic ampoules 72
plastic substrate 10-11, 35, 155
PLGA 155
Pluronic F-68 (polyglycol) 22
polystyrene substrate 10, 35
Ponten Plates 295
population doublings 23, 281
porous carriers 48, 62-6
positional cloning 254-5
pre-hybridization 92
primary cultures 11-12
probes 222-3
  denaturation 227
  detection 228-31
  labelling 223
  precipitation 223-4
  Web sites 232-3
proliferative heterogeneity 260-5
prostate epithelial cells 262
protein content determination 188, 189, 202-3

quiescence 282

radiolabelling 187-8, 204-6
redox potential 29
respiration measurement 187
retinal development 140
ring cloning 264-5
roller bottles 35-7, 48
Roux bottles 26, 27, 32, 48

sampling 21-2
scaling-up cultures 19-67
  of anchorage-dependent cells 32-3
  immobilized culture 58-66
  microcarrier culture 41-8
  monolayer culture 32-48
  oxygenation 28
  suspension culture 48-58
seeding 163-5

INDEX

senescence 281-2, 284, 288-90, 295-6
serum 105-6
   and cell differentiation 3
serum-free media 7-8, 105-21
shear force models 162
short tandem repeats (STR) 95-7, 100-1
Siran spheres 62, 63-4
skin 2-3, 140-3, 152, 157, 163-4
sodium carboxymethylcellulose 22
sodium chloride 110, 112
solvents 193
sparging 27
species verification 78-81
specific growth rate 23
specific utilization/respiration rate 21
speed of cells 160
spheroids 123-39
   automated dissociation 133
   co-cultures 133-9
     with endothelial cells 136-9
     with fibroblasts 135-6
     with monocytes 134-5
   drug sensitivity testing 177-8, 191-2, 198-200
   liquid-overlay cultures 131-2, 134-6
   monocultures 124-33
   proliferation status 132
   spinner flask culture 126-30
spinner flasks 51, 52-3, 126-30
SpiraCel 36, 48
stack plates 41, 48
stainless steel substrate 35
stem cells 259-79
   cell separation 265-73
   characterization 274-8
   clinical applications 259-60
   identification 260
   maintenance 273-4
   proliferative heterogeneity 261-5

purification 260
stirred bioreactors 54-6, 65-6
stirrers 53-4
stirring rate 22
subculture 5-7, 12
substrates 10-11, 21, 34-5, 154-7
sulforhodamine B 202-3
SuperSpinner 52-3
suppliers 303-10
survival assays 191-2, 194-200
suspension cultures 7, 24, 48-58
   adaptation to 49-50
   for drug sensitivity testing 178
   scaling-up 53-4
   small scale 50-3
   static 50

TdT-mediated dUTP nick end-labelling (TUNEL) 290-1
Tecnomouse 60
telomerase activity 296, 297-9
telomere length 296-7, 299-301
tensional forces 162
terminal restriction fragment length analysis 296-7, 299-301
tetrazolium dye reduction (MTT) 190, 207
thymidine labelling 187
thyroid cells 140
'tissue culture' 9
tissue engineering 149-73
   cell orientation 159-63
   cell sources 150, 157-8
   culture methods 158-9
   design stages 150-4
   gene therapy 158
   protocols 163-8
   support materials 154-7
tissue modelling 139-43

tissue repair 149-50
tissue sources 4-5, 150, 157-8
titanium substrate 35
tracheal epithelial cells, see NHBE
transfection 235-45
   calcium phosphate-DNA co-precipitation 237-8
   electroporation 240-3
   lipid-mediated 238-40
TRAP assay 296, 297-9
trypan blue 20, 75, 287
tubing 21
tumour spheroids, see spheroids
tumour tissue 4, 5
TUNEL 290-1

uridine labelling 187, 205-6
urothelial cells 152-3
   seeding protocol 165

variable number of tandem repeats 95
velocity of cells 160
Verax system 64-5
viability testing 20-1, 75-6, 183-91, 206-10
Vibro-mixer 54
viral contamination tests 87-93
vitality measurement 77-8

Web sites 232-3

xenogeneic cells 150
XTT 210

yield 21, 22-3